Credit Scoring and Its Applications

SIAM Monographs on Mathematical Modeling and Computation

About the Series

In 1997, SIAM began a new series on mathematical modeling and computation. Books in the series develop a focused topic from its genesis to the current state of the art; these books

- present modern mathematical developments with direct applications in science and engineering;
- describe mathematical issues arising in modern applications;
- develop mathematical models of topical physical, chemical, or biological systems;
- present new and efficient computational tools and techniques that have direct applications in science and engineering; and
- illustrate the continuing, integrated roles of mathematical, scientific, and computational investigation.

Although sophisticated ideas are presented, the writing style is popular rather than formal. Texts are intended to be read by audiences with little more than a bachelor's degree in mathematics or engineering. Thus, they are suitable for use in graduate mathematics, science, and engineering courses.

By design, the material is multidisciplinary. As such, we hope to foster cooperation and collaboration between mathematicians, computer scientists, engineers, and scientists. This is a difficult task because different terminology is used for the same concept in different disciplines. Nevertheless, we believe we have been successful and hope that you enjoy the texts in the series.

Mark Holmes

Johnny T. Ottesen, Mette S. Olufsen, and Jesper K. Larsen, *Applied Mathematical Models in Human Physiology*

Ingemar Kaj, *Stochastic Modeling in Broadband Communications Systems*

Peter Salamon, Paolo Sibani, and Richard Frost, *Facts, Conjectures, and Improvements for Simulated Annealing*

Lyn C. Thomas, David B. Edelman, and Jonathan N. Crook, *Credit Scoring and Its Applications*

Frank Natterer and Frank Wübbeling, *Mathematical Methods in Image Reconstruction*

Per Christian Hansen, *Rank-Deficient and Discrete Ill-Posed Problems: Numerical Aspects of Linear Inversion*

Michael Griebel, Thomas Dornseifer, and Tilman Neunhoeffer, *Numerical Simulation in Fluid Dynamics: A Practical Introduction*

Khosrow Chadan, David Colton, Lassi Päivärinta, and William Rundell, *An Introduction to Inverse Scattering and Inverse Spectral Problems*

Charles K. Chui, *Wavelets: A Mathematical Tool for Signal Analysis*

Credit Scoring and Its Applications

Lyn C. Thomas
University of Southampton
Southampton, United Kingdom

David B. Edelman
Royal Bank of Scotland
Glasgow, United Kingdom

Jonathan N. Crook
University of Edinburgh
Edinburgh, United Kingdom

Society for Industrial and Applied Mathematics
Philadelphia

10 9 8 7 6 5 4 3

Library of Congress Cataloging-in-Publication Data
Thomas, L. C.
Credit Scoring and its Applications / Lyn C. Thomas, David B. Edelman, Jonathan N. Crook.
p.cm. – (SIAM Monographs on mathematical modeling and computation)
Includes bibliographical referencesand index.
ISBN-10: 0-89871-483-4 (pbk.)
ISBN-13: 978-0-898714-83-8 (pbk.)

1. Credit scoring systems. I.Edelman, David B. II. Crook, Jonathan N. III. Title IV. Series

HG3751.5.T47 2002
65.8'8–dc21

2001049846

To Margery, Matthew, Beth, and Stephen

To Kate, Rachael, and Becky, who taught me
which decisions really matter

To Monica, Claudia, and Zoë, who provide the light in my life,
and to Susan, who keeps that light shining brightly

Contents

Preface **xiii**

1 The History and Philosophy of Credit Scoring **1**
1.1 Introduction: What is credit scoring? 1
1.2 History of credit 2
1.3 History of credit scoring 3
1.4 Philosophical approach to credit scoring 4
1.5 Credit scoring and data mining 6

2 The Practice of Credit Scoring **9**
2.1 Introduction 9
2.2 Credit assessment before scoring 9
2.3 How credit scoring fits into a lender's credit assessment 11
2.4 What data are needed? 13
2.5 The role of credit-scoring consultancies 13
2.6 Validating the scorecard 14
2.7 Relation with information system 14
2.8 Application form 15
2.9 Role of the credit bureau 15
2.10 Overrides and manual intervention 16
2.11 Monitoring and tracking 17
2.12 Relationship with a portfolio of lender's products 18

3 Economic Cycles and Lending and Debt Patterns **19**
3.1 Introduction 19
3.2 Changes in credit over time 19
3.3 Microeconomic issues 21
3.3.1 Present value 21
3.3.2 Economic analysis of the demand for credit 22
3.3.3 Credit constraints 25
3.3.4 Empirical evidence 27
3.4 Macroeconomic issues 28
3.4.1 A simplified Keynesian-type model of the economy 29
3.4.2 The money channel 32
3.4.3 The credit channel 33
3.4.4 Empirical evidence 37
3.5 Default behavior 39

4 Statistical Methods for Building Credit Scorecards **41**
4.1 Introduction 41
4.2 Discriminant analysis: Decision theory approach 42
4.2.1 Univariate normal case 45
4.2.2 Multivariate normal case with common covariance 46
4.2.3 Multivariate normal case with different covariance matrices 46
4.3 Discriminant analysis: Separating the two groups 47
4.4 Discriminant analysis: A form of linear regression 48
4.5 Logistic regression 50
4.6 Other nonlinear regression approaches 52
4.7 Classification trees (recursive partitioning approach) 53
4.7.1 Kolmogorov–Smirnov statistic 55
4.7.2 Basic impurity index $i(v)$ 56
4.7.3 Gini index 57
4.7.4 Entropy index 58
4.7.5 Maximize half-sum of squares 58
4.8 Nearest-neighbor approach 60
4.9 Multiple-type discrimination 61

5 Nonstatistical Methods for Scorecard Development **63**
5.1 Introduction 63
5.2 Linear programming 64
5.3 Integer programming 68
5.4 Neural networks 70
5.4.1 Single-layer neural networks 70
5.4.2 Multilayer perceptrons 72
5.4.3 Back-propagation algorithm 73
5.4.4 Network architecture 76
5.4.5 Classification and error functions 78
5.5 Genetic algorithms 79
5.5.1 Basic principles 79
5.5.2 Schemata 81
5.6 Expert systems 84
5.7 Comparison of approaches 86

6 Behavioral Scoring Models of Repayment and Usage Behavior **89**
6.1 Introduction 89
6.2 Behavioral scoring: Classification approaches 89
6.3 Variants and uses of classification approach–based behavioral scoring systems 90
6.4 Behavioral scoring: Orthodox Markov chain approach 92
6.5 Markov decision process approach to behavioral scoring 96
6.6 Validation and variations of Markov chain models 98
6.6.1 Estimating parameters of a stationary Markov chain model 98
6.6.2 Estimating parameters of nonstationary Markov chains . . 98
6.6.3 Testing if $p(i, j)$ have specific values $p^0(i, j)$ 99
6.6.4 Testing if $p_t(i, j)$ are stationary 99
6.6.5 Testing that the chain is Markov 100
6.6.6 Mover-stayer Markov chain models 101

6.7 Behavioral scoring: Bayesian Markov chain approach 102

7 Measuring Scorecard Performance 107
7.1 Introduction 107
7.2 Error rates using holdout samples and 2×2 tables 108
7.3 Cross-validation for small samples 110
7.4 Bootstrapping and jackknifing 111
7.5 Separation measures: Mahalanobis distance and Kolmogorov–Smirnov statistics 112
7.6 ROC curves and Gini coefficients 115
7.7 Comparing actual and predicted performance of scorecards: The delta approach 117

8 Practical Issues of Scorecard Development 121
8.1 Introduction 121
8.2 Selecting the sample 121
8.3 Definitions of good and bad 123
8.4 Characteristics available 124
8.5 Credit bureau characteristics 126
8.5.1 Publicly available information 126
8.5.2 Previous searches 127
8.5.3 Shared contributed information 128
8.5.4 Aggregated information 128
8.5.5 Fraud warnings 129
8.5.6 Bureau-added value 129
8.6 Determining subpopulations 131
8.7 Coarse classifying the characteristics 131
8.7.1 χ^2-statistic 132
8.7.2 Information statistic 133
8.7.3 Somer's D concordance statistic 134
8.7.4 Maximum likelihood monotone coarse classifier 138
8.8 Choosing characteristics 139
8.9 Reject inference 141
8.9.1 Define as bad 141
8.9.2 Extrapolation 142
8.9.3 Augmentation 142
8.9.4 Mixture of distributions 143
8.9.5 Three-group approach 143
8.10 Overrides and their effect in the scorecards 144
8.11 Setting the cutoff 145
8.12 Aligning and recalibrating scorecards 147

9 Implementation and Areas of Application 151
9.1 Introduction 151
9.2 Implementing a scorecard 151
9.3 Monitoring a scorecard 152
9.4 Tracking a scorecard 156
9.5 When are scorecards too old? 161
9.6 Champion versus challenger 163

10 Applications of Scoring in Other Areas of Lending **169**
10.1 Introduction 169
10.2 Prescreening 169
10.3 Preapproval 171
10.4 Fraud prevention 171
10.5 Mortgage scoring 172
10.6 Small business scoring 173
10.7 Risk-based pricing 174
10.8 Credit extension and transaction authorization 175
10.9 Debt recovery: Collections scoring and litigation scoring 176
10.10 Provisioning for bad debt 177
10.11 Credit reference export guarantees 177

11 Applications of Scoring in Other Areas **179**
11.1 Introduction 179
11.2 Direct marketing 179
11.3 Profit scoring 180
11.4 Tax inspection 183
11.5 Payment of fines and maintenance payments 184
11.6 Parole 185
11.7 Miscellany 185

12 New Ways to Build Scorecards **187**
12.1 Introduction 187
12.2 Generic scorecards and small sample modeling 188
12.3 Combining two scorecards: Sufficiency and screening 189
12.4 Combining classifiers 193
12.5 Indirect credit scoring 195
12.6 Graphical models and Bayesian networks applied to credit scoring . . 196
12.7 Survival analysis applied to credit scoring 203

13 International Differences **209**
13.1 Introduction 209
13.2 Use of credit 209
13.2.1 Consumer credit 209
13.2.2 Credit cards 212
13.3 International differences in credit bureau reports 212
13.3.1 The U.S. 213
13.3.2 Other countries 213
13.4 Choice of payment vehicle 214
13.5 Differences in scorecards 215
13.6 Bankruptcy 215

14 Profit Scoring, Risk-Based Pricing, and Securitization **219**
14.1 Introduction 219
14.2 Profit-maximizing decisions and default-based scores 220
14.3 Holistic profit measure 224
14.4 Profit-scoring systems 225
14.5 Risk-based pricing 227

14.6 Securitization 229
14.7 Mortgage-backed securities 231

References **233**

Index **245**

Preface

Credit scoring is one of the most successful applications of statistical and operations research modeling in finance and banking, and the number of scoring analysts in the industry is constantly increasing. Yet because credit scoring does not have the same glamour as the pricing of exotic financial derivatives or portfolio analysis, the literature on the subject is very limited. However, credit scoring has been vital in allowing the phenomenal growth in consumer credit over the last 40 years. Without an accurate and automatically operated risk assessment tool, lenders of consumer credit could not have expanded their loan books in the way they have.

Credit scoring was one of the earliest financial risk management tools developed. Its use by U.S. retailers and mail-order firms in the 1950s is contemporary with the early applications of portfolio analysis to manage and diversify the risk inherent in investment portfolios. Also, credit scoring could claim to be the grandfather of data mining because it was one of the earliest uses of data on consumer behavior. In fact, the commonest techniques used in data mining—segmentation, propensity modeling, and clustering—are also techniques that have been used with considerable success in credit scoring.

This book outlines the basic theory underpinning the development of credit-scoring systems. It also looks at the practical issues that arise in the building and monitoring of such systems. The areas where scoring is used and the possible developments in scoring usage are also discussed. It is hoped that such an introduction will be of use to those studying scoring as part of courses in operations research and statistics and to practitioners who might find it useful as a reference book. The need for such a book was brought home to us in the series of international conferences on credit scoring and credit control run by the Credit Research Centre at the University of Edinburgh over the past decade. At these conferences, there was a plethora of new uses and new ideas for scoring but a dearth of literature where these ideas were expounded.

The book developed out of a course in credit scoring run jointly for master's degrees in operations research and financial mathematics. The book could also be used as part of applications modules in degree and master's courses in statistics and operations research. In that case, one might concentrate on the chapters that outline the basic problems and the techniques used to develop risk assessment systems: Chapters 1, 2, 4–8, and 12. Those who want to look only at current areas of application and avoid the modeling aspects should look at Chapters 1, 2, 9–11, 13, and 14, while Chapter 3 gives an introduction to the economic theory underlying models of consumer indebtedness.

Thus the book as a whole should give practitioners a background in which to place the everyday decisions in which they will be involved when they commission, build, operate, and monitor credit and behavioral scoring systems. Because decisions must be made about what to include and exclude in such a text, we concentrate on consumer credit and behavioral scoring and do not investigate in great detail the risk-assessment tools used for company and trade credit, although there are some strong overlaps. Similarly, in the last few years, it has become clear that there are some interesting connections in the ways of dealing with credit risk in the pricing of bonds and other financial instruments and the techniques that are used or are being developed for consumer credit scoring. Again, we decided not to investigate such connections in this book, nor do we go deep into the relationship between credit scoring and the securitization of the loans that have been scored. We tried to focus primarily on assessing the risk involved in lending to consumers.

One way to understand the ideas in the text is to try them out on real data. For practitioners this is not a problem, but for those who do not have access to data sets, we have included a small data set.

Credit scoring is a fascinating subject. It affects almost everyone—in the U.S. or Europe, a typical person will be credit scored on average at least twice a month. Yet it is also an area where many techniques in statistics and operations research can be used. Finally, it is an area in which even a small improvement in performance can mean a tremendous increase in profit to the lender because of the volume of lending made by using scoring. Relevant, ubiquitous, profitable—credit scoring is all of these, and we hope this book convinces the reader that it is also interesting and thought provoking.

In writing this book we benefited considerably from comments and suggestions from many people. We especially thank Phil Bowers, David Hand, Alan Lucas, and Bob Oliver for their helpful suggestions and their encouragement. All the errors are, of course, ours.

Chapter 1

The History and Philosophy of Credit Scoring

1.1 Introduction: What is credit scoring?

Credit scoring is the set of decision models and their underlying techniques that aid lenders in the granting of consumer credit. These techniques decide who will get credit, how much credit they should get, and what operational strategies will enhance the profitability of the borrowers to the lenders.

Credit-scoring techniques assess the risk in lending to a particular consumer. One sometimes hears it said that credit scoring assesses the creditworthiness of the consumer, but this is an unfortunate turn of phrase. Creditworthiness is not an attribute of individuals like height or weight or even income. It is an assessment by a lender of a borrower and reflects the circumstances of both and the lender's view of the likely future economic scenarios. Thus some lenders will assess an individual as creditworthy and others will not. One of the longer-term dangers of credit scoring is that this may cease to be the case, and there will be those who can get credit from all lenders and those who cannot. Describing someone as uncreditworthy causes offense. It is better for the lender to state the reality, which is that the proposition of lending to this consumer represents a risk that the lender is not willing to take.

A lender must make two types of decision—first, whether to grant credit to a new applicant and, second, how to deal with existing applicants, including whether to increase their credit limits. The techniques that aid the first decision are called credit scoring, and these are outlined in Chapters 4 and 5, while the techniques that assist the second type of decision are called behavioral scoring. The latter are described in detail in Chapter 6.

In both cases, whatever the techniques used, the vital point is that there is a very large sample of previous customers with their application details and subsequent credit history available. All the techniques use the sample to identify the connections between the characteristics of the consumers and how "good" or "bad" their subsequent history is. Many of the methods lead to a scorecard, where the characteristics are given a score and the total of these scores says whether the risk of a consumer being bad is too great to accept. Other techniques described in this book do not lead to such scorecards but instead indicate directly the chance that the consumer is good and so whether accepting the account is worthwhile. Although these approaches do not lead to a scorecard, they go by the name of credit and behavioral scoring.

Although most of this text, with the exception of Chapter 11, concentrates on the credit applications of scoring, scoring has been used in a number of other contexts in the last decade.

In particular, it is proving very useful for targeting customers in direct mailing and other marketing techniques. With the advent of data warehousing, many organizations, especially in the financial and retail sectors, find that they have the information needed to apply scoring techniques. Similarly, data mining, the other much-hyped advance in using information systems, has as one of its most successful application areas response classification, which is essentially scoring applied in a different context.

1.2 History of credit

As soon as humans started to communicate, it is certain that they started borrowing and repaying. One could speculate that a hunter would take the best club belonging to the group and then share the meat killed with that club with the rest of the group. The first recorded instance of credit comes from ancient Babylon. Lewis (1992) records that on a stone tablet dated to around 2000 BC is the inscription "Two shekels of silver have been borrowed by Mas-Schamach, the son of Adadrimeni, from the Sun priestess Amat-Schamach, the daughter of Warad-Enlil. He will pay the Sun-God's interest. At the time of the harvest he will pay back the sum and the interest upon it." Thus farmers were already dealing with their cash flow problems by borrowing at planting to pay back at harvest.

By the time of the Greek and Roman empires, banking and credit institutions were well advanced, although the idea of a credit card offering XVIII.IX% APR might have proved difficult to promote. The next thousand years, the "Dark Ages" of European history, saw little development in credit, but by the time of the Crusades in the thirteenth century, pawn shops had been developed. Initially, these were charities that charged no interest, but merchants quickly saw the possibilities, and by 1350 commercial pawn shops charging interest were found throughout Europe. These pawn shops, which lend on almost anything left on deposit, and their three-ball sign can still be found in most European and South American countries. During the Middle Ages there was an ongoing debate on the morality of charging interest on loans—a debate that continues today in Islamic countries. The outcome of the debate in Europe was that if the lender levied small charges, this was interest and was acceptable, but large charges were usury, which was bad. Even Shakespeare got into this debate with his portrait of the Merchant of Venice. Also at this time, kings and potentates began to have to borrow in order to finance their wars and other expenses. Lending at this level was more politics than business, and the downside for poor lending could be severe. The abbots who dealt with Henry VIII were given the choice, "Your monasteries or your life."

The rise of the middle classes in the 1800s led to the formation of a number of private banks, which were willing to give bank overdrafts to fund businesses and living expenses. However, this start of consumer credit was restricted to a very small proportion of the population. Mail-order and other enterprises began as clubs, mainly in Yorkshire. If an item costing £1 was desired by each of 20 people, then they would form a club and each pay in 5p (1 shilling in old money) a week. After one week they had enough to buy one of the items and they drew lots to determine whom should receive the item. This continued each week until week 20, when the last nonrecipient got the item. If they had all saved separately, then none of them would have had the item until week 20 without credit being extended. Therefore, in some sense this was a credit union with credit extended by ordinary people to other people, which allowed the lucky person to get his item when 19 further payments were due.

The real revolution started in the 1920s, when consumers started to buy motor cars. Here for the first time was an article that most consumers wanted and that was very mobile and so could not be deposited for security (like pawn shop lending) or even used as a security (like land and property, whose location the lender knows at all times). Finance companies were developed to respond to this need and experienced rapid growth before World War II. At the same time, mail-order companies began to grow as consumers in smaller towns demanded the clothes and household items that were available only in larger population centers. These were advertised in catalogues, and the companies were willing to send the goods on credit and allow customers to pay over an extended period.

Over the last half of the twentieth century, lending to consumers has exploded. Consumer credit has had one of the highest growth rates in any sector of business. The advent of credit cards in the 1960s was one of the most visible signs of this growth, and it is now difficult to function in society without one. In many cases purchases essentially can be made only if one uses a credit card—for example, over the Internet or on the telephone. It is important to remember, however, that credit cards account for less than 15% of consumer credit; far more is being borrowed through personal loans, hire purchases, overdrafts, and, of course, mortgages.

1.3 History of credit scoring

While the history of credit stretches back 5000 years, the history of credit scoring is only 50 years old. Credit scoring is essentially a way to identify different groups in a population when one cannot see the characteristic that defines the groups but only related ones. The first approach to solving this problem of identifying groups in a population was introduced in statistics by Fisher (1936). He sought to differentiate between two varieties of iris by measurements of the physical size of the plants and to differentiate the origins of skulls using their physical measurements. In 1941, Durand (1941) was the first to recognize that one could use the same techniques to discriminate between good and bad loans. His was a research project for the U.S. National Bureau of Economic Research and was not used for any predictive purpose.

During the 1930s, some mail-order companies had introduced numerical scoring systems to try to overcome the inconsistencies in credit decisions across credit analysts (Weingartner 1966, Smalley and Sturdivant 1973). With the start of the World War II, all the finance houses and mail-order firms began to experience difficulties with credit management. Credit analysts were being drafted into military service, and there was a severe shortage of people with this expertise. Hence the firms had the analysts write down the rules of thumb they used to decide to whom to give loans (Johnson 1992). Some of these were the numerical scoring systems already introduced; others were sets of conditions that needed to be satisfied. These rules were then used by nonexperts to help make credit decisions—one of the first examples of expert systems.

It did not take long after the war ended for some folks to connect the automation of credit decisions and the classification techniques being developed in statistics and to see the benefit of using statistically derived models in lending decisions (Wonderlic 1952). The first consultancy was formed in San Francisco by Bill Fair and Earl Isaac in the early 1950s, and their clients were mainly finance houses, retailers, and mail-order firms.

The arrival of credit cards in the late 1960s made the banks and other credit card issuers realize the usefulness of credit scoring. The number of people applying for credit cards each day made it impossible in both economic and manpower terms to do anything but automate the lending decision. The growth in computing power made this possible. These organizations found credit scoring to be a much better predictor than any judgmental scheme, and default rates dropped by 50% or more—see Myers and Forgy (1963) for an early report on such success, and see Churchill et al. (1977) for one from a decade later. The only opposition came from those like Capon (1982), who argued that "the brute force empiricism of credit scoring offends against the traditions of our society." He believed that there should be more dependence on credit history and it should be possible to explain why certain characteristics are needed in a scoring system and others are not. The event that ensured the complete acceptance of credit scoring was the passage of the Equal Credit Opportunity Acts and its amendments in the U.S. in 1975 and 1976. These outlawed discrimination in the granting of credit unless the discrimination "was empirically derived and statistically valid." It is not often that lawmakers provide long-term employment for anyone but lawyers, but this ensured that credit-scoring analysis was to be a growth profession for the next 25 years. This is still the case, and this growth has spread from the U.S. across the world so that the number of analysts in the U.K. has doubled in the last four years.

In the 1980s, the success of credit scoring in credit cards meant that banks started using scoring for other products, like personal loans, while in the last few years, scoring has been used for home loans and small business loans. In the 1990s, growth in direct marketing led to the use of scorecards to improve the response rate to advertising campaigns. In fact, this was one of the earliest uses in the 1950s, when Sears used scoring to decide to whom to send its catalogues (Lewis 1992). Advances in computing allowed other techniques to be tried to build scorecards. In the 1980s, logistic regression and linear programming, the two main stalwarts of today's card builders, were introduced. More recently, artificial intelligence techniques, like expert systems and neural networks, have been piloted.

At present, the emphasis is on changing the objectives from trying to minimize the chance a customer will default on one particular product to looking at how the firm can maximize the profit it can make from that customer. Moreover, the original idea of estimating the risk of defaulting has been augmented by scorecards that estimate response (How likely is a consumer to respond to a direct mailing of a new product?), usage (How likely is a consumer to use a product?), retention (How likely is a consumer to keep using the product after the introductory offer period is over?), attrition (Will the consumer change to another lender?), debt management (If the consumer starts to become delinquent on the loan, how likely are various approaches to prevent default?), and fraud scoring (How likely is that application to be fraudulent?).

There is a fairly limited literature on credit scoring. Ted Lewis, one of the founders of the industry, wrote a monograph on the practical aspects of credit scoring (Lewis 1992). The proceedings of an international conference on credit scoring appeared in the same year (Thomas et al. 1992), and there are several chapters on credit scoring in a textbook (Hand and Jacka 1998). Mays edited a book with chapters on different aspects of credit and mortgage scoring (Mays 1998), while there are a number of textbooks that look at classification problems in general (Hand 1981, 1997). A series of review articles in journals addressed the techniques used in credit scoring (Rosenberg and Gleit 1994, Hand and Henley 1997, Thomas 1998).

1.4 Philosophical approach to credit scoring

The philosophy underlying credit scoring is pragmatism and empiricism. The aim of credit scoring and behavioral scoring is to predict risk, not to explain it. For most of the last 50 years, the aim has been to predict the risk of a consumer defaulting on a loan. More recently, the approach has been to predict the risk that a consumer will not respond to a mailing for a new product, the risk that a consumer will not use a credit product, or even the risk that a consumer will move an account to another lender. Whatever the use, the vital point is that credit scoring is a predictor of risk, and it is not necessary that the predictive model also explain why some consumers default and others do not. The strength of credit scoring is that its methodology is sound and that the data it uses are empirically derived.

Thus credit-scoring systems are based on the past performance of consumers who are similar to those who will be assessed under the system. This is usually done by taking a sample of past customers who applied for the product as recently as it is possible to have good data on their subsequent performance history. If that is not possible because it is a new product or only a few consumers have used it in the past, then systems can be built on small samples or samples from similar products (see Chapter 12), but the resultant system will not be as good at predicting risk as a system built on the performance of past customers for that product.

There is a parallel development to credit scoring in using scoring approaches to predict the risk of companies going bankrupt (Altman 1968). Although this has provided some interesting results connecting accounting ratios to subsequent bankruptcy, because samples are so much smaller than in consumer credit and because accounting information is open to manipulation by managers, the predictions are less accurate than for the consumer credit case.

The pragmatism and empiricism of credit scoring implies that any characteristic of the consumer or the consumer's environment that aids prediction should be used in the scoring system. Most of the variables used have obvious connections with default risk. Some give an idea of the stability of the consumer—time at address, time at present employment; some address the financial sophistication of the consumer—having a current or checking account, having credit cards, time with current bank; others give a view of the consumer's resources—residential status, employment, spouse's employment; while others look at the possible outgoings—number of children, number of dependents. Yet there is no need to justify the case for any variable. If it helps the prediction, it should be used. There are apocryphal stories of the first letter of the consumer's surname being used—although this is probably a surrogate for racial origin. In the U.K., there has been an angry battle with the data protection registrar over whether information on people who have lived at the same address as the consumer could be used. Such information certainly enhanced the prediction of the scorecard; otherwise, the battle would not have taken place.

This battle makes the point that it is illegal to use some characteristics like race, religion, and gender in credit-scoring systems. (This is discussed in more detail in section 8.8.) A number of studies (Chandler and Ewert 1976) have showed that if gender were allowed to be used, then more women would get credit than is the case. This is because other variables, like low income and part-time employment, are predictors of good repayment behavior in females but poor repayment behavior in the whole population. Yet legislators will not allow use of gender because they believe it will discriminate against women.

Other characteristics, although not legally banned, are not used to predict default risk because they are culturally unacceptable. Thus a poor health record or lots of driving convictions are predictors of an increasing risk to default, but lenders do not use them because they fear the approbation of society. However, checking whether the consumer has taken out insurance protection on credit card debts in the case of loss of employment is used by some lenders and is found to be positively related to chance of default. Thus what are acceptable characteristics is very subjective.

This led to a debate in the early 1980s about the ethics of credit scoring between those (Nevin and Churchill 1979, Saunders 1985) who advocated it and those (Capon 1982) who were critical of its philosophy and implementation compared with the subjective judgmental systems based on credit analysts and underwriters' opinions. The former described the advantages of credit scoring as its ability to maximize the risk-reward trade-off, that it gave managerial control of this trade-off, and that it is efficient at processing applications. They argued that credit scoring reduced the need for credit investigations, but this is not really the case. Seeking a bank reference was on the decline anyway since this allowed the applicant's bank the opportunity to make a counteroffer. Using a credit bureau has become a standard fraud check, so the credit bureau might as well give the information on the consumer's credit history and check the validity of the address. The proponents of credit scoring also advocated its consistency over applicants and stated that it improved the information available on the accounts and the quality of the overall portfolio of accounts.

Those against credit scoring attacked its philosophy and the soundness of its methodology. They criticized the fact that it did not give any explanation of the links between the characteristics it found important and the subsequent credit performance. They argued that there was a complex chain of interacting variables connecting the original characteristics and the performance. It is interesting to note that some of the recent developments in credit scoring, like graphical modeling, outlined in section 12.6, do seek to model such chains. The soundness of the statistical methodology was criticized because of the bias in the sample used, which does not include those who were previously rejected. The appropriate size of a sample was questioned, as was the problem of overriding the systems decisions. Other problems highlighted were the collinearity between the variables and the discontinuities that coarse classifying introduces into continuous variables. Eisenbeis (1977, 1978) also noted these criticisms. The credit-scoring industry has long been aware of these criticisms and either has found ways to overcome any deficiency or allows for them in its decision making. These issues are all addressed in Chapter 8.

The overwhelming point is that in the 20 years since this debate took place, scoring systems have been introduced for many different consumer lending products by many different types of lending organization all over the world, and the results have almost always given significant improvements in the risk-return trade-off. The proof of the pudding is in the eating. That is not to say the debate has gone away completely. For example, the introduction of credit scoring into mortgage lending in the U.K. is even now sparking a debate on the role of the underwriters vis-à-vis the mortgage scorecard in the lending institutions.

1.5 Credit scoring and data mining

Data mining is the exploration and analysis of data in order to discover meaningful patterns and relationships. As with mining for minerals, to be successful you have to know where to look and recognize what is important when you find it. In this case, one is looking for information that helps answer specific business problems. Over the last decade, organizations, especially retailers and banks, have recognized the importance of the information they have about their customers. This is because with the advent of electronic funds transfer at point of sale (EFTPOS) and loyalty cards, they are able to gather information on all customer transactions. Moreover, the growth in computer power makes it feasible to analyze the large numbers of transaction data collected. Telecommunications companies, for example, are able to analyze each night all the calls made by their customers in the previous 24 hours. Moreover, new competitors, substitute products, and easier communication channels like the Internet mean that it is much easier for customers to move to rival firms. Thus it is essential for firms to understand and target their customers. This is why they are willing to spend so much money developing data warehouses to store all this customer information and on data mining techniques for analyzing it.

When one looks at the main techniques of data mining, not surprisingly one finds that they are the ones that have proved so successful in credit scoring. In his review of data mining, Jost (1998) acknowledged this fact. The basic data-mining techniques are data summary, variable reduction, observation clustering, and prediction and explanation. The standard descriptive statistics like frequencies, means, variances, and cross-tabulations are used for data summary. It is also useful to categorize or "bin" the continuous variables into a number of discrete classes. This is exactly the technique of coarse classifying (see section 8.8) that has proved so useful in credit scoring. Trying to find which variables are most important, so one can reduce the number of variables one has to consider, is standard in many statistical applications, and the methods used in credit scoring are also the ones that are successful in the other data-mining application areas. Clustering the customers into groups who can be targeted in different ways or who buy different products is another of the data-mining tools. Credit scoring also clusters the consumers into different groups according to their behavior and builds different scorecards for each group. This idea of segmenting into subpopulations and then having a portfolio of scorecards—one for each subpopulation—is discussed in detail later in the text.

The prediction uses of data mining—who will respond to this direct mailing or who will buy this financial product in the next year—are discussed in Chapter 11, since the techniques used are exactly those developed earlier for credit scoring. Jost (1998) described explanation analysis as a major use of data mining but then went on to make the point that rarely are satisfactory explanations possible. Instead, data mining uses segmentation analysis to indicate which segment is most likely to exhibit a particular type of behavior. The philosophical difficulties that Capon (1982) had with credit scoring still occur in this context in that it is very hard to explain why that segment behaves in that way. Human behavior is complex and it seems all one can do is point out the correlations between the segments and their subsequent purchase and repayment behavior. The idea of causality between the two should not be attempted.

Thus it is clear that data mining has at its heart the techniques and methodologies of credit scoring but these are applied in a wider context. It is be hoped that those who use and advocate data mining study the successes and developments of credit scoring since credit scoring proves a useful guide to the pitfalls to be avoided and the ideas that will work in these other application areas.

Chapter 2
The Practice of Credit Scoring

2.1 Introduction

In this chapter, we introduce the basic operation and ideas of scoring, at least from a business and lending perspective. Many of these ideas are referred to in later chapters.

First, we set the scene of how credit assessment was carried out before the widespread use of credit scoring. Then we examine at a high level what a scorecard is and how it fits into the lender's overall operation. We touch on the required data and how they are managed. We also look at some of the other parties involved—the credit bureaus and credit-scoring consultancies.

By the end of this chapter, the reader should have a grasp of the basic business framework in which scoring is used so that, as issues and challenges are discussed in later chapters, the context in which these arise can be understood.

2.2 Credit assessment before scoring

Not so long ago—certainly in the 1970s in the U.K. and the U.S. and perhaps, for a few lenders, still in the late 1990s—credit scoring was not used. Traditional credit assessment relied on "gut feel" and an assessment of the prospective borrower's character, ability to repay, and collateral or security. What this meant was that a prospective borrower did not approach a bank manager or building society manager until they had been saving or using other services for several years. Then, with some trepidation, an appointment would be made and, wearing Sunday best, the customer would ask to borrow some money.

The manager would consider the proposition and, despite the length of relationship with the customer, would ponder the likelihood of repayment and assess the stability and honesty of the individual and their character. He—the manager was invariably male—would also assess the proposed use of the money, and then he might ask for an independent reference from a community leader or the applicant's employer. He might arrange for a further appointment with the customer and then perhaps reach a decision and inform the customer. This process was slow and inconsistent. It suppressed the supply of credit because the prospective borrower would have such a relationship with only one lender. If one asked the bank manager about this, he would answer that such a task required many years of training and experience. (Of course, one could not learn from one's mistakes because the process was so cautious that these were rarely made.)

These disadvantages had existed for some time, so what happened to change things? During the 1980s in the U.K., many changes occurred to the lending environment. Some of these changes were as follows:

- Banks changed their market position considerably and began to market their products. This in turn meant that they had to sell products to customers not only whom they hardly knew but whom they had enticed.

- There was phenomenal growth in credit cards. Sales authorizations of this product meant that there had to be a mechanism for making a lending decision very quickly and around the clock. Also, the volumes of applications were such that the bank manager or other trained credit analyst could would not have the time or opportunity to interview all the applicants. Clearly there would be insufficient numbers of experienced bank managers to handle the volume. During the 1980s, a handful of U.K. operations were dealing with several thousand applications each day.

- Banking practice changed emphasis. Previously, banks had focused almost exclusively on large lending and corporate customers. Now consumer lending was an important and growing part of the bank. It would still be a minority part by value but was becoming significant. Banks could not control the quality across a branch network of hundreds or thousands of branches, and mistakes were made. With corporate lending, the aim was usually to avoid any losses. However, banks began to realize that with consumer lending, the aim should not be to avoid any losses but to maximize profits. Keeping losses under control is part of that, but one could maximize profits by taking on a small controlled level of bad debts and so expand the consumer lending book.

There were many other reasons why scoring came to be introduced in the U.K. during the 1980s, although, as we said, some lenders continue to ignore scoring. Most of these are small lenders who cannot or choose not to expand their lending book in a controlled way, and so they do have the luxury of a few well-trained and experienced managers who can make the critical underwriting decisions. For the vast majority of the market, scoring is used in one form or another.

In consumer lending, it is probably used most in credit cards, because of the volumes and 24-hour coverage. Also, once the decision has been taken to grant a credit card, the lending process does not end. Decisions may be required, not only on authorizations but on whether to amend the customer's credit limit or even on whether to reissue a card and for how long.

For installment loans, the scoring decision is a little simpler. The basic decision is whether to grant a loan. In some situations, it may be more complex as there are additional parameters to consider. These might include the interest margin or interest rate to charge, or the security to demand, or the term for which the money is to be borrowed. Other parameters are possible, but the key factor with most loans is that once the lender has agreed to lend the money, provided the borrower makes payments when due, there is no further decision to make.

For overdrafts, the situation is similar to that for credit cards. A decision is required regarding what the overdraft limit should be. If customers operate their accounts within that limit, there is little else to do. However, if a check is presented or an electronic payment request is made that would take the account over the current overdraft limit, then there is a decision on whether to pay the check or to return it unpaid. If the customer requests an increase in the overdraft limit, a decision is required on whether to grant or decline this.

While credit card limits are reviewed from time to time to see if an increase is warranted, this has not yet happened on a widespread basis with overdrafts.

Once we begin to consider some other types of lending, other factors are introduced. Two of the main additional factors are ownership and security. With leasing or hire purchase, the item concerned—typically a car—may not be actually owned by the consumer until the final payment has been made. This may have an effect on the likelihood of the consumer continuing to make payments. On the one hand, it may be easier for the lender to reclaim the outstanding balance by repossessing the car. This is easier than having to go through court action to establish the debt and then have the court order the consumer to repay the debt, often at an even slower schedule than was in the original credit agreement. On the other hand, the consumer may also find it easier once they get into arrears to return the asset. Of course, the matter is not that simple in practical terms. For example, if the consumer has made, say, 27 of 36 payments, there may be no entitlement to a refund when the car is repossessed. For example, the lender may find that when repossessing the car, it is no longer worth the outstanding balance, and so the consumer may still have to be pursued for money. Also, it may be difficult to repossess the car because it cannot be located.

Of course, if repossession does occur, the lender now has to concern itself not only with lending money but also with the value of second-hand cars.

With mortgages, the consumer owns the property. The borrower will take a legal charge over the property. This prevents the property being sold without the debt being repaid. It also gives the lender a right to repossess the property when the arrears situation has deteriorated, provided a prescribed legal process is followed. As with leasing, a key issue here is whether the property will be worth enough on repossession to extinguish the debt. Falling property values can be caused by the external market or by the residents failing to look after the property. On the other hand, in most cases, the mortgage is the last of the consumer's payments to go into arrears as there is a clear wish to protect the home. (This is also referred to in section 13.6.) After all, people need to live somewhere.

2.3 How credit scoring fits into a lender's credit assessment

At a high level, the potential borrower presents a proposition to the lender. The lender considers the proposition and assesses the related risk. Previously, bankers wove some magic which allowed them to gauge whether the risk was of an acceptably low level. With scoring, the lender applies a formula to key elements of the proposition, and the output of this formula is usually a numeric quantification of the risk. Again, the proposition will be accepted if the risk is suitably low.

How credit scoring fits into the lender's assessment may vary a little from product to product. Let us consider, as an example, an application for a personal loan. Nowadays, the applicant will complete an application form. This could either be a paper form or on a computer screen. An application may be made onscreen if the application is made over the Internet or is made within the branch, where data may be entered by either the customer or the branch staff. Screen-based entry may also arise if the lender is a telephone-based lender and the applicant provides details over the telephone, which are then entered into a screen by a telesales operator.

Typically, the application data will be scored. Not all the application data are used in calculating a credit score. However, as we shall see below, other pieces of information are needed for a variety of purposes, including identification, security, and future scorecard development.

The credit score calculation may also include some information from a credit bureau (see section 2.9). In many cases and in many environments, the result of the credit-scoring process makes a recommendation or decision regarding the application. The role of subjective human assessment has been reduced to a small percentage of cases where there is a genuine opportunity for the human lender to add value.

To make matters more concrete, let us consider a simple scoring operation. Suppose that we have a scorecard with four variables (or characteristics): residential status, age, loan purpose, and value of county court judgements (CCJs); see Table 2.1.

Table 2.1. *Simple scorecard.*

Residential Status		**Age**	
Owner	36	18–25	22
Tenant	10	26–35	25
Living with parents	14	36–43	34
Other specified	20	44–52	39
No response	16	53+	49

Loan Purpose		**Value of CCJs**	
New car	41	None	32
Second-hand car	33	£1–£299	17
Home improvement	36	£300–£599	9
Holiday	19	£600–£1199	–2
Other	25	£1200+	–17

A 20-year-old, living with his or her parents, who wishes to borrow money for a second-hand car and has never had a CCJ, will score 101 ($14 + 22 + 33 + 32$). On the other hand, a 55-year-old house owner, who has had a £250 CCJ and wishes to borrow money for a daughter's wedding, would score 127 ($36 + 49 + 25 + 17$).

Note that we are not saying that someone aged 53+ scores 27 points more than someone aged 18–25. For the characteristic age, this 27-point differential is true. However, as we can clearly see, there are correlations involved. For example, someone aged 53+ may also be more likely to own their house than someone aged 18–25, while it might be quite rare to find someone aged 53+ living with their parents. Thus what we might find is that someone in the older age category may score, on average, 40 or 50 or 60 points more once the other characteristics have been taken into account.

In setting up a credit-scoring system, a decision will be made on what is the pass mark. This is a simple thing to implement but not necessarily a simple thing to decide on. This is touched on in Chapter 9.

Let us suppose that in the example above, the pass mark is 100. Thus any application scoring 100 or more would carry a recommendation for approval. This would be the case whatever the answer to the four questions. What scoring allows, therefore, is a trade-off, so that a weakness in one factor can be compensated for a strength in other factors.

In assessing an application for credit, the lender collects information on the applicant. This can be from a variety of sources, including the applicant, a credit bureau, and the lender's files of the applicant's other accounts. Credit bureau reports are usually available electronically but paper reports are still available (and are still quite common in business and commercial lending).

The lender will examine the information available and calculate a score. There are many ways to use this score.

Some lenders operate a very strict cutoff policy. If the score is greater than or equal to the cutoff, the application is approved. If it is less than the cutoff, the application is declined.

Some lenders operate a simple variation to this. A referral band or gray area is created. This might be 5 or 10 points on one or both sides of the cutoff. Applications falling into such a gray area are referred for a closer look. This closer look might involve some subjective underwriting or might involve seeking further information that is still assessed objectively.

Some lenders operate policy rules that force potentially accepted cases into a referral band. For example, this might be where the application achieves the cutoff but there is some adverse event in the credit bureau information, e.g., a bankruptcy. In other words, we would not be permitting the strength of the rest of the application to compensate automatically for a weakness.

Some lenders operate what are called super-pass and super-fail categories. We discuss the role of the credit bureau below, but it is acknowledged that getting a credit bureau report incurs a cost. Therefore, certainly there are cases that score so poorly that even the best credit bureau report will not raise the score to the cutoff. These would be classed as super-fails. At the other extreme, we might have cases that score so well that even the worst credit bureau report will not reduce the score to below the cutoff. These are super-passes. In effect, these lenders are operating two or three cutoffs: one to define a super-pass, one to define a super-fail, and, in between, a cutoff to be used once the credit bureau information has been used. An alternative use of this might be where the cost of the credit bureau report is low relative to the cost of capturing the applicant information. This might occur in a telephone-based operation: the credit bureau report might be obtained early on, and if a case scores very poorly the application could be curtailed quickly.

Some lenders operate risk-based pricing or differential pricing. Here, we may no longer have a simple fixed price. Rather, the price is adjusted according to the risk (or the profit potential) the proposition represents. Instead of having one cutoff, the lender may have several. There might be a high cutoff to define the best applicants who might be offered an upgraded product, another cutoff for the standard product at a lower interest rate, a third cutoff for the standard product at the standard price, and a fourth cutoff for a downgraded product. In commercial lending, to some extent, the assessment of price takes the risk into account, although other issues, such as competition and the customer relationship, will also have a bearing.

2.4 What data are needed?

Requirements for data fit several categories. Table 2.2 presents listed some ideas of what data may be required. However, in most credit environments, much of the data collected are needed for more than one purpose.

2.5 The role of credit-scoring consultancies

Once again, things have changed over the past 10 or 15 years. Until the mid 1980s, a lender who wanted a scorecard would approach an external scorecard developer and contract with it. Much of the process of actual scorecard development is covered in later chapters. At a high level, this approach involved the lender providing a sample of its data and the scoring developer producing a model with the data.

There were few scoring developers in each market but they provided a useful service. They tended to have very large computers that had enough space to carry out enormous

Table 2.2. *Reasons for data collection.*

Purpose	Examples
To identify customer	Name, address, date of birth
To be able to contract with customer	Name, address, date of birth, loan amount, repayment schedule
To process/score the application	Scorecard characteristics
To get a credit bureau report	Name, address, date of birth, previous address
To assess marketing effectiveness	Campaign code, date of receipt of application, method of receipt—post, telephone, Internet
To effect interbank transfers of money	bank account number, bank branch details
To develop scorecards	Any information legally usable in a scorecard. (Law may vary from country to country.)

calculations. Also, many of them either had direct links to credit bureaus or were part of a larger organization which itself had a credit bureau.

Things have changed but not completely. There are still credit-scoring companies who develop scoring models. They also advise on strategies and implementation and provide training where required. Some of the advantages of having an external developer remain—the credit bureau's links and also the fact that developing models for many lenders allows the external developer an opportunity to identify trends across an industry or portfolio type.

On the other hand, many sizable lenders have built up their own internal analytic teams who can develop scorecards at a fraction of the cost. The growth in power and the reduction in cost of computers have facilitated this internal development. As well as the benefit of lower costs, internal teams may understand the data better and should be better placed to anticipate some of the implementation challenges. Scoring consultancies have arisen to bridge the gap between internal and external development. These do not do the development work but advise internal development teams and keep them away from many pitfalls, both practical and analytic.

At the same time, the traditional scoring vendors have greatly changed their position. Many of them are now much more willing for the development to be a cooperative venture. They will openly discuss methodology and are willing to receive input from the lender, who is, after all, the customer. This cooperation might produce a better scorecard—better at least in the sense that it will generate fewer implementation challenges. It might also enhance the customer's understanding. And it could generate different types of problem for the consultancy and so advance the knowledge of the industry as a whole.

2.6 Validating the scorecard

In Chapter 8, we discuss a typical scorecard development. However, at the simplest level, when a scorecard is built, it is by necessity built on historical cases. This is because we need to allow these cases some time to mature so that we know their performance. Before using or implementing a scorecard, it is important to validate it. One validation check that is carried out is to compare the profile of the cases used in the scorecard development with the profile of current cases. Differences will occur. These may be due to differences in branding, in marketing, or in action by a competitor, and they must be investigated and understood.

2.7 Relation with information system

Not all data used appear on an application form. Some are obtained from a credit bureau (see section 2.9). However, the bank may extract data from its own files. For example, in assessing a personal loan application, a strong scorecard variable could be the performance of the applicant's previous loan. Even more likely is that for an application for an increased credit card limit or overdraft limit, the recent performance on the credit card account or current (checking) account is highly predictive and will indicate whether the customer's finances are in or out of control.

Also, in developing a bank of data for scorecard monitoring and tracking or future scorecard development, details of the customer's performance on the loan or credit card must be saved. The more detail stored, the greater the opportunity to identify niches and opportunities or to build a better scorecard.

2.8 Application form

In section 2.3 and Table 2.2, we looked at the data we might need to process a lending application. Clearly, the more data we need to get from the applicant, the longer we need to make the application form. The longer the application form, the less likely the applicant is either to submit the application or to complete the details. Therefore, there is often pressure to make the process as simple as possible for the applicant.

One way to do this is to reduce the form to its bare minimum. Unfortunately, this sometimes makes future scorecard development difficult. For example, if we do not capture details on a characteristic such as Time with Bank, we cannot analyze it to discover if it would help to predict future performance.

Another way to do this is to find alternative sources for the same or approximately equivalent information. The credit bureau may be able to supply information rather so the applicant need not be asked. This is particularly the case with lenders that share account information (see section 2.9). It may also be the case that we can use the length of time that someone is registered at an address to vote as a proxy for the length of time they have resided at the address.

2.9 Role of the credit bureau

Credit bureaus (or credit reference agencies) are well established in the U.S. and in the U.K. In other countries in Western Europe and in the major developed countries, they are in varying stages of development. Eastern Europe, for example, has only recently begun to tackle the issue of how to develop them. Where they are well established, they are state owned or there is a small number of very large players in the market. In the U.S., there are currently three major bureaus, while the U.K. has two.

To understand the role of credit bureaus, we can examine how they arose. Prior to their widespread use, when one considered a lending application, one might write to an employer or a bank for a reference. Certainly, in the U.K., these references became more guarded and less useful. In addition, if Bank Ours, for example, became aware that Mr. Brown is applying to Bank Other for a credit card or a mortgage, before replying to the reference, Bank Ours could offer him one of its credit cards or mortgage packages. Of course, the bank reference would reveal only, at best, details of the performance of the bank account, a general indication of the prospective borrower's character, and any adverse information of which the bank or the branch at which Mr. Brown banked was aware. As credit availability was expanded, bad debts began to appear. What was galling was that these bad debts were easily avoidable as indications that they could occur were available at the time of the application.

Before describing how credit bureaus operate in these two nations, we should recognize the very different position that U.S. and U.K. credit bureaus occupy. A principal piece of legislation governing credit bureaus in the U.K. is the Data Protection Act. In the U.S., one of the key pieces of legislation is the Freedom of Information Act. Therefore, the two regimes start from opposite ends of a spectrum. In very rough terms, in the U.S., information is available unless there is a good reason to restrict it. Also, in very rough terms, in the U.K., information is restricted or protected unless there is a good reason to make it available. Thus U.S. credit bureaus have a greater wealth of information on consumers.

In both environments, the credit reference agencies first began to accumulate publicly available information and put it in a central point. Even in the 1980s, this might be achieved with a large room full of index card cabinets. On receipt of an inquiry, an agent would put the inquirer on hold and run up and down the room accessing relevant cards. Obviously, over time, this became computerized. This means that the inquiry can also be made electronically, by a human operator dialing up the inquiry service or, more commonly, two computers talking to each other. This public information might now include electoral roll information and information on public court debts. Using the computing power, the bureaus are also able to link addresses so that, when a consumer moves house, the debt and the consumer do not become detached.

The bureaus also act as agents for the lenders. Lenders contribute to the bureau details of the current status of their borrowers' accounts. These statuses can be viewed and used by other lenders when considering a credit application. They can also be viewed and used by other lenders in marketing although with some increased greater restrictions on the use of the data. (In the U.K., this agency arrangement works for the lenders on a reciprocal basis. Roughly speaking, if the lender contributes only details on their defaulting customers, they get to see only details on other lenders' defaulting customers.)

Another service that bureaus offer is to accumulate details of all inquiries and try to assess mismatches and potential fraudulent applications. Clearly, the greater the level of detail with which to work, the better. Major bureaus also offer a variety of application processing services.

A further service used by many lenders is a generic score. This score is calculated from a scorecard built by the bureau based on its experience with millions of applications and millions of credit history records. It is particularly useful in cases where the lender is not large enough to develop scorecards for their own portfolios or in the early year or two of a new product. It is also used to get an up-to-date view of the borrower's credit position as it will incorporate the borrower's recent credit performance with all contributing lenders and any inquiries being made as a result of fresh credit applications. Indeed, some lenders, especially in credit card portfolios, buy a score for each of their cardholders every month and use these to assess how to deal with cases which miss payments or go overlimit, or when and by how much to increase the customers' credit limit.

Other discussions of the credit bureaus and of the data available from them appears in sections 8.5 and 13.3.

2.10 Overrides and manual intervention

For a variety of reasons, the decision or recommendation offered as a result of credit scoring an application for credit is not always the one followed. First, there is the fact that the applicant may appeal. In the U.K., the *Guide to Credit Scoring* (Finance and Leasing Association 2000) encourages lenders to have a process that allows this. Lenders need to consider these appeals, especially as there may have been a wrong decision, e.g., due to the incorrect keying of application details or incorrect information loaded at the credit bureau.

Second, we talked earlier about the fact that some lenders will operate a gray area or referral band, above or below or surrounding the optimal cutoff. Therefore, case which reach a final decision to accept or decline having gone through an intermediate step of referral might also be appropriate for consideration as overrides.

A possible third reason may arise when we develop a credit scorecard for applications for a product, but then, in making a decision on whether to lend, we take a customer rather than account view. For example, we may consider a loan application as a marginal decline but recognize that the applicant also has a large amount of money tied up in a savings account as well as family and business connections. Further, if these other connections generate profit and we consider that these connections might be in jeopardy were the loan application declined, we may overturn the decline decision. Now such claims are quite common in large lending organizations. However, it needs to be recognized that few institutions have the data and systems to be able to accurately assess the profitability or otherwise of a connection, especially when we must project that connection's worth to the lender into the future. The subject of assessing customer and connection profitability is too large a topic for this book. Indeed, it could occupy a text on its own. However, the reader must recognize it as an issue to be addressed.

For whatever reason, overrides are created. They can be of two types, i.e., in two directions. First, we may have cases that are acceptable to the scorecard but that are now declined. These usually will be of marginal profitability and so they will not hugely affect the profitability of the portfolio. However, they may affect our ability to assess the accuracy of the scorecard.

We may also have cases that were not acceptable on the scorecard but that have been approved. These are more serious because they may be accounts on which we expect to lose money and we are adding them to the portfolio. This topic is further explored in section 8.10.

2.11 Monitoring and tracking

Monitoring and tracking are covered in later chapters, especially Chapter 9. However, it is worthwhile to introduce the item now and to establish some key concepts.

Many practitioners use these two terms—tracking and monitoring—almost identically and interchangeably. In this text, we differentiate between these two activities. Monitoring is passive, rather like traffic census takers who sit beside a road and mark off on sheets the number of different types of vehicle that pass a particular point in discrete intervals of time. On the other hand, tracking is active, following one's quarry until either they are captured or the tracker discovers their base.

In scoring, monitoring a scorecard is a set of activities involved in examining the current batch of applications and new accounts and assessing how close they are to some benchmark. Often this benchmark is the development sample, although it need not be. In scoring, tracking involves following groups of accounts to see how they perform and whether the scorecard's predictions come to pass.

Most monitoring activity for a tranche of business can be carried out soon after the lending decisions have been made. It may include a forecast itself, but this forecast, of provisions, attrition, write-off, etc., is made soon after the point of application.

Tracking activity may take up to two years after this point. At a variety of points in the life of a tranche of accounts, we should assess whether this particular group of accounts is performing as the scorecard suggests that it should. If we can identify early on that they are performing worse, we may be able to take corrective action and mitigate any losses. On the other hand, if the business is performing better, there is the opportunity to expand credit assessment and to take on additional profitable business.

Earlier, we introduced the idea of credit-scoring consultancies. A service that they can provide is to do the lender's monitoring and tracking. This might be attractive for a small organization where senior management wish to benefit from a scoring approach but without the overhead of an analytic team.

2.12 Relationship with a portfolio of lender's products

As has been touched on, in reaching a lending decision, we may include details of a customer's other holdings. These may be because they are credit related. An example is the case in which the customer has an existing credit card and has applied for a personal loan. We may also include details because they are important for the management of the customer relationship and its profitability. Examples would be cases in which the customer has applied for a personal loan and already has a mortgage or a substantial deposit.

There is also the issue of defining the customer and the boundaries of the customer relationship. We may be trying to assess a small business loan application from a small accountancy partnership that repeatedly refers good potential new customers. Similarly, we may be about to decline a loan application from someone whose extended family represents highly profitable business to the lender but may consider moving their connections or their business. When stepping beyond a simple consideration of one product for one consumer—as we should in managing our business—many issues require careful consideration if we are to strike out in the general direction, at least, of a position of optimum profitability.

Chapter 3

Economic Cycles and Lending and Debt Patterns

3.1 Introduction

In this chapter, we explain how changes in macroeconomic factors affect the demand and supply of credit and how changes in the demand and supply of credit combine with other factors to affect an economy's output. This will allow us to understand the profound role that credit has in the workings of a nation's economy. From a strategic point of view, we are then able to understand how and when the demand for credit is likely to change. We also explain how the output of an economy affects default rates over time. In this chapter, we first describe the cyclical variations in credit over time. Second, we explain the relationships between the volume of credit and variations in output of an economy. Third, we explain how the state of the economy affects default behavior.

3.2 Changes in credit over time

Figure 3.1 shows quarterly data for the U.K. for net consumer credit extended (new credit extended less payments) and gross domestic product (GDP), a measure of the output of an economy and the total income of its participants, for the period 1965Q1–1999Q2. Figure 3.2 shows figures for consumer credit outstanding between 1976Q1 and 1999Q2. All series are measured at constant (1990) prices. Between 1963 and 1988, the growth rate of net credit extended per quarter was much greater than that of GDP, and the same is true for debt outstanding between 1976 and 1990. However, between 1988Q3 and 1992Q2, net credit extended per quarter plummeted and then rose again between 1992Q3 and 1994Q3. The corresponding changes in the stock of debt outstanding can be seen with a decrease in its growth rate from mid 1988 until 1992.

However, these figures do not clearly show cyclical variations in the volumes. To achieve this, we need to remove the trend. This has been done, and the results are shown in Figure 3.3, which is reproduced from Crook (1998). Crook argued that a number of conclusions are suggested. First, by identifying the dates of peaks and troughs it seems that on average those in net credit extended lead those in GDP by +2.9 quarters. That is, people take more credit (net of payments) approximately nine months before the domestic economy makes the items or provides the services they wish to buy and, conversely, people take less credit (net of payments) approximately nine months before the home economy reduces production of the goods and services they wish to buy. Second, Crook found that

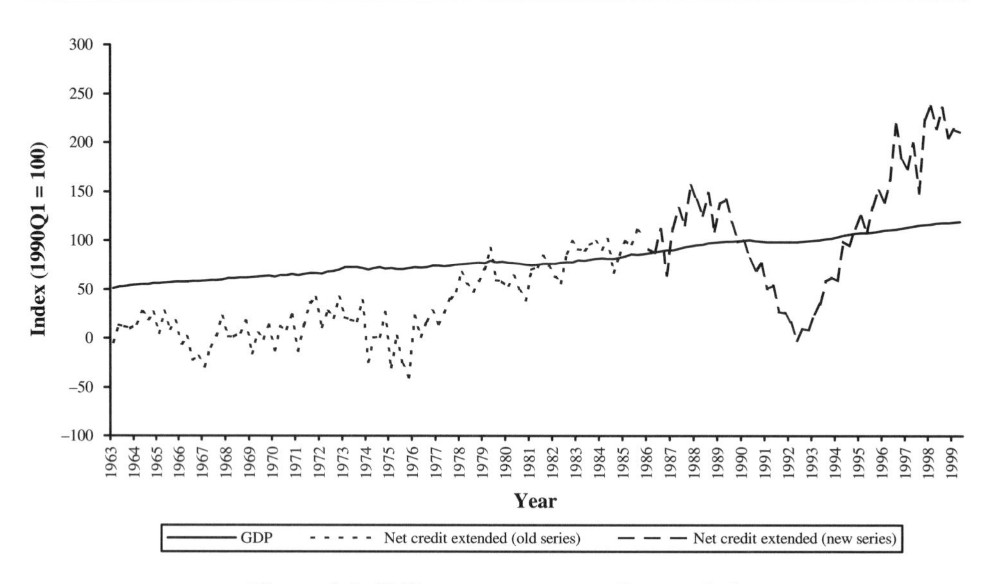

Figure 3.1. *U.K. net consumer credit extended.*

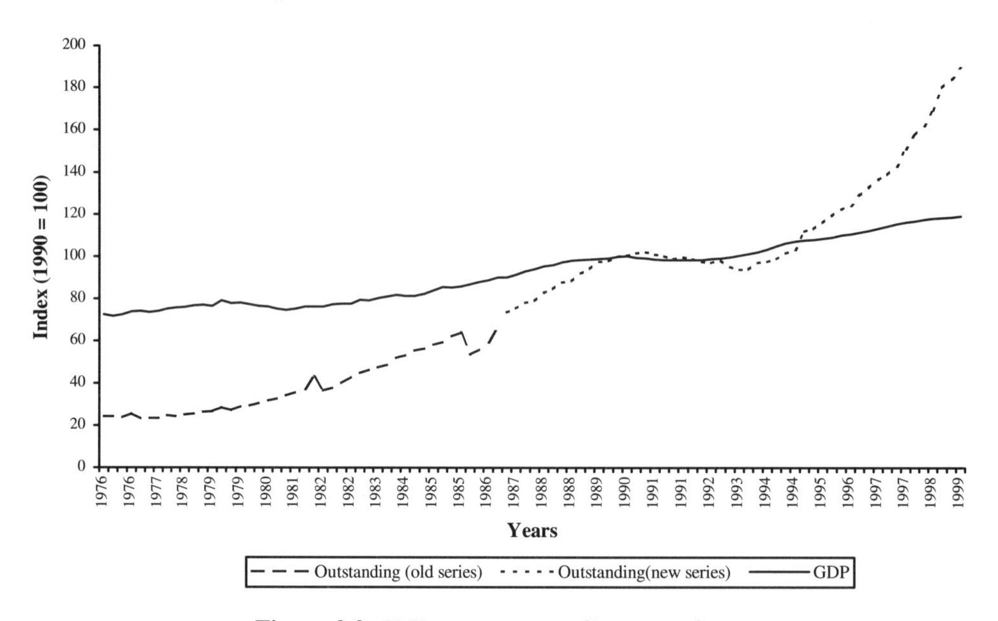

Figure 3.2. *U.K. consumer credit outstanding.*

the average length of the GDP and net credit extended cycles was almost identical at 18.67 and 18.25 quarters, respectively.

Figure 3.4 shows similar figures for the U.S. The data have been manipulated to reveal turning points rather than magnitudes. Figure 3.4 suggests that, using data that when detrended cover the period 1949Q4–1996Q2, the turning points in consumer credit debt outstanding follow those in GDP. The average number of quarters by which outstandings lead GDP is 1.8, and the range of leads is from -1 quarter to $+7$ quarters. Figure 3.4 suggests that the turning points in net credit extended lead those in GDP by typically one or two quarters, the average being $+1.25$ quarters.

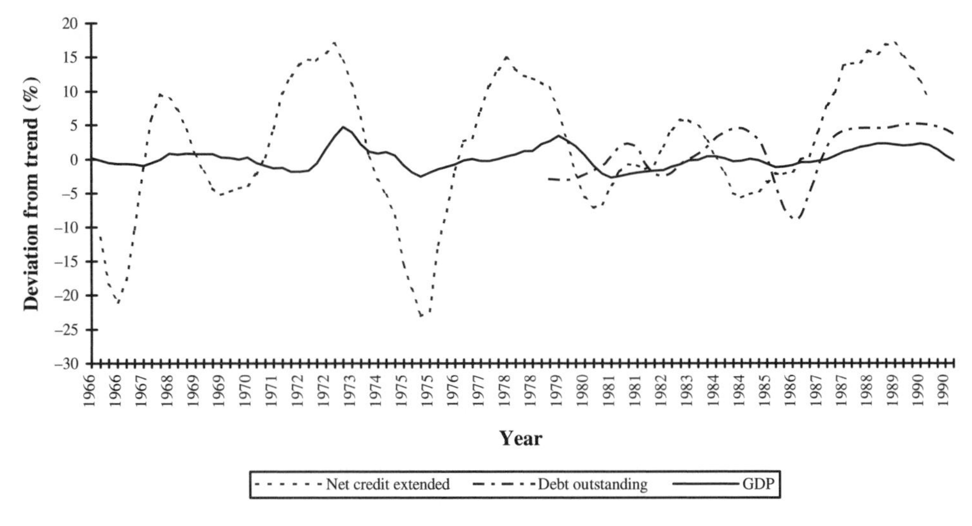

Figure 3.3. *U.K. cyclical behavior of GDP, net consumer credit extended, and consumer debt outstanding. (Reproduced from Crook* (1998) *with permission.)*

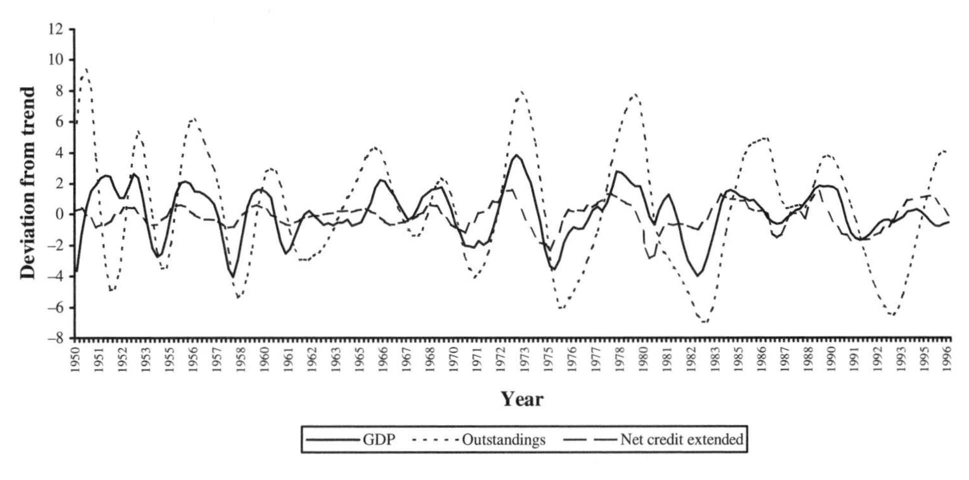

Figure 3.4. *U.S. cyclical behavior of GDP, net consumer credit extended, and consumer debt outstanding.*

Thus far, we have considered some apparent correlations in credit data. To understand how the volume of credit and output relate to each other over time, we need to consider some economic explanations of the relationship and we need to know the results of far more detailed correlation analyses that have tested these explanations. This is our task in the remainder of this chapter.

3.3 Microeconomic issues

3.3.1 Present value

We use the concept of present value often in this book—the first time in this section. Consider an individual investing, today (period 0), $\$R_0$ in a savings account that gives an interest rate of r per year. In year 1 the individual would receive $R_1 = R_0(1 + r)$. When we discount we ask the question, How much does one need to invest today to receive R_1 in year 1, given the interest rate of r? We know the answer is R_0, which can be derived simply as $\frac{R_1}{(1+r)}$. Now the individual may gain receipts in each of several years, so to calculate the amount the person would need to invest today to receive all these amounts, we simply multiply each future receipt by $\frac{1}{(1+r)^t}$, where t is the number of years into the future in which the receipt is gained. (The term $\frac{1}{(1+r)^t}$ is known as the discount factor.) Therefore, the amount a person would need to invest today (year 0) to gain R_t, for each of $t = 1, 2, \dots T$, which is called the present value of these future receipts, is

$$PV = \frac{R_1}{(1+r)} + \frac{R_2}{(1+r)^2} + \cdots + \frac{R_T}{(1+r)^T}. \tag{3.1}$$

If, in order to gain these future receipts, we need to spend C_0 in the current period, the present value of these future receipts, net of the current outlay, called the net present value (NPV), is

$$\text{NPV} = -C_0 + \frac{R_1}{(1+r)} + \frac{R_2}{(1+r)^2} + \cdots + \frac{R_T}{(1+r)^T}. \tag{3.2}$$

The discount rate, r, is the rate of return on the best alternative use of funds. If instead of investing funds we borrow them, then the discount rate is the lowest borrowing rate instead of the highest lending rate. Notice that discounting has nothing to do with inflation. When we discount we simply remove the interest we would have earned had the future receipt been available for investment today.

3.3.2 Economic analysis of the demand for credit

To understand the effect of the level of output (and thus income) of an economy on the demand for credit we need to consider how the income of an individual borrower would affect his demand for credit. We consider an individual who, for simplicity, is considering how much to consume in each of two periods: periods 0 and 1. We assume the individual wishes to maximize his satisfaction over the two periods but is constrained by his income. First we consider the individual's income constraint.

Suppose the individual receives a certain income in each time period, Y_0 in period 0 and Y_1 in period 1. Suppose the individual can save an amount S_0 in period 0. If we denote his consumption in period 0 as C_0, we can write

$$S_0 = Y_0 - C_0. \tag{3.3}$$

In period 1 the amount of income a consumer has available to consume is his income in period 1, Y_1, plus his savings from period 0, including any interest on them:

$$C_1 = S_0(1+r) + Y_1, \tag{3.4}$$

where r is the rate of interest. Substituting (3.3) into (3.4), we can see that $C_1 = (Y_0 - C_0)(1+r) + Y_1$, which, after manipulation, gives

$$C_0 + \frac{C_1}{(1+r)} = Y_0 + \frac{Y_1}{(1+r)}. \tag{3.5}$$

In other words, the present value of consumption over the two periods equals that of income. Equation (3.5) is known as the budget constraint and can be represented as line AB in Figure 3.5.

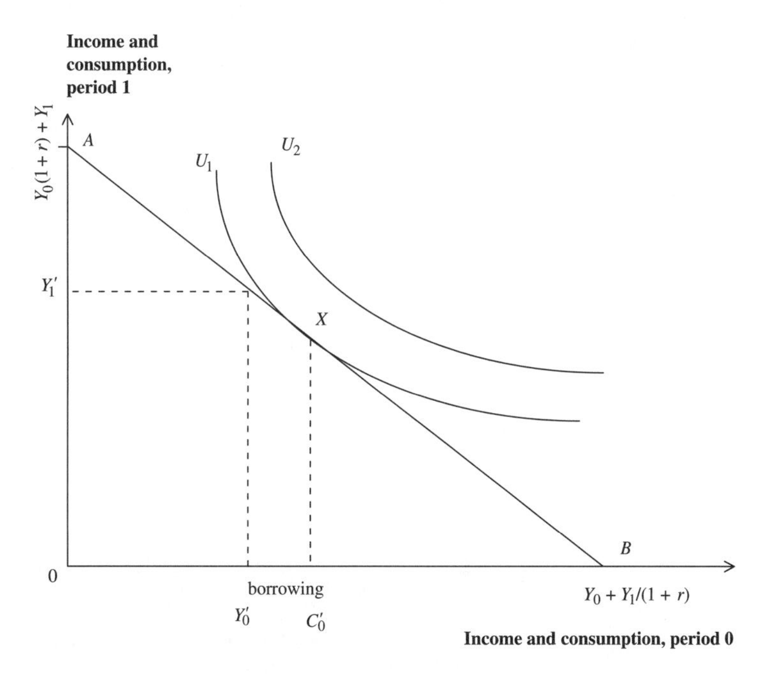

Figure 3.5. *Utility maximization.*

If the individual saved all of his income in period 0, he would receive income of $Y_0(1+r)+Y_1$ in period 1 and would be able to consume this amount then. This is point A in Figure 3.5. Alternatively, the individual might choose to receive income of $Y_0 + \frac{Y_1}{(1+r)}$ in period 0, where $\frac{Y_1}{(1+r)}$ is the amount the individual could borrow in period 0 such that his payment in period 1 just equalled his income, then Y_1. For each dollar by which our consumer increases his consumption in period 0, he must reduce consumption in period 1 by the payment: $(1+r)$. Thus the slope of AB is $-(1+r)$.

Now consider the wishes of the consumer. Suppose he requires additional consumption in period 0 to compensate for loss of consumption in period 1. Then a line representing combinations of consumption in the two periods, each giving equal satisfaction, would slope downward. If we also assume that as the person gains more consumption in period 0 he is willing to give up less consumption in period 1 to compensate, then the line is convex downward, as in the figure. Such a line is called an indifference curve and we have drawn several in the figure. Curves further from the origin represent greater satisfaction because they represent more consumption in period 1 given a level of consumption in period 0.

The combination of consumption in both periods that maximizes the consumer's satisfaction is shown by point X because this is on the highest indifference curve possible, given that the point must also be on the budget line constraint. Given our consumer's incomes of Y_0^1 and Y_1^1, point X indicates that he wishes to consume more than Y_0^1 in period 0 and less than Y_1^1 in period 1. In short, he wishes to borrow an amount $Y_0^1 C_0^1$ in period 0.

Now we consider the effect of an increase in income. If the income of the consumer increases, a budget line shifts out parallel to itself because its slope depends only on the interest rate, which has not changed. This is shown in Figure 3.6.

Suppose the additional income is received only in period 0 and the new budget line is $A'B'$ with the new equilibrium at Z. Whether the new amount of desired borrowing, B_2, is larger or smaller than the original amount clearly depends on the shape and position of the

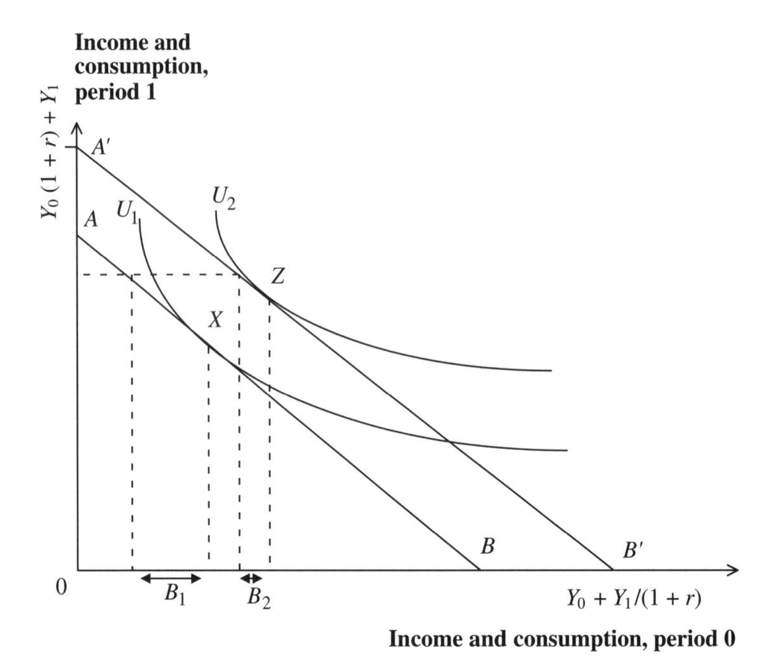

Figure 3.6. *A change in income.*

indifference curves. It may be higher or lower—we cannot say. If the income is received in period 1 this conclusion is reinforced. We conclude that if over time the income of a household rises, the demand for credit may rise or fall.

Equation (3.5) showed that the present value of consumption in the two periods equals an individual's present value of income. It can be extended to as many periods as a person expects to live. Modigliani (1986) argued that given that this is true and that people can borrow or lend as much as they wish, given their lifetime profile of income, each person will typically borrow early in their adult life (when expenditure exceeds income), then save in middle age, and finally, in retirement, run down their savings. This is known as the life cycle theory of consumption. According to this theory we would expect that an increase in the proportion of the population in younger age groups (but above the legal age necessary for acquiring debt) would be associated with increases in the stock of debt. Park (1993) suggests that over the period 1960–1990 the percentage of the U.S. population in the 20–34 years age group increased, especially in 1960–1980. He suggests this may explain the rise in debt outstanding especially in the 1970s and the stabilization of the volume of debt in the 1980s and early 1990s when the percentage of the population in these age groups reached a plateau. Of course, such demographic changes cannot explain short-term changes in demand.

Now consider an increase in the interest rate. This case is shown in Figure 3.7. The slope of the budget line becomes steeper (remember that the slope is $-(1+r)$) and changes from AB to CD. Notice that the new and the old budget line both go through point E, the combination of incomes in periods 0 and 1. This combination is still available after the change.

The initial consumption was at W and the final point is at X. The consumer's desired stock of debt changes from B_1 to B_2, an increase. Economists can show that this is the normal case: when the interest rate rises, the consumer will choose a lower stock of debt.

It has also been argued (see Park 1993) that many individuals have both financial

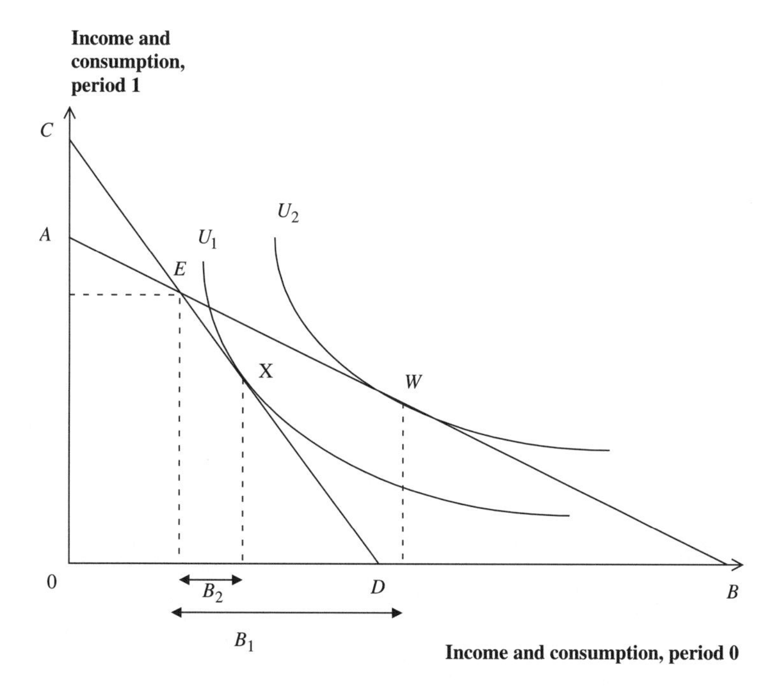

Figure 3.7. *A change in interest rate.*

assets and debt. Although the interest rate on debt exceeds that on assets, people prefer to borrow to pay for some purchases because of the flexibility that having financial assets provides. However, if the difference between the interest rate on assets and that on liabilities changes, so will the proportion of people who take credit rather than run down their assets to buy goods. By examining visually the time series of the difference between these two interest rates and the ratio of net credit extended to consumption expenditure over the period 1972–1992, Park found support for this view.

3.3.3 Credit constraints

If the consumer credit market was very competitive, one might think of representing demand and supply as shown in Figure 3.8.

Given all the factors that affect supply, one might think that the higher the interest rate the more net credit banks would be willing to supply (it is more profitable) and, as we explained above, the less net credit consumers would demand. We would actually observe the equilibrium amount of net credit extended of C_e at an interest rate determined by supply and demand, of r_e. If the interest rate is above r_e suppliers of credit are willing to supply more than consumers wish to borrow, and thus they would offer credit at lower interest rates so that the amount they wish to supply equals the amount consumers wish to borrow. The reverse would happen if the rate was below r_e. Changes in income over the business cycle would shift the demand curve and so affect the equilibrium interest rate and the observed amount of net credit extended. Unfortunately, the market for credit is not so simple.

First, there is evidence that many consumers do not receive all the credit they desire (see, for example, Crook 1996, Jappelli 1990, Crook 2000). Evidence suggests that approximately 18% of households in the U.S. have applied for credit in any five-year period and have been

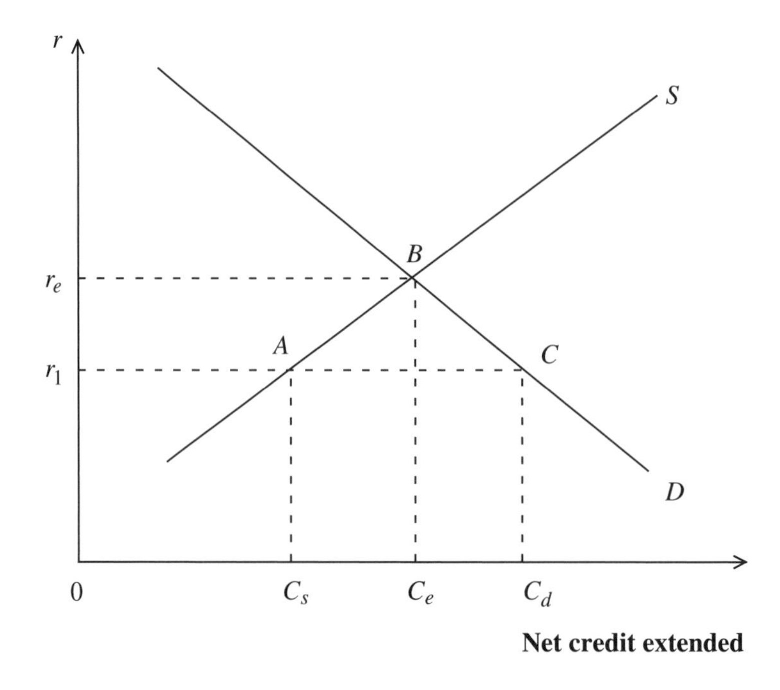

Figure 3.8. *Demand and supply of credit.*

unable to receive all they wished. This proportion varies slightly over time but on the whole is quite stable.

This might be incorporated into our diagram by suggesting that the interest rate is not determined by the intersection of demand and supply but is instead determined by other factors, perhaps the central bank, and the amount of net credit extended that we actually observe is the amount banks wish to offer at this rate. This amount and rate would be r_1 and c_s in the figure.

However, Stiglitz and Weiss (1981) considered four definitions of credit rationing. The first is the type above. Second, some applicants and banks differ in their assessment of the appropriate interest rate to be received by the bank to compensate it for its perception of the riskiness of lending to the applicant. Third, there may be no difference in the rates described in the second definition, but the agreed risk is so high that there is no interest rate that would compensate the lender and at which the applicant would wish to borrow any amount. Fourth, some people may receive credit when others who are identical do not. This may arise as follows. Suppose a bank chose a higher interest rate. This would increase the revenue from each volume of loans. But it also makes each loan more risky because it increases the chance of default. Increasing the interest rate further would result in low-risk applicants turning to other suppliers or not borrowing at all, and so only high-risk borrowers would remain and they yield lower profits. Given that the level of profits and the amount the bank wishes to supply are positively related, a backward bending supply curve is deduced, as shown in Figure 3.9.

Credit rationing can now be seen. If the demand curve is D_0, the equilibrium interest rate is r_0. If the demand curve is D_1 the optimal rate for the bank is r_1, which maximizes its profits, with supply C_1. But the volume of net credit demanded is C_2 at this rate and C_1C_2 is unsupplied. If the business cycle leads to fluctuations in the position of the demand curve, the amount of rationing will also vary over time.

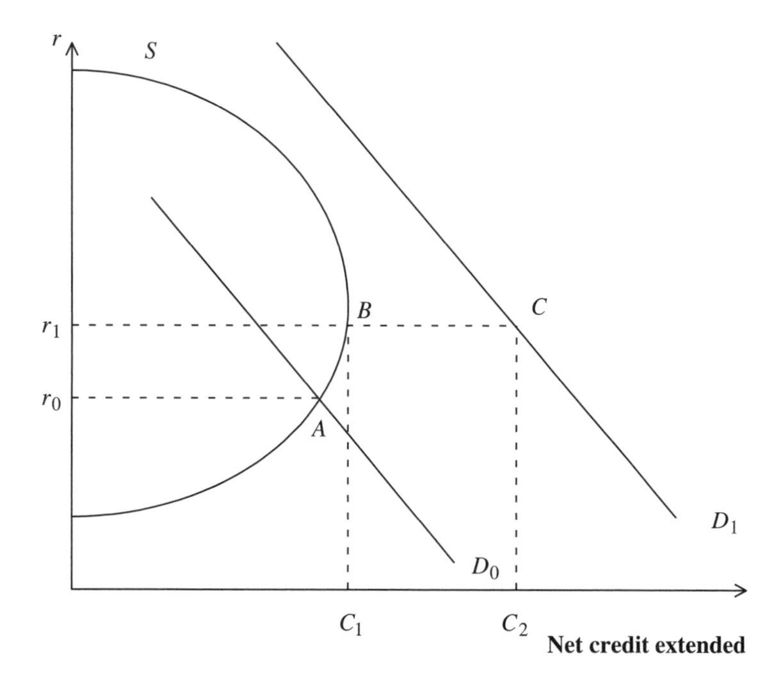

Figure 3.9. *Credit rationing.*

Stiglitz and Weiss proposed their argument as applying to borrowing by firms. However, Drake and Holmes (1995, 1997) found evidence for its applicability to the U.K. mortgage market and in the U.K. market for consumer bank loans, while Martin and Smyth (1991) did likewise for the U.S. mortgage market.

3.3.4 Empirical evidence

We divide the empirical evidence into, first, that relating to the demand for nonmortgage loans and, second, that relating to the demand for mortgage finance.

Table 3.1, relating to nonmortgage loans, shows the results of econometric studies that have related the demand for either net credit extended or new credit extended to various explanatory variables. The elasticity of demand is the percentage change in credit extended when the explanatory variable changes by 1%. Mathematically, the elasticity of y with respect to x is $\frac{\partial y}{\partial x} \cdot \frac{x}{y}$, where $y = f(x)$. Short-run elasticities, that is, elasticities that are estimated from an equation relating quarterly changes in demand to quarterly changes in the explanatory variables, are positive and, generally speaking, have a value approximately equal to one. Thus a 1% increase (decrease) in real personal disposable income (PDI) results in a 1% increase (decrease) in credit demanded. Long-run elasticity, that is, the elasticity estimated from an equation that represents the stable relationship between demand and its explanatory variables, is a little larger. All these studies take the existence of credit rationing into account to identify the demand function. Of particular interest to us is the lagging structures of the affect of income on demand. One of the most sophisticated of the studies is that by Drake and Holmes (1995) using data for the U.K. They established that current values of net credit extended are correlated with current values of income (and other variables) and from this deduced the long-run elasticities. Second, they estimate the relationship between quarterly

Table 3.1. *Elasticities of demand for nonmortgage consumer loans.* (Note: *All four studies relate to the U.K.*)

	Drake and Holmes (1975)	Garganas (1995)	Crook (1989)	Hartropp (1992)
	Net credit extended	New credit extended	New credit extended	Net credit extended
	1977–1992	1957–1971	1977–1985	1968–1987
PDI	+1.349	+0.48 to +1.41	+1.73 to +1.02	positive
Real interest rate				−0.75
Nominal interest rate		−0.20 to −0.22		−0.58
Real net financial wealth	+1.099			
Vacancies		+0.26 to 0.37	+0.07 to +0.04	
Real housing wealth	+1.621			

changes in net credit extended about this long-run relationship and lagged net changes in income. After deleting insignificant variables they find that the change in demand is related to the current change in PDI, the change in housing wealth lagged two quarters, and the change in financial wealth lagged four quarters. They also found evidence of a backward-bending supply curve.

Table 3.2 shows estimated elasticities of demand for mortgage loans. Unfortunately, no clear conclusions concerning the sizes of elasticities can be drawn because the estimates differ markedly between the two studies cited. Possible reasons for the difference include differences in the econometric methodology, the different markets considered with associated differences in tax regimes and other regulations (U.K. versus U.S.), and different methods for taking account of rationing.

Table 3.2. *Elasticities of demand for mortgage debt.*

	Drake and Holmes (1997)	Martin and Smyth (1991)
	(U.K.)	(U.S.)
	1980–1992	1968–1989
PDI	+1.555	+4.860
Nominal interest rate	–0.007	–4.254
House price inflation	+0.031	
Nonprice rationing	+0.056	

3.4 Macroeconomic issues

In the previous section we considered the market for credit alone. Such an analysis is know as a partial equilibrium analysis. In this section we consider the relationship between credit and aggregate output of an economy taking into account the interrelationships between different parts of the economy. When we consider many markets together we are undertaking a general equilibrium analysis.

Most economists would agree that changes in the money supply result in changes in the total output of an economy in the short run. There is a considerable amount of evidence to support this, beginning with Friedman and Schwartz (1963). However, there is considerable debate about *how* a change in the money supply effects changes in real variables (such as output and employment) and the role of credit in this relationship. The mechanism by which changes in the money supply effect changes in real variables is known as the transmission mechanism. While there is no universally accepted mechanism, there appear to be three major views: the money channel, the exchange rate channel, and the credit channel. The role of credit differs between the first two and the third. In this section, we first explain a very simplified Keynesian-type model of the economy. Then we explain each channel, and finally we evaluate the empirical evidence to see if it supports the existence on one channel rather than the others.

3.4.1 A simplified Keynesian-type model of the economy

For simplicity we assume that the price level in an economy is fixed and that the economy is closed: it has no trade with other countries. While these assumptions are unrealistic, they help us to explain the mechanism most simply.

Such an economy consists of many markets, but we assume that they can be represented by two: the market for goods and services and the market for money. In the goods market the total demand for goods and services consists of the planned (or desired) consumption expenditure by households, the planned investment expenditure by firms, and planned government expenditure. We assume that firms always have spare capacity so that the total output of an economy is determined by the volume of goods and services demanded.

What determines planned investment and planned consumption expenditures? We assume that planned investment by firms depends (negatively) on the interest rate they would forego if they used their funds in this way, or the interest rate they would have to pay if they had to borrow funds. This can be justified by assuming that firms invest in those items—plant, equipment, advertising campaigns, etc.—that give a positive net present value (NPV). If the interest rate they forego rises, the NPV of the planned expenditure on plant decreases, and if the rate rises sufficiently the NPV will become negative and the firm would not make the investment. Second, we assume that planned consumption expenditure depends positively on income but negatively on the interest rate consumers have to pay if they borrow funds to spend, for example, on consumer durables or houses. Again, the higher the rate the lower the present value of future satisfaction in each year. Also, the more expensive credit is, the less credit people will wish to take, as we explained in section 3.3. Third, we assume that planned government expenditure is determined by political considerations and is unrelated to output or interest rate. We can represent these relationships by the following equations:

$$I_p = a - br, \tag{3.6}$$

$$C_p = c - dr + eY, \tag{3.7}$$

$$G_p = \overline{G}, \tag{3.8}$$

$$\text{AD} = I_p + C_p + \overline{G}_p, \tag{3.9}$$

where I_p denotes planned investment, C_p denotes planned consumption expenditure, G_p denotes planned government expenditure, Y denotes income and value of output, r denotes the interest rate, AD denotes aggregate demand, and a, b, c, d, e, and $\overline{G}$ all denote constants. In equilibrium, aggregate demand equals actual output:

$$\text{AD} = Y. \tag{3.10}$$

Therefore, we can deduce that

$$Y = \frac{\left(a + c + \overline{G}\right) - (b + d)r}{(1 - e)}, \tag{3.11}$$

which can be written as

$$Y = \theta - \phi r, \tag{3.12}$$

where

$$\theta = \frac{\left(a + c + \overline{G}\right)}{(1 - e)}$$

and

$$\phi = \frac{(b + d)}{(1 - e)},$$

which is known as the IS (investment and saving) curve.

We now consider the market for money. There are several measures of the supply of money. M0 consists of notes and coins in the banking system as reserves held by banks, plus cash outside the banks, such as that held by individuals and firms. M1 consists of cash outside the banks plus sight deposits. Sight deposits are deposits against someone's name, which they can withdraw whenever they wish without any warning, for example, deposits in checking accounts. At this stage we suppose that the supply of money is determined by the central bank, such as the Federal Reserve.

Returning to the demand for money, we suppose that there are only two types of asset in this economy: money and bonds. The term *bonds* is generic and refers to all assets that may give a rate of return: government securities, houses, debt that you are owed, and so on. Given the wealth that each person has, he will keep some of it as bonds and some as money. People, firms, and organizations demand a certain amount of money for several reasons. First, salaries are not perfectly synchronized with expenditures, so each of us wishes to keep some of our wealth as money rather than bonds to enable us to buy things between the times we are paid. The more things we buy, the more money we need. Hence economists assume that the demand for money for this, the transactions motive for holding it, is positively related to output in the economy, Y. But since keeping cash and noninterest bearing bank accounts (checking accounts) gives no (or minimal) interest, whereas bonds do give interest, the higher the interest rate on bonds, the less money we wish to hold to facilitate transactions. A second reason people hold money is to enable them to make purchases they were not planning to make: to repair the car that broke down, to buy an item on sale. Again the amount of money demanded for this precautionary motive is assumed to be related to output. Third, people hold money as an asset. For example, people may put money into a low-risk bank savings account because holding a bond involves a risk that the return will be less than expected. Holding a mix of bonds and money enables each of us to reduce the risk we incur compared with that incurred if we held only bonds. The greater the return on bonds, relative to that in money accounts, the less money people will wish to hold to reduce their risk. In summary we can say that the demand for money is positively related to output and negatively related to the interest rate.

The equilibrium level of the interest rate is determined by the demand and supply of money. Given the level of output, Y, Figure 3.10 plots both the amount of money demanded and the amount supplied at each interest rate. The amount demanded equals the amount supplied at interest rate r_e. To see how the interest rate would adjust if it were not r_e, suppose the interest rate is r_0. At this rate, the supply of money is greater than the amount demanded. Thus, since the alternative to holding money is to hold bonds, at r_0 the demand for bonds must exceed its supply. The price of bonds will rise and the interest rate on them fall.[1] This will continue until there is no more excess demand for bonds and so no more excess supply of money. Thus the interest rate in Figure 3.10 will fall until r_e.

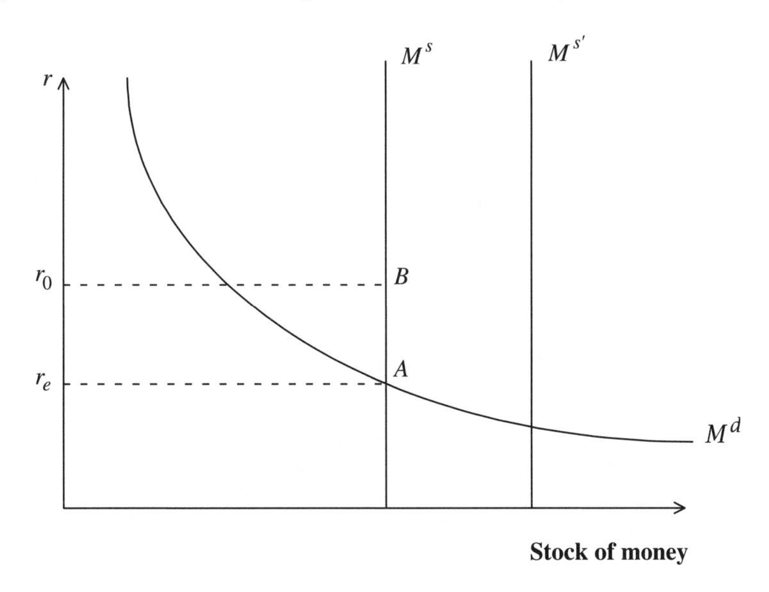

Figure 3.10. *Determination of the interest rate.*

Now suppose that the number of transactions in the economy increased because the level of output increased. The demand for money at each interest rate would increase to facilitate the additional transactions. Thus the demand for money curve would shift to the right and the equilibrium level of interest rate would increase. Thus we can represent the relationship between the level of output, Y, and the interest rate, r, given the money supply as

$$Y = \delta + \lambda r. \tag{3.13}$$

This is known as the LM curve (where LM stands for liquidity preference—money).

We now put the two markets in our economy together, as in Figure 3.11. For the economy to be in equilibrium there must be equilibrium in both the goods and services market and in the money market. The same interest rate and output must exist in both markets. Therefore, the economy is in equilibrium when both curves cross; that is, they are simultaneous equations. The equilibrium values in the figure are r_e and Y_e. In the next sections we explain how changes in monetary variables are related to changes in real variables like output in the economy: the transmission mechanism. The role of credit in a macroeconomy is explained by its role in this mechanism.

[1] For a bond with given payments of R_t in year t, the amount the marginal buyer is just willing to pay for it, its price P, is the present value of these receipts over the life (T years) of the bond:

$$P = \sum_{t=0}^{T} \frac{R_t}{(1+r)^t},$$

where r is the interest rate. Thus r and P are inversely related.

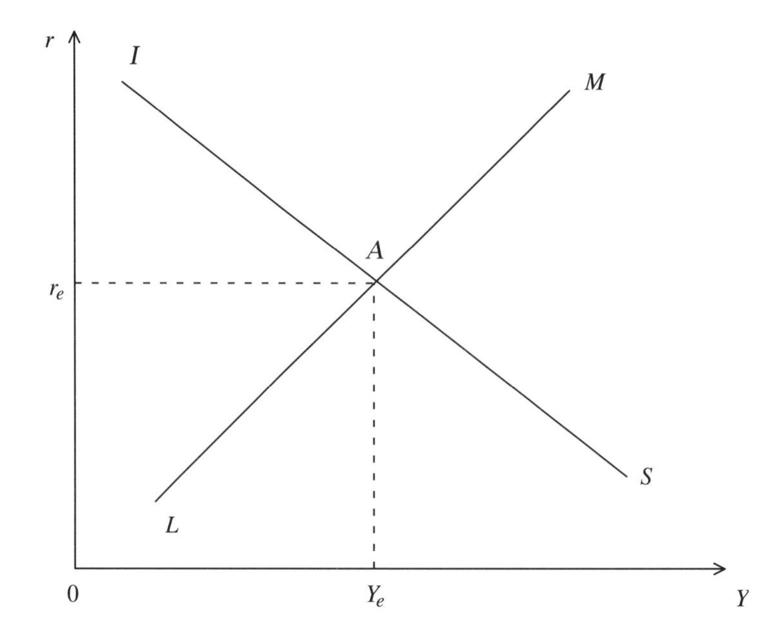

Figure 3.11. *Equilibrium in the goods and money markets.*

3.4.2 The money channel

The money channel of the transmission mechanism, sometimes called the interest rate channel, is the standard Keynesian mechanism explained in most macroeconomics textbooks. Suppose the central bank, the Fed, buys bonds and so pays cash to the private sector—citizens and firms. The private sector will deposit this cash into commercial banks. Each bank's balance sheet consists of assets and liabilities, each of which must sum to the same value. The banks' assets have increased because the banks have received the cash, and their liabilities have increased because the banks have created deposits against the names of those who paid the cash in. If banks choose a cash:deposits ratio of, say, α (such as 3%), they will create deposits of $\frac{\text{cash}}{\alpha}$. This will be done by lending to people who wish to borrow. The borrowers will spend their loan and the recipients of these expenditures will pay their receipts back into banks. In turn, the banks will create deposits against the depositors' names. These receipts will be lent again, and deposited again, and so on. The process will stop when the chosen cash:deposit ratio is reached, but with more cash now in the banks (the original amount plus the additional deposit that came from the Fed). Thus on the assets side of banks' balance sheets is the original cash paid in, plus the value of a number of loans. On the liabilities side are total deposits created, as money is paid in at each round of the process.

Remember that M1 is cash outside the banks plus the sight deposits created. After the cash was paid in it was repeatedly lent again. When the process stopped, all the cash was inside banks and new sight deposits equal to $\frac{\text{additional cash}}{\alpha}$ had been created. Thus M1 had increased, and by considerably more than the additional cash.

When the deposits increased, at the original interest rate, the supply of money exceeded the demand for it and so the demand for bonds exceeded their supply. The price of bonds rose and, by inference, their interest rate decreased. The decrease in interest rate would increase both planned investment and planned consumption. In the former case, the NPVs of additions to plant and machinery would increase so firms wished to invest more. In the latter case, the PV of each consumer's expected future income and their current wealth would increase. Thus each consumer would choose to increase his or her expenditure. If their desired expenditure exceeds their current income, they may reduce their stock of savings or borrow. Therefore, aggregate demand increases and output rises.

Notice that the creation of loans is the result of an increase in the demand for them by firms and households. The increase in demand is due to the fall in interest rates due to the creation of additional deposits by banks.

3.4.3 The credit channel

According to Bernanke and Blinder (1988), the credit channel acts to exaggerate the effect of the money supply on output. There are at least two ways in which the mechanism works: through the balance sheet channel and through the bank lending channel. We consider the balance sheet channel first.

Borrowers (who act as agents) are more aware than lenders (who act as principals) of their ability to repay. To reduce the risk of nonpayment, lenders incur costs in assessing the risks of lending and also of monitoring borrowers. To compensate for the risk, banks may require a higher return than if there was no risk. These costs would not be borne by a firm (or consumer) that used its own funds to finance investment. We call these higher costs moral hazard costs. Therefore, if a firm uses external finance, it would expect to pay a higher interest rate than if it used retained profits. Bernanke and Blinder argue that the difference between these rates, the "external finance premium," is itself positively related to the level of the rates. Thus a change in, say, banks' reserves effects a change in interest rates (as in the interest rate channel) but also in the premium.

In the case of the balance sheet channel, a reduction in banks' reserves makes the financial position of firms worse, the moral hazard costs rise, and so does the premium charged by banks. At the original interest rate, banks supply fewer loans and, provided firms cannot gain loans from others at a lower interest rate than from the banks, firms will invest less than if the channel did not operate. The financial position of firms is made worse in several ways. When banks' reserves fall and the interest rate rises, the interest that firms have to pay on outstanding debt increases. In addition, the value of collateral that a borrower may have offered when he took out the loan decreases. (Think of the value as being the expected present value of the future profits the collateral would generate.) Therefore, the risk for the bank associated with outstanding loans and any new loans would be higher and the interest rate premium charged would be higher. Thus banks would supply fewer loans and firms would plan to invest less. Aggregate demand would fall and so therefore would output.

Similar arguments apply to households' expenditure on durables and housing. For example, when interest rates rise, the interest rate that consumers have to pay on outstanding debt increases and the value of collateral falls. The costs of moral hazard rise, and at the original rate, banks supply fewer loans to consumers. Therefore, the consumers who cannot perfectly substitute loans from other sources reduce their investment expenditure and output decreases.

In the case of the bank lending channel, a reduction in banks' reserves reduces the volume of bank loans directly. This is because, given the reserve ratio α, $\frac{\text{cash reserves}}{\text{deposits}}$, chosen by commercial banks, a reduction in reserves causes banks to reduce deposits. Since α is less than 1, the reduction in reserves is less than the reduction in deposits. Since the total liabilities (that is, deposits) must equal total assets for commercial banks, the banks must reduce the nonreserve elements of their assets also. These nonreserve elements of their assets consist of loans and securities. If banks wish to preserve the ratio of loans to securities (to maintain the same risk-return mix), then they will reduce their supply of loans. The high-risk firms will then have to borrow from nonbank lenders. Banks are assumed to be the most efficient lenders at assessing the riskiness of firms. Therefore, the high-risk firms will have to apply to lenders who will incur more risk assessment costs than banks, and they will have to incur higher search costs than if banks had lent to them. These additional costs increase the financing premium that firms would have to pay and so reduce the NPV of investments that firms wish to make. Thus investment and output would fall. The same process would reduce the supply of loans to consumers for durables and housing purchases.

We now explain Bernanke and Blinder's formal representation of the bank lending channel mechanism. Their model predicts how changes in the money supply or demand for credit, as well as changes in bank reserves, affect output. First, assume that instead of bonds and money being the only assets in an economy, as in the conventional LM model, borrowers and lenders can spread their wealth among bonds, money, and securities. Bernanke and Blinder develop models for the money market (the standard LM model as explained above), the credit market, and the goods market. They assume that the demand for loans depends negatively on their interest rate (they are less expensive) and positively on the return on bonds (bonds are more attractive to buy and can be bought using credit). Demand for credit is also positively related to the number of transactions made to enable purchases, as proxied by output. Now consider the supply of loans. Bernanke and Blinder argue that this depends positively on the loan interest rate because, everything else constant, the return earned by a bank by making a loan is greater, relative to all other uses of its funds, when this rate is higher. The supply is also negatively related to the interest rate on bonds because, holding everything else constant, the lower the rate on bonds, the greater the relative return from making loans. Third, the supply depends positively on the level of deposits above the level banks retain to be able to return cash to customers who have deposits. These relationships are shown in Figure 3.12. Holding everything constant except C_s, C_d, and i_1, the equilibrium will occur at A.

Now consider the market for money. The demand for deposits depends positively on the number of transactions, proxied by output, and negatively on the return on bonds for reasons given earlier. The supply of deposits equals the money multiplier (which is $\frac{1}{\alpha}$; see above) times reserves R. The money multiplier is assumed to depend positively on the return on bonds because the higher this return, the lower the proportion of deposits banks wish to keep in reserve (they prefer to buy bonds) and the multiplier is the inverse of this fraction. The demand and supply of money is represented by Figure 3.13. Using the same explanation as earlier (see section 3.4.1), we can derive the LM curve: combinations of i_b and output at which the demand for deposits equals supply.

Finally, Bernanke and Blinder assume the traditional IS model of the goods market, explained earlier. Thus equilibrium output in this market depends negatively on the interest rate on bonds (a rise in this rate reduces the NPV of investment) and negatively on the interest rate on loans (the greater the interest rate on loans, the fewer loans are demanded and firms invest less and consumer demand decreases also). As aggregate demand decreases, so does output. This relationship is shown in Figure 3.14.

To explain the equilibrium, remember that we require equilibrium in the money market giving the LM curve, which is plotted in Figure 3.15. We require equilibrium in the credit market giving the interest rate on loans. We also require equilibrium in the goods market given the interest rate on loans; this is represented by the line in Figure 3.14, which corresponds to this rate. Call this line XX. Thus conditions represented by LM and XX must hold, so the overall equilibrium levels of i_b and Y must be where the two curves intersect.

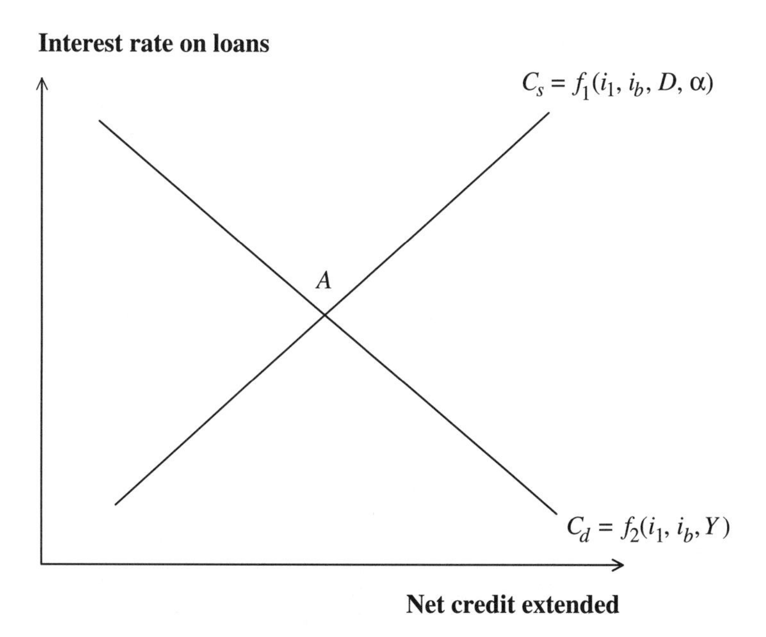

Figure 3.12. *The demand and supply of loans.*

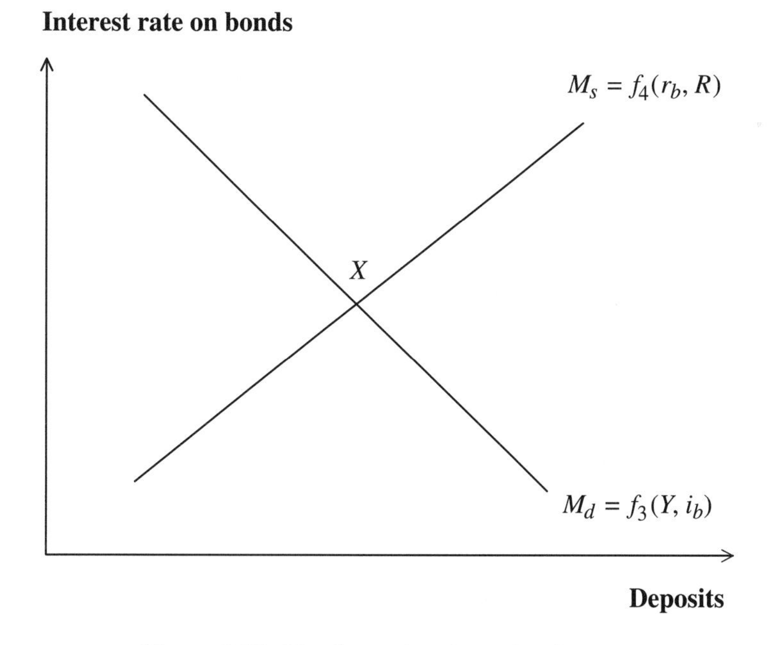

Figure 3.13. *The demand and supply of money.*

This model can be used to show the effects of an increase in the supply of credit at every loan interest rate. This might be the result of banks decreasing their assessment of the level of risk associated with loans. This would shift the supply curve in Figure 3.12 to the right, thus reducing the loan interest rate and increasing the amount of loans. Consumer demand and investment demand would increase and so would output: the XX line in Figure 3.14

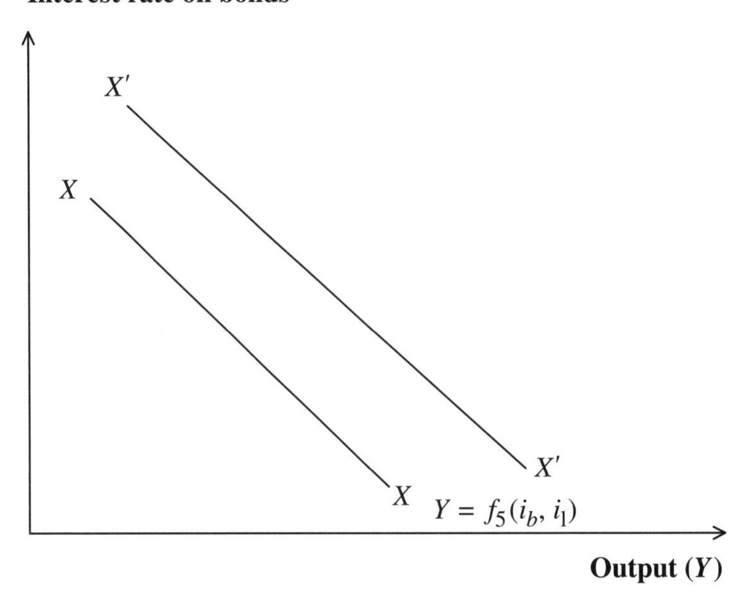

Figure 3.14. *Augmented IS curve.*

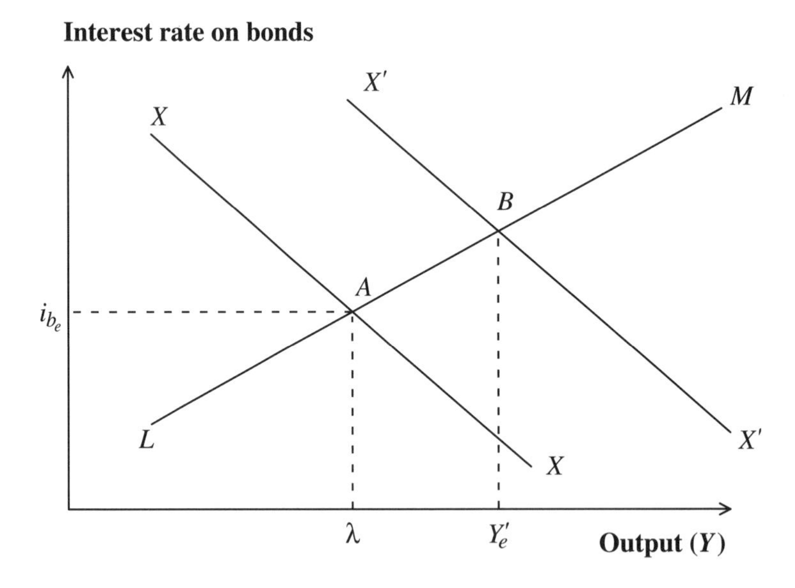

Figure 3.15. *Overall equilibrium.*

shifts to the right to $X'X'$. In Figure 3.15, the new $X'X'$ line is shown, and it can be seen that output would rise from Y_e to Y_e'. Bernanke and Blinder comment that exactly this pattern of adjustments occurred (in reverse) when, in the 1920s, banks believed that loans were more risky. The supply of credit decreased, the loan rate rose, the XX curve moved to the left, and output fell: the Great Depression.

3.4.4 Empirical evidence

A large number of studies have tried to discover whether the credit channel operates as well as the money channel. The conclusions of the studies differ and this is an active area of research. These studies often estimated systems of simultaneous time series regression equations. In each case, a variable was related to lagged values of itself and lagged values of other variables (typically up to six lagged terms each). These are known as vector autoregression (VAR) models and, with some simplification, have the form

$$\mathbf{y_t} = \mathbf{a} + \beta' \mathbf{y_{t-1}} + \mathbf{e_t},$$

where $\mathbf{y_t}$ is a column vector of variables, β is a matrix of coefficients, and $\mathbf{e_t}$ is a column vector of residuals. Each set of equations allows the researcher to examine the effects of changes in $\mathbf{e_t}$ for one variable on the future values of all of the other variables in $\mathbf{y_t}$.

Using this method, an early study by King (1986) examined the relationships between the growth rates of GNP, deposits, and loans to different sectors of the economy for the U.S. between 1950 and 1979. King found that deposits explained a larger proportion of the variations in future output than did loans, which he interpreted as indicating support for the money channel and little support for the credit channel. However, Bernanke (1986) argued that King imposed ad hoc constraints on the values of the coefficients in the matrix of βs in his equations. When constraints that are economically plausible are applied, both channels appear to be equally important.

The methodology of King and of Bernanke (1986) was criticized by Bernanke and Blinder (1992) as being sensitive to the choice of imposed constraints and specifications of the VARs. Instead, Bernanke and Blinder used a direct measure of monetary policy: the federal funds interest rate. Having estimated the appropriate VARs, they examined the time series paths of bank deposits, bank loans and bank securities, and unemployment to a shock (that is, not explained by other variables in the model) in the funds rate. The funds rate is directly controlled by the Federal Reserve and an increase in it can be seen in Figure 3.10, to correspond to a decrease in the money supply and so a tightening of monetary policy. Bernanke and Blinder's results show that an increase in the funds rate leads to an immediate reduction in both total bank deposits and total bank assets (both values are equal, being on opposite sides of the bank balance sheet). But total bank assets consist of loans and securities, and for the first six months after the shock, it is securities, not loans, that are reduced. After six months, banks' holdings of securities began to rise and they reduce their loans. This continued, and by two years after the shock, banks' holdings of securities had returned to their original levels and loans had decreased by the same amount as deposits.

Similar results were found in other studies. For example, using U.S. data, Kashyap et al. (1993) observed that after each of several specific dates on which monetary contraction occurred, commercial paper held by banks increased and bank loans did not decrease until two years later. Bachetta and Ballabriga (1995), using data for each of 14 European countries, found that following a shock rise in the federal funds rate, the decrease in deposits was initially greater than that in loans (that is, banks initially reduced securities), but after several months, the reverse happened.

However, authors disagree over the interpretation. Bernanke and Blinder (1992) argued that the initial stability of loans occurs because, unlike deposits, they involve a contract and cannot be easily repaid quickly. In the long run, banks can reallocate their portfolio between loans and securities. Therefore, the initial constancy of loans is consistent with a credit channel. Alternatively, Romer and Romer (1990) argued that if a monetary contraction involves banks losing reserves and so demand deposits, banks may respond by issuing deposits (certificates of deposit) whose quantity is not linked to reserves rather than reducing total liabilities. Since total deposits fall hardly at all, neither do total assets and thus loans. Bernanke and Gertler (1995) accepted that this may be valid for the U.S. after 1980, when legal reserve requirements for banks were removed.

A second test of money versus the credit channel is to examine the change in the fraction $\frac{\text{bank loans}}{\text{commercial paper plus bank loans}}$ after monetary tightening. Kayshap et al. (1993) argued that a decrease in this ratio is consistent with a reduction in the supply of debt rather than a reduction in the demand for debt because if there were a decrease in demand, commercial paper and bank loans would decrease by the same proportion and the ratio would remain constant. Kayshap, using U.S. business loan data for 1963–1989, found that the ratio did indeed decrease after monetary tightenings. Ludvigson (1998) found the same results for consumer loans for automobiles between 1965 and 1994.

A third test of the credit channel is to see if there is support for the assumption of the theory that those firms and consumers whose supply of bank debt is reduced are unable to gain debt from other sources. If they can raise debt from nonbank sources, then a reduction in bank debt will not affect aggregate demand and so will not affect output. Kayshap et al. tested this by seeing whether the above ratio significantly explained firms' investment and inventories. They found that it did. A similar result was found by Ludvigson. In addition, Gertler and Gilchrist (1993) argued that small firms were less able to substitute paper for bank credit because of the higher-risk assessment costs to issuers of paper than to banks. Thus the credit channel would predict that an increase in the funds rate would cause small firms to have not only less bank credit but also less of other types of credit and so would cut back investment more than large firms. This is indeed what they found.

Bernanke and Blinder (1992) found that following monetary contraction, eventually unemployment rises at the same time as loans decrease. Romer and Romer (1990) found a consistent result for output. However, this correlation on its own is consistent with both the supply of loans causing a fall in output and with the fall in output reducing the demand for loans: the money channel and the bank lending version of the credit channel. It is also consistent with the balance sheet version of the credit channel, whereby the rise in interest rates makes potential borrowers more risky and so many do not receive credit. Ludvigson (1998) tried to identify which credit channel was operating. He argued that the balance sheet channel implies that given a monetary contraction, riskier borrowers would face a greater reduction in credit supply than less risky borrowers. His data did not support this prediction. Increases in the funds rate led to larger proportionate increases in commercial paper, which is given to riskier borrowers, than in bank credit, which is given to less risky borrowers. Ludvigson argued that, because there is evidence in favor of a credit channel, if the balance sheet channel is not operating, then the bank lending channel must be the mechanism in operation.

A further interesting relationship was found by Dale and Haldane (1995). Using U.K. data for 1974–1992, they found that an increase in interest rates results in an increase in business borrowings in the short term with business lending becoming lower than its original level a year later, after a decrease in output. For the personal sector, a positive shock to interest rates results in an immediate decrease in borrowing, and two years later, the level of deposits becomes less than their original value at the time of the shock. The decrease in lending to the personal sector preceded the decrease in demand, which is consistent with a credit channel mechanism. But the credit channel did not operate for businesses. Furthermore, they found the significance of the credit channel to be small.

3.5 Default behavior

In this section, we consider how default rates vary with the output of an economy. Figure 3.16 shows the GDP (at constant prices) and the default rate on all consumer loans for the U.S. Both series have been detrended so we are observing fluctuations around their long-term trend values. The immediate impression is that the default rate is negatively related to output and that the timing of the peaks (troughs) in output occur, in most cases, in the same quarter as the troughs (peaks) in the default rate.

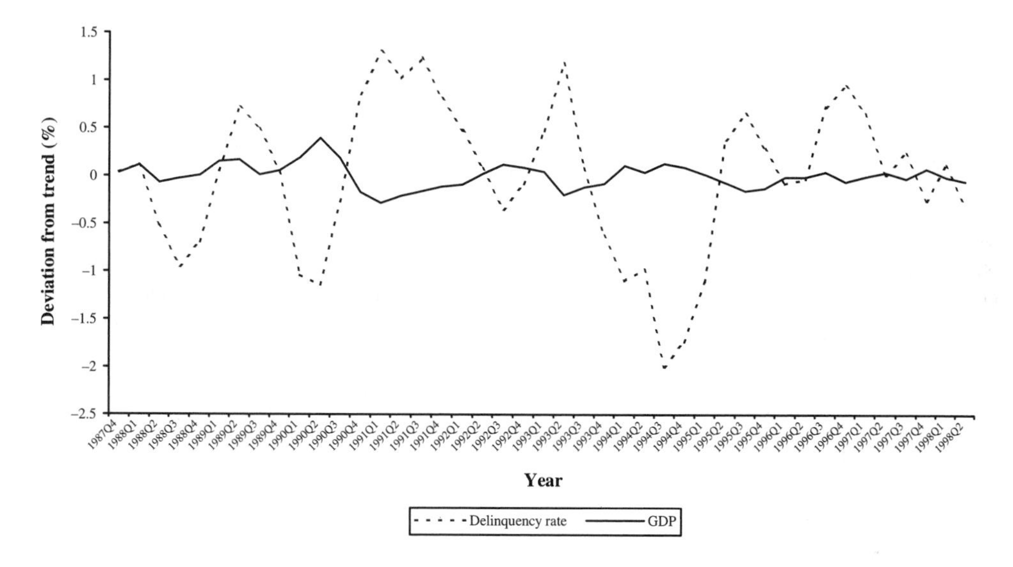

Figure 3.16. *U.S. cyclical behavior of delinquency and business cycles.*

There is no corresponding data for the U.K. The most aggregated data relates to mortgage arrears and is collected by the Council of Mortgage Lenders. Figure 3.17 shows the proportion of mortgages between 6 months and 12 months in arrears and more than 12 months in arrears as well as the number of homes repossessed and GDP. Although the frequency of the data is annual, the same pattern as was observed for the U.S. for all consumer loans is evident.

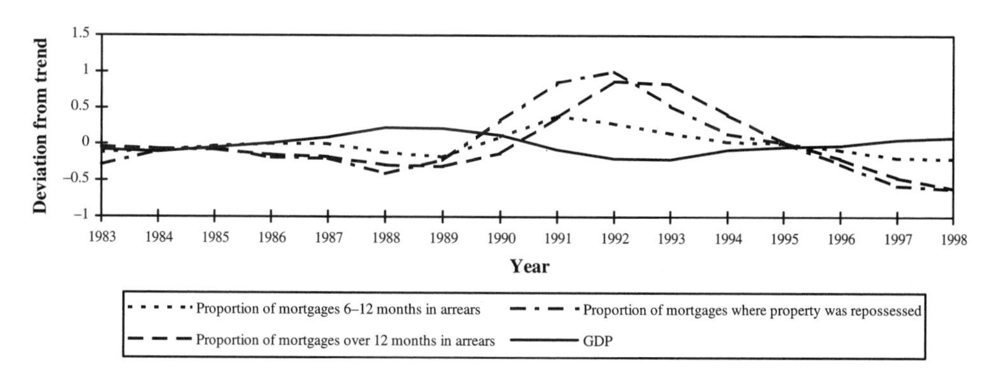

Figure 3.17. *U.K. cyclical behavior of mortgage arrears. (Data from the Council of Mortgage Lenders.)*

Little work has been done to explain patterns of default over time. Sullivan (1987) made the most thorough study and we discuss her results here. Sullivan argued that the default rate varies over time because of changes in the ability of borrowers to repay and because of changes in lenders' credit granting and collection policies. She argued that a borrower will default when the costs of maintaining the loan payments are greater than the costs of not doing so. The costs of the payments are their present value. The costs of not keeping up to date include the repayment costs, λ, which the lender may be able to force onto the borrower, the effect of default on the ability of the borrower to gain credit in the future, and its price. She argued that the costs of maintaining the loan will exceed the costs of default when a borrower's ability to repay deteriorates due to, for example, a decrease in real income. Using aggregate U.S. time series data for the period 1975–1986, she found that the delinquency rate (delinquent for at least 30 days) on consumer installment loans at commercial banks is positively related to total consumer debt outstanding as a proportion of total PDI, and the share that commercial banks had of the consumer loan market, and it is negatively related to the growth rate of this debt. Other indicators of consumers' ability to repay their loans were not significantly related to the delinquency rate. She also found that the factors affecting the default rate varied between types of lenders. For example, for indirect auto loans borrowed from auto finance companies, she found that the delinquency rate was higher in periods of high unemployment and in periods when the debt-to-income ratio was high. She also found that delinquencies increased in periods of high inflation but decreased as the share of the loan market taken by such finance companies increased. Alternatively, for indirect loans borrowed from commercial banks, unemployment and inflation had no effect. The higher the debt-to-income ratio and the market share of banks, the higher the delinquency rate. These differences may reflect the different types of borrower from each sector.

Chapter 4

Statistical Methods for Building Credit Scorecards

4.1 Introduction

When credit scoring was first developed in the 1950s and 1960s, the only methods used were statistical discrimination and classification methods. Even today statistical methods are by far the most common methods for building credit scorecards. Their advantage is that they allow one to use knowledge of the properties of sample estimators and the tools of confidence intervals and hypothesis testing in the credit-scoring context. Thus one is able to comment on the likely discriminating power of the scorecard built and the relative importance of the different characteristics (variables) that make up the scorecard in its discrimination. These statistical techniques then allow one to identify and remove unimportant characteristics and ensure that all the important characteristics remain in the scorecard. Thus one is able to build lean and mean scorecards. One can also use this information when one is looking at what changes might be needed in the questions asked of prospective borrowers.

Although statistical methods were the first to be used to build scoring systems and they remain the most important methods, there have been changes in the methods used during the intervening 40 years. Initially, the methods were based around the discrimination methods suggested by Fisher (1936) for general classification problems. This led to a linear scorecard based on the Fisher linear discriminant function. The assumptions that were needed to ensure that this was the best way to discriminate between good and bad potential customers were extremely restrictive and clearly did not hold in practice, although the scorecards produced were very robust. The Fisher approach could be viewed as a form of linear regression, and this led to an investigation of other forms of regression that had less restrictive assumptions to guarantee their optimality and still led to linear scoring rules. By far the most successful of these is logistic regression, which has taken over from the linear regression–discriminant analysis approach as the most common statistical method. Another approach that has found favor over the last 20 years is the classification tree or recursive partitioning approach. With this, one splits the set of applicants into a number of different subgroups depending on their attributes and then classifies each subgroup as satisfactory or unsatisfactory. Although this does not give a weight to each of the attributes as the linear scorecards does, the result is the same—a method for deciding whether a new applicant will be classified as satisfactory or unsatisfactory. All these methods have been used in practice to devise scorecards for commercial organizations, but there is still plenty of experimentation in using statistical methods. The final statistical method discussed in this chapter is one of these experimental methods—the nonparametric approach based on nearest neighbors.

In this chapter, we review each of these methods and the statistical background that underpins the methods. We begin with discriminant analysis. The next three sections look at how the linear discriminant function is arrived at as a classifier by three different approaches to the problem. The first approach is decision making, where one looks to find the rule that minimizes the expected cost in deciding whether to accept a new customer. The second approach is that which motivated Fisher's original work—namely, seeking to find a function that best differentiates between the two groups of satisfactory (good) and unsatisfactory (bad) customers. The third approach is to consider a regression equation that tries to find the best estimate of the likelihood of a client being good. Each of these approaches arrives at the same linear scorecard method for determining good from bad clients. Sections 4.5 and 4.6 build on the regression approach by looking at regression approaches where the dependent variable is some nonlinear function of the probability of a client being good. One such function leads to logistic regression; other functions lead to probit and tobit analysis. Section 4.7 deals with the classification tree approach and section 4.8 with the nearest-neighbor approach, which are quite different statistical formulations of the original classification problem. Finally, there is a discussion of how some of the methods could be extended to allow for credit-scoring systems where one wants to classify clients into more than the two categories of good and bad.

4.2 Discriminant analysis: Decision theory approach

The credit-granting process leads to a choice between two actions—give this new applicant credit or refuse this applicant credit. Credit scoring tries to assist this decision by finding what would have been the best rule to apply on a sample of previous applicants. The advantage of doing this is that we know how these applicants subsequently performed. If there are only two actions possible—accept or reject—then there is no advantage in classifying this performance into more than two classes—good and bad. Good is any performance that is acceptable to the lending organization, while bad is performance that means the lender wishes he had rejected the applicant. This is looked at more closely in section 8.3. In some organizations, bad performance is taken as missing a given number of consecutive payments, while in others it is the total number of missed payments that matters.

There is an inherent bias in this approach in that the sample is of previous applicants who were granted credit and there is no information on the performance of those applicants who were rejected in the past. Thus the sample is representative of those accepted in the past and is not representative of those who applied in the past. How organizations deal with this bias is looked at in section 8.9.

There is an argument that there are more than the two choices of accept and reject in the credit granting process. For example, one might decide to request further information on the applicant or that this applicant should be considered by a credit analyst for a manual decision. However, these variations are more to do with the ways a lender organizes decision making, and the final decision in every case will be accept or reject. It is not worthwhile to try to classify which applicants will be in a particular group when the decision on whether they are in that group is wholly that of the lender. Thus even in these multiple action processes it is sensible to classify the applicants into only two groups—good and bad—since the final decision will result in one of two actions.

Let $X = (X_1, X_2, \ldots, X_p)$ be the set of p random variables that describe the information available on an applicant for credit both from the application form and through a credit reference bureau check. Of course, nowadays there may not be a physical form, as the details could be captured on a screen via the Internet or they could be written down by a member of the lender staff in a telephone call. We use the words *variable* and *characteristic* interchangeably to describe a typical X_i: the former when we want to emphasize the random nature of this information between applicants and the latter when we want to recall what sort of information it is. The actual value of the variables for a particular applicant is denoted $\mathbf{x} = (x_1, x_2, \ldots, x_p)$. In credit-scoring terminology, the different possible values or answers, x_i, to variable X_i are called the attributes of that characteristic. So if a typical characteristic is the applicant's residential status, then its attributes might be owner, rent unfurnished, rent furnished, living with parents, or other. Different lenders may have different groups of attributes of the same characteristic. Thus another lender may decide to classify residential status into owner with no mortgage, owner with mortgage, renting unfurnished property, renting furnished property, subletting property, mobile home, accommodation provided, living with parents, living with other than parents, or other status. It is not uncommon for attribute and characteristic to be mixed up. An easy way to remember which is which is that the Attribute is the Answer to the application form question and the Characteristic is the Cw(qu)estion that was asked.

Returning to the decision that has to be made by the lending organization, suppose A is the set of all possible values that the application variables $\mathbf{X} = (X_1, X_2, \ldots, X_p)$ can take, i.e., all the different ways the application form can be answered. The objective is to find a rule that splits the set A into two subsets A_G and A_B so that classifying applicants whose answers are in A_G as "goods" and accepting them while classifying those whose answers are in A_B as "bads" and rejecting them minimizes the expected cost to the lender. The two types of cost correspond to the two types of error that can be made in this decision. One can classify someone who is good as a bad and hence reject the person. In that case the potential profit from that applicant is lost. Assume for now that the expected profit is the same for each applicant and is L. The second error is to classify a bad as a good and so accept the applicant. In that case a debt will be incurred when the customer defaults on the loan. We assume that the expected debt incurred is the same for all customers and is set at D.

Assume that p_G is the proportion of applicants who are goods.

Similarly, let p_B be the proportion of applicants who are bads.

Assume the application characteristics have a finite number of discrete attributes so that A is finite and there are only a finite number of different attributes $\mathbf{x}$. This is like saying there is only a finite number of ways of filling in the application form. Let $p(\mathbf{x}|G)$ be the probability that a good applicant will have attributes $\mathbf{x}$. This is a conditional probability and represents the ratio

$$p(\mathbf{x}|G) = \frac{\text{Prob(applicant is good and has attributes } \mathbf{x})}{\text{Prob(applicant is good)}}. \tag{4.1}$$

Similarly, define $p(x|B)$ to be the probability that a bad applicant will have attributes $\mathbf{x}$.

If $q(G|\mathbf{x})$ is defined to be the probability that someone with application attributes $\mathbf{x}$ is a good, then

$$q(G|\mathbf{x}) = \frac{\text{Prob(applicant has attributes } \mathbf{x} \text{ and is good)}}{\text{Prob(applicant has attributes } \mathbf{x})}, \tag{4.2}$$

and if $p(x) = $ Prob(applicant has attributes $\mathbf{x}$), then (4.1) and (4.2) can be rearranged to read

$$\text{Prob(applicant has attributes } \mathbf{x} \text{ and is good)} = q(G|\mathbf{x})p(\mathbf{x}) = p(\mathbf{x}|G)p_G. \tag{4.3}$$

Hence we arrive at Bayes's theorem, which says

$$q(G|\mathbf{x}) = \frac{p(\mathbf{x}|G)p_G}{p(\mathbf{x})}. \tag{4.4}$$

A similar result holds for $q(B|\mathbf{x})$, the probability that someone with application attributes $\mathbf{x}$ is a bad, namely,

$$q(B|\mathbf{x}) = \frac{p(\mathbf{x}|B)p_B}{p(\mathbf{x})}. \tag{4.5}$$

Note that from (4.4) and (4.5), it follows that

$$\frac{q(G|\mathbf{x})}{q(B|\mathbf{x})} = \frac{p(\mathbf{x}|G)p_G}{p(\mathbf{x}|B)p_B}. \tag{4.6}$$

The expected cost per applicant if we accept applicants with attributes in A_G and reject those with attributes in A_B is

$$L\sum_{\mathbf{x}\in A_B} p(\mathbf{x}|G)p_G + D\sum_{\mathbf{x}\in A_G} p(\mathbf{x}|B)p_B = L\sum_{\mathbf{x}\in A_B} q(G|\mathbf{x})p(\mathbf{x}) + D\sum_{\mathbf{x}\in A_G} q(B|\mathbf{x})p(\mathbf{x}). \tag{4.7}$$

The rule that minimizes this expected cost is straightforward. Consider what the two costs are if we categorize a particular $\mathbf{x} = (x_1, x_2, \ldots, x_p)$ into A_G or A_B. If it is put into A_G, then there is only a cost if it is a bad, in which case the expected cost is $Dp(\mathbf{x}|B)p_B$. If $\mathbf{x}$ is classified into A_B, there is only a cost if it is a good, and so the expected cost is $Lp(\mathbf{x}|G)p_G$. Thus one classifies $\mathbf{x}$ into A_G if $Dp(\mathbf{x}|B)p_B \le Lp(\mathbf{x}|G)p_G$. Thus the decision rule that minimizes the expected costs is given by

$$\begin{aligned} A_G &= \{\mathbf{x}|Dp(\mathbf{x}|B)p_B \le Lp(\mathbf{x}|G)p_G\} = \left\{\mathbf{x}|\frac{D}{L} \le \frac{p(\mathbf{x}|G)p_G}{p(\mathbf{x}|B)p_B}\right\} \\ &= \left\{\mathbf{x}|\frac{D}{L} \le \frac{q(G|\mathbf{x})}{q(B|\mathbf{x})}\right\}, \end{aligned} \tag{4.8}$$

where the last expression follows from (4.6).

One criticism of the above criterion is that it depends on the expected costs D and L, which may not be known. So instead of minimizing the expected cost, one could seek to minimize the probability of committing one type of error while keeping the probability of committing the other type of error at an agreed level. In the credit-granting context, the obvious thing to do is to minimize the level of default while keeping the percentage of applicants accepted at some agreed level. The latter requirement is equivalent to keeping the probability of rejecting good applicants at some fixed level.

Suppose one wants the percentage of applicants accepted (the acceptance rate) to be a. Then A_G must satisfy

$$\sum_{\mathbf{x}\in A_G} p(\mathbf{x}|G)p_G + \sum_{\mathbf{x}\in A_G} p(\mathbf{x}|B)p_B = a \tag{4.9}$$

while at the same time minimizing the default rate

$$\sum_{\mathbf{x}\in A_G} p(\mathbf{x}|B)p_B.$$

If we define $b(\mathbf{x}) = p(\mathbf{x}|B)p_B$ for each $x \in A$, then one wants to find the set A_G so that we can

$$\text{minimize} \sum_{\mathbf{x} \in A_G} b(\mathbf{x}) = \sum_{\mathbf{x} \in A_G} \left(\frac{b(\mathbf{x})}{p(\mathbf{x})} \right) p(\mathbf{x}) \quad \text{subject to} \sum_{\mathbf{x} \in A_G} p(\mathbf{x}) = a. \tag{4.10}$$

Using Lagrange multipliers (or common sense using the greedy principle), one can see that this must be the set of attributes $\mathbf{x}$, where $\frac{b(\mathbf{x})}{p(\mathbf{x})} \leq c$, where c is chosen so that the sum of the $p(\mathbf{x})$ that satisfy this constraint is equal to a. Hence

$$\begin{aligned} A_G &= \left\{ \mathbf{x} | \frac{b(\mathbf{x})}{p(\mathbf{x})} \leq c \right\} = \{ \mathbf{x} | q(B|\mathbf{x}) \leq c \} \\ &= \left\{ \mathbf{x} | \frac{1-c}{c} \leq \frac{p(\mathbf{x}|G)p_G}{p(\mathbf{x}|B)p_B} \right\}, \end{aligned} \tag{4.11}$$

where the second inequality follows from the definitions of $p(\mathbf{x})$ and $b(\mathbf{x})$.

Thus the form of the decision rule under this criterion is the same as (4.8), the decision rule under the expected cost criterion for some suitable choice of costs D and L.

The whole analysis could be repeated assuming the application characteristics are continuous and not discrete random variables. The only difference would be that the conditional distribution functions $p(\mathbf{x}|G)$, $p(\mathbf{x}|B)$ are replaced by conditional density functions $f(\mathbf{x}|G)$, $f(\mathbf{x}|B)$ and the summations are replaced by integrals. So the expected cost if one splits the set A into sets A_G and A_B and accepts only those in A_G becomes

$$L \int_{\mathbf{x} \in A_B} f(\mathbf{x}|G)p_G d\mathbf{x} + D \int_{\mathbf{x} \in A_G} f(\mathbf{x}|B)p_B d\mathbf{x}, \tag{4.12}$$

and the decision rule that minimizes this is the analogue of (4.8), namely,

$$A_G = \{ \mathbf{x} | Df(\mathbf{x}|B)p_B \leq Lf(\mathbf{x}|G)p_G \} = \left\{ \mathbf{x} | \frac{Dp_B}{Lp_G} \leq \frac{f(\mathbf{x}|G)}{f(\mathbf{x}|B)} \right\}. \tag{4.13}$$

4.2.1 Univariate normal case

Consider the simplest possible case where there is only one continous characteristic variable X and its distribution among the goods $f(x|G)$ is normal with mean μ_G and variance σ^2, while the distribution among the bads is normal with mean μ_B and variance σ^2. Then

$$f(x|G) = (2\pi\sigma^2)^{-\frac{1}{2}} \exp\left(\frac{-(x-\mu_G)^2}{2\sigma^2} \right),$$

and so the rule of (4.13) becomes

$$\begin{aligned} \frac{f(x|G)}{f(x|B)} &= \frac{\exp\left(\frac{-(x-\mu_G)^2}{2\sigma^2}\right)}{\exp\left(\frac{-(x-\mu_B)^2}{2\sigma^2}\right)} = \exp\left(\frac{-(x-\mu_G)^2 + (x-\mu_B)^2}{2\sigma^2} \right) \geq \frac{Dp_B}{Lp_G} \\ &\Rightarrow x(\mu_G - \mu_B) \geq \frac{\mu_G^2 - \mu_B^2}{2} + \sigma^2 \log\left(\frac{Dp_B}{Lp_G} \right). \end{aligned} \tag{4.14}$$

Thus the rule becomes "accept if the x value is large enough."

4.2.2 Multivariate normal case with common covariance

A more realistic example is when there are p characteristics (variables) in the application information and the outcomes of these among the goods and the bads both form multivariate normal distributions. Assume the means are $\boldsymbol{\mu}_G$ among the goods and $\boldsymbol{\mu}_B$ among bads with common covariance matrix $\boldsymbol{\Sigma}$. This means that $E(X_i|G) = \mu_{G,i}$, $E(X_i|B) = \mu_{B,i}$, and $E(X_iX_j|G) = E(X_iX_j|B) = \Sigma_{ij}$.

The corresponding density function in this case is

$$f(\mathbf{x}|G) = (2\pi)^{-\frac{p}{2}}(\det \Sigma)^{-\frac{1}{2}} \exp\left(\frac{-(\mathbf{x}-\boldsymbol{\mu}_G)\boldsymbol{\Sigma}^{-1}(\mathbf{x}-\boldsymbol{\mu}_G)^T}{2}\right), \tag{4.15}$$

where $(\mathbf{x}-\boldsymbol{\mu}_G)$ is a vector with 1 row and p columns and $(\mathbf{x}-\boldsymbol{\mu}_G)^{\mathrm{T}}$ is its transpose, which is the same numbers represented as a vector with p rows and 1 column. Following the calculation in (4.14), we get

$$\begin{aligned}&\frac{f(\mathbf{x}|G)}{f(\mathbf{x}|B)} \geq \frac{Dp_B}{Lp_G}\\ &\Rightarrow \mathbf{x}\cdot\boldsymbol{\Sigma}^{-1}(\boldsymbol{\mu}_G-\boldsymbol{\mu}_B)^{\mathrm{T}} \geq \frac{\boldsymbol{\mu}_G\cdot\boldsymbol{\Sigma}^{-1}\boldsymbol{\mu}_G^{\mathrm{T}} - \boldsymbol{\mu}_B\cdot\boldsymbol{\Sigma}^{-1}\boldsymbol{\mu}_B^{\mathrm{T}}}{2} + \log\left(\frac{Dp_B}{Lp_G}\right).\end{aligned} \tag{4.16}$$

The left-hand side of (4.16) is a weighted sum of the values of the variables, namely, $x_1w_1 + x_2w_2 + \cdots + x_pw_p$, while the right-hand side is a constant. Hence (4.16) leads to a linear scoring rule, which is known as the linear discriminant function.

The above example assumed that the means and covariances of the distributions were known. This is rarely the case, and it is more normal to replace them by the estimates, namely, the sample means $\mathbf{m}_G$ and $\mathbf{m}_B$ and the sample covariance matrix $\mathbf{S}$. The decision rule (4.16) then becomes

$$\mathbf{x}\cdot\mathbf{S}^{-1}(\mathbf{m}_G-\mathbf{m}_B)^T \geq \frac{\mathbf{m}_G\cdot\mathbf{S}^{-1}\mathbf{m}_G^{\mathrm{T}} - \mathbf{m}_B\cdot\mathbf{S}^{-1}\mathbf{m}_B^{\mathrm{T}}}{2} + \log\left(\frac{Dp_B}{Lp_G}\right). \tag{4.17}$$

4.2.3 Multivariate normal case with different covariance matrices

Another obvious restriction on the previous case is that the covariance matrices are the same for the population of goods and of bads. Suppose the covariance matrix on the population of the goods is $\boldsymbol{\Sigma}_G$ for the goods and $\boldsymbol{\Sigma}_B$ for the bads. In this case, (4.16) becomes

$$\begin{aligned}&\frac{f(\mathbf{x}|G)}{f(\mathbf{x}|B)} \geq \frac{Dp_B}{Lp_G}\\ &\quad\Rightarrow \exp\left\{-\frac{1}{2}((\mathbf{x}-\boldsymbol{\mu}_G)\boldsymbol{\Sigma}_G^{-1}(\mathbf{x}-\boldsymbol{\mu}_G)^{\mathrm{T}} - (\mathbf{x}-\boldsymbol{\mu}_B)\boldsymbol{\Sigma}_B^{-1}(\mathbf{x}-\boldsymbol{\mu}_B)^{\mathrm{T}})\right\} \geq \frac{Dp_B}{Lp_G}\\ &\quad\Rightarrow (\mathbf{x}(\boldsymbol{\Sigma}_G^{-1}-\boldsymbol{\Sigma}_B^{-1})\mathbf{x}^{\mathrm{T}} + 2\mathbf{x}\cdot(\boldsymbol{\Sigma}_G^{-1}\boldsymbol{\mu}_G^{\mathrm{T}} - \boldsymbol{\Sigma}_B^{-1}\boldsymbol{\mu}_B^{\mathrm{T}}) \geq (\boldsymbol{\mu}_G\boldsymbol{\Sigma}_G^{-1}\boldsymbol{\mu}_G^{\mathrm{T}} + \boldsymbol{\mu}_B\boldsymbol{\Sigma}_B^{-1}\boldsymbol{\mu}_B^{\mathrm{T}})\\ &\quad\quad + 2\log\left(\frac{Dp_B}{Lp_G}\right).\end{aligned} \tag{4.18}$$

The left-hand side here is a quadratic in the values $x_1, x_2, \ldots, x_p$. This appears to be a more general decision rule and so one would expect it to perform better than the linear rule. In practice, however, one has to estimate double the number of parameters $\boldsymbol{\Sigma}_B$ and $\boldsymbol{\Sigma}_G$. The extra uncertainty involved in these estimates makes the quadratic decision rule less robust than the linear one, and in most cases it is not worth trying to get the slightly better accuracy that may come from the quadratic rule. This is confirmed by the work of Reichert, Cho, and G. M. Wagner, who made such comparisons (Reichert et al. 1983).

4.3 Discriminant analysis: Separating the two groups

In Fisher's original work (1936), which introduced the linear discriminant function, the aim was to find the combination of variables that best separated two groups whose characteristics were available. These two groups might be different subspecies of a plant and the characteristics are the physical measurements, or they might be those who survive or succumb to some traumatic injury and the characteristics are the initial responses to various tests. In the credit-scoring context, the two groups are those classified by the lender as goods and bads and the characteristics are the application form details and the credit bureau information.

Let $Y = w_1X_1 + w_2X_2 + \cdots + w_pX_p$ be any linear combination of the characteristics $\mathbf{X} = (X_1, X_2, \ldots, X_p)$. One obvious measure of separation is how different are the mean values of Y for the two different groups of goods and bads in the sample. Thus one looks at the difference between $E(Y|G)$ and $E(Y|B)$ and chooses the weights w_i with $\sum_i w_i = 1$, which maximize this difference. However, this is a little naive because it would say the two groups in Figure 4.1 were equally far apart. What that example shows is that one should also allow for how closely each of the two groups cluster together when one is discussing their separation.

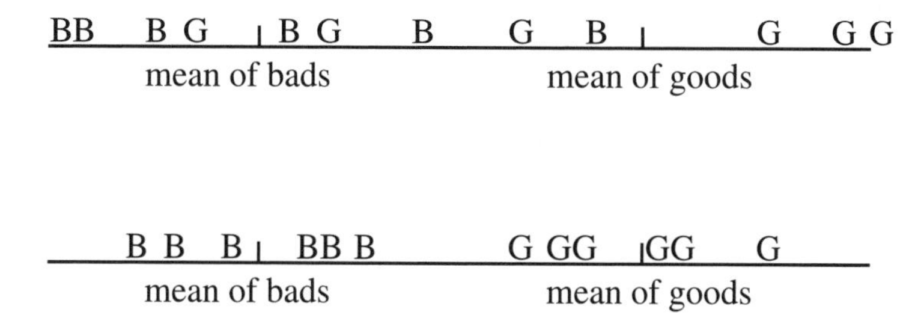

Figure 4.1. *Two examples where means of goods and bads are the same.*

Fisher suggested that if we assume the two groups have a common sample variance, then a sensible measure of separation is

$$M = \frac{\text{distance between sample means of two groups}}{(\text{sample variance of each group})^{\frac{1}{2}}}.$$

One divides by the square root of the sample variance so as to make the measure scale independent. If one changes the variable from Y to cY, then the measure M does not change.

Assume sample means are $\mathbf{m}_G$ and $\mathbf{m}_B$ for the goods and the bads, respectively, and $\mathbf{S}$ is the common sample variance. If $Y = w_1X_1 + w_2X_2 + \cdots + w_pX_p$, then the corresponding separating distance M would be

$$M = \mathbf{w}^{\mathrm{T}} \cdot \frac{\mathbf{m}_G - \mathbf{m}_B}{(\mathbf{w}^{\mathrm{T}} \cdot \mathbf{S} \cdot \mathbf{w})^{\frac{1}{2}}}. \tag{4.19}$$

This follows since $E(Y|G) = \mathbf{w}\cdot\mathbf{m}_G^{\mathbf{T}}$, $E(Y|B) = \mathbf{w}\cdot\mathbf{m}_B^{\mathbf{T}}$, and $\mathrm{Var}(Y) = \mathbf{w}\cdot\mathbf{S}\cdot\mathbf{w}^{\mathbf{T}}$. Differentiating this with respect to $\mathbf{w}$ and setting the derivative equal to 0 shows that this value M is maximized when

$$\frac{\mathbf{m}_G-\mathbf{m}_B}{(\mathbf{w}\cdot\mathbf{S}\cdot\mathbf{w}^{\mathbf{T}})^{\frac{1}{2}}} - \frac{(\mathbf{w}\cdot(\mathbf{m}_G-\mathbf{m}_B)^{\mathbf{T}})(\mathbf{S}\mathbf{w}^{\mathbf{T}})}{(\mathbf{w}\cdot\mathbf{S}\cdot\mathbf{w}^{\mathbf{T}})^{\frac{3}{2}}} = 0,$$
$$(\mathbf{m}_G-\mathbf{m}_B)(\mathbf{w}\cdot\mathbf{S}\cdot\mathbf{w}^{\mathbf{T}}) = (\mathbf{S}\mathbf{w}^{\mathbf{T}})(\mathbf{w}\cdot(\mathbf{m}_G-\mathbf{m}_B)^{\mathbf{T}}). \tag{4.20}$$

In fact, all this shows is that it is a turning point, but the fact that the second derivatives of M with respect to $\mathbf{w}$ form a positive definite matrix guarantees that it is a minimum. Since $\frac{\mathbf{w}\cdot\mathbf{S}\cdot\mathbf{w}^{\mathbf{T}}}{(\mathbf{w}\cdot(\mathbf{m}_G-\mathbf{m}_B)^{\mathbf{T}})}$ is a scalar λ, this gives

$$\mathbf{w}^T \propto (\mathbf{S}^{-1}(\mathbf{m}_G-\mathbf{m}_B)^{\mathbf{T}}). \tag{4.21}$$

Thus the weights are the same as those obtained in (4.17), although there has been no assumption of normality this time. It is just the best separator of the goods and the bads under this criterion no matter what their distribution. This result holds for all distributions because the distance measure M involves only the mean and variance of the distributions and thus gives the same results for all distributions with the same mean and variance.

Figure 4.2 shows graphically what the scorecard (4.21) seeks to do. The $\mathbf{S}^{-1}$ term standardizes the two groups so they have the same dispersion in all directions. $\mathbf{w}$ is then the direction joining the means of the goods and the bads after they have been standardized. Thus a line perpendicular to this line joins the two means. The cutoff score is then the midpoint of the distance between the means of the standardized groups.

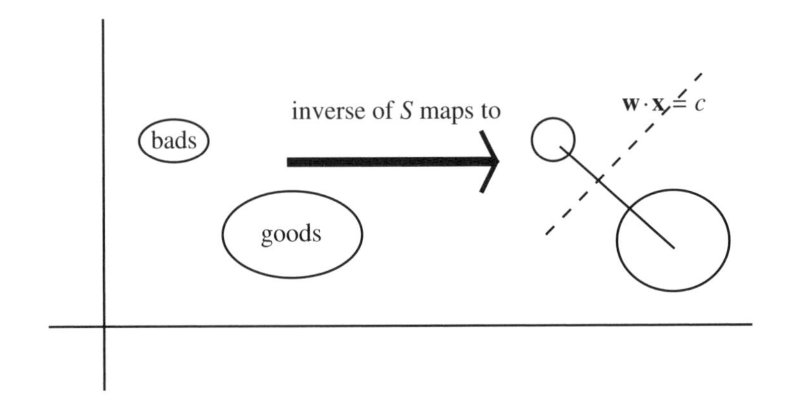

Figure 4.2. *Line corresponding to scorecard.*

4.4 Discriminant analysis: A form of linear regression

Another approach to credit scoring, which again arrives at the linear discriminant function, is linear regression. In this, one tries to find the best linear combination of the characteristics

$$w_0 + w_1X_1 + w_2X_2 + \cdots + w_pX_p = \mathbf{w}^*\cdot\mathbf{X}^{*^{\mathrm{T}}},$$

where

$$\mathbf{w}^* = (w_0, w_1, w_2, \ldots, w_p), \qquad \mathbf{X}^* = (1, X_1, X_2, \ldots, X_p),$$

which explains the probability of default. So if p_i is the probability that applicant i in the sample has defaulted, one wants to find $\mathbf{w}^*$ to best approximate

$$p_i = w_0 + x_{i1}w_1 + x_{i2}w_2 + \cdots + x_{ip}w_p \quad \text{for all } i. \tag{4.22}$$

Suppose n_G of the sample are goods; then for ease of notation, we assume these are the first n_G in the sample and so $p_i = 1$ for $i = 1, \ldots, n_G$. The remaining n_B of the sample $i = n_G + 1, \ldots, n_G + n_B$ are bad, so for them $p_i = 0$, where $n_G + n_B = n$.

In linear regression, we choose the coefficient that minimizes the mean square error between the left- and right-hand sides of (4.22). This corresponds to minimizing

$$\sum_{i=1}^{n_G}\left(1 - \sum_{j=0}^{p} w_j x_{ij}\right)^2 + \sum_{i=n_G+1}^{n_G+n_B}\left(\sum_{j=0}^{p} w_j x_{ij}\right)^2. \tag{4.23}$$

In vector notation, (4.22) can be rewritten as

$$\begin{pmatrix} 1 & \mathbf{X}_G \\ 1 & \mathbf{X}_B \end{pmatrix}\begin{pmatrix} w_0 \\ \mathbf{w} \end{pmatrix} = \begin{pmatrix} \mathbf{1}_G \\ \mathbf{0} \end{pmatrix} \quad \text{or} \quad \mathbf{Y}\mathbf{w}^{\mathbf{T}} = \mathbf{b}^{\mathbf{T}}, \tag{4.24}$$

where

$$\mathbf{Y} = \begin{pmatrix} \mathbf{1}_G & \mathbf{X}_G \\ \mathbf{1}_B & \mathbf{X}_B \end{pmatrix}$$

is an n-row $\times$ $(p + 1)$-column matrix,

$$\mathbf{X}_G = \begin{pmatrix} x_{11} & \cdots & \cdots & x_{1p} \\ x_{21} & \cdots & \cdots & x_{2p} \\ \vdots & \vdots & \vdots & \vdots \\ x_{n_G 1} & \cdots & \cdots & x_{n_G p} \end{pmatrix}$$

is an $n_G \times p$ matrix,

$$\mathbf{X}_B = \begin{pmatrix} x_{n_G+11} & \cdots & \cdots & x_{n_G+1p} \\ \vdots & \cdots & \cdots & \vdots \\ \vdots & \vdots & \vdots & \vdots \\ x_{n_G+n_B 1} & \cdots & \cdots & x_{n_G+n_B p} \end{pmatrix}$$

is an $n_B \times p$ matrix, and

$$\mathbf{b}^{\mathbf{T}} = \begin{pmatrix} \mathbf{1}_G \\ 0 \end{pmatrix},$$

where $\mathbf{1}_G$ $(\mathbf{1}_B)$ is the $1 \times n_G$ (n_B) vector with all entries 1.

Finding the coefficients of the linear regression corresponds as in (4.23) to

$$\text{Minimize } (\mathbf{Y}\mathbf{w}^{\mathbf{T}} - \mathbf{b}^{\mathbf{T}})^{\mathbf{T}}(\mathbf{Y}\mathbf{w}^{\mathbf{T}} - \mathbf{b}^{\mathbf{T}}). \tag{4.25}$$

Differentiating with respect to $\mathbf{w}$ says this is minimized when the derivative is zero; i.e.,

$$\mathbf{Y}^{\mathbf{T}}(\mathbf{Y}\mathbf{w}^{\mathbf{T}} - \mathbf{b}^{\mathbf{T}}) = \mathbf{0} \quad \text{or} \quad \mathbf{Y}^{\mathbf{T}}\mathbf{Y}\mathbf{w}^{\mathbf{T}} = \mathbf{Y}^{\mathbf{T}}\mathbf{b}^{\mathbf{T}},$$

$$\mathbf{Y}^{\mathbf{T}} \cdot \mathbf{b}^{\mathbf{T}} = \begin{pmatrix} \mathbf{1} & \mathbf{1} \\ \mathbf{X}_G & \mathbf{X}_B \end{pmatrix} \cdot \begin{pmatrix} \mathbf{1}_G \\ \mathbf{0} \end{pmatrix} = \begin{pmatrix} n_G \\ n_G \mathbf{m}_G \end{pmatrix},$$

$$\text{and} \quad \mathbf{Y}^{\mathbf{T}}\mathbf{Y} = \begin{pmatrix} \mathbf{1} & \mathbf{1} \\ \mathbf{X}_G & \mathbf{X}_B \end{pmatrix}\begin{pmatrix} 1 & \mathbf{X}_G \\ 1 & \mathbf{X}_B \end{pmatrix} = \begin{pmatrix} n & n_G\mathbf{m}_G + n_B\mathbf{m}_B \\ n_G\mathbf{m}_G^{\mathbf{T}} + n_B\mathbf{m}_B^{\mathbf{T}} & \mathbf{X}_G^{\mathbf{T}}\mathbf{X}_G + \mathbf{X}_B^{\mathbf{T}}\mathbf{X}_B \end{pmatrix}. \tag{4.26}$$

If for explanatory purposes we denote sample expectations as actual expectations, we get

$$\mathbf{X}_G^{\mathbf{T}}\mathbf{X}_G + \mathbf{X}_B^{\mathbf{T}}\mathbf{X}_B = nE\{X_iX_j\} = n\,\mathrm{Cov}(X_i, X_j) + n_G\mathbf{m}_G\mathbf{m}_G^{\mathbf{T}} + n_B\mathbf{m}_B\mathbf{m}_B^{\mathbf{T}}.$$

If S is the sample covariance matrix, this gives

$$\mathbf{X}_G^{\mathbf{T}}\mathbf{X}_G + \mathbf{X}_B^{\mathbf{T}}\mathbf{X}_B = n\mathbf{S} + n_G\mathbf{m}_G\mathbf{m}_G^{\mathbf{T}} + n_B\mathbf{m}_B\mathbf{m}_B^{\mathbf{T}}. \tag{4.27}$$

Expanding (4.26) and using (4.27) gives

$$\begin{aligned} nw_0 + (n_G\mathbf{m}_G + n_B\mathbf{m}_B)\mathbf{w}^{\mathbf{T}} &= n_G, \\ (n_G\mathbf{m}_G^{\mathbf{T}} + n_B\mathbf{m}_B^{\mathbf{T}})w_0 + (n\mathbf{S} + n_G\mathbf{m}_G\mathbf{m}_G^{\mathbf{T}} + n_B\mathbf{m}_B\mathbf{m}_B^{\mathbf{T}})\mathbf{w}^{\mathbf{T}} &= n_G\mathbf{m}_G^{\mathbf{T}}. \end{aligned} \tag{4.28}$$

Substituting the first equation in (4.28) into the second one gives

$$\begin{aligned} &((n_G\mathbf{m}_G^{\mathbf{T}} + n_B\mathbf{m}_B^{\mathbf{T}})(n_G - (n_G\mathbf{m}_G + n_B\mathbf{m}_B)\mathbf{w}^{\mathbf{T}})/n) \\ &\quad + (n_G\mathbf{m}_G\mathbf{m}_G^{\mathbf{T}} + n_B\mathbf{m}_B\mathbf{m}_B^{\mathbf{T}})\mathbf{w}^{\mathbf{T}} + n\mathbf{S}\mathbf{w}^{\mathbf{T}} = n_G\mathbf{m}_G^{\mathbf{T}}, \\ &\text{so } \left(\frac{n_Gn_B}{n}\right)(\mathbf{m}_G - \mathbf{m}_B)\mathbf{w}^{\mathbf{T}} + n\mathbf{S}\mathbf{w}^{\mathbf{T}} = \left(\frac{n_Gn_B}{n}\right)(\mathbf{m}_G - \mathbf{m}_B)^{\mathbf{T}}; \\ &\text{thus } \mathbf{S}\mathbf{w}^{\mathbf{T}} = c(\mathbf{m}_G - \mathbf{m}_B)^{\mathbf{T}}. \end{aligned} \tag{4.29}$$

Thus (4.29) gives the best choice of $\mathbf{w} = (w_1, w_2, \ldots, w_p)$ for the coefficients of the linear regression. This is the same $\mathbf{w}$ as in (4.21), namely, the linear discriminant function. This approach shows, however, that one can obtain the coeffieients of the credit scorecard by the least-squares approach beloved of linear regression.

We have taken the obvious left-hand sides in the regression equation (4.22), where goods have a value 1 and bads have a value 0. These have given a set of constants, which we label $\mathbf{w}(\mathbf{1}, \mathbf{0})^*$. If one takes any other values so that the goods have a left-hand side of g and the bads have a left-hand side of b, then the coefficients in the regression—$\mathbf{w}(\mathbf{g}, \mathbf{b})^*$—differ only in the constant term w_0 since

$$\mathbf{w}(\mathbf{a}, \mathbf{b})^* = b + (g - b)\mathbf{w}(\mathbf{1}, \mathbf{0})^*. \tag{4.30}$$

4.5 Logistic regression

The regression approach to linear discrimination has one obvious flaw. In (4.22), the right-hand side could take any value from $-\infty$ to $+\infty$, but the left-hand side is a probability and so should take only values between 0 and 1. It would be better if the left-hand side were a function of p_i, which could take a wider range of values. Then one would not have the difficulty that all the data points have very similar values of the dependent variables or that the regression equation predicts probabilities that are less than 0 or greater than 1. One such function is the log of the probability odds. This leads to the logistic regression approach, on which Wiginton (1980) was one of the first to publish credit-scoring results. In logistic regression, one matches the log of the probability odds by a linear combination of the characteristic variables; i.e.,

$$\log\left(\frac{p_i}{1 - p_i}\right) = w_0 + w_1x_1 + w_2x_2 + \cdots + w_px_p = \mathbf{w} \cdot \mathbf{x}^{\mathbf{T}}. \tag{4.31}$$

Since $\frac{p_i}{1-p_i}$ takes values between 0 and ∞, $\log(\frac{p_i}{1-p_i})$ takes values between $-\infty$ and $+\infty$. Taking exponentials on both sides of (4.31) leads to the equation

$$p_i = \frac{e^{\mathbf{w}\cdot\mathbf{x}}}{1+e^{\mathbf{w}\cdot\mathbf{x}}}. \tag{4.32}$$

This is the logistic regression assumption. It is interesting to note that if we assume that the distribution of the characteristic values of the goods and of the bads is multivariate normal, as was suggested in section 4.2, then this example satisfies the logistic regression assumption. Again assume that the means are $\boldsymbol{\mu}_G$ among the goods and $\boldsymbol{\mu}_B$ among the bads with common covariance matrix $\boldsymbol{\Sigma}$. This means that $E(X_i|G) = \mu_{G,i}$, $E(X_i|B) = \mu_{B,i}$, and $E(X_iX_j|G) = E(X_iX_j|B) = \Sigma_{ij}$.

The corresponding density function in this case as given in (4.15) is

$$f(\mathbf{x}|G) = (2\pi)^{-\frac{p}{2}}(\det\Sigma)^{-\frac{1}{2}}\exp\left(\frac{-(\mathbf{x}-\boldsymbol{\mu}_G)\boldsymbol{\Sigma}^{-1}(\mathbf{x}-\boldsymbol{\mu}_G)^{\mathbf{T}}}{2}\right), \tag{4.33}$$

where $(\mathbf{x}-\boldsymbol{\mu}_G)$ is a vector with 1 row and p columns while $(\mathbf{x}-\boldsymbol{\mu}_G)^{\mathbf{T}}$ is its transpose, which is the same numbers represented as a vector with p rows and 1 column. If p_B is the proportion of the population who are bad and p_G is the proportion of the population who are good, then the log of the probability odds for customer i who has characteristics $\mathbf{x}$ is

$$\begin{aligned}\log\left(\frac{p_i}{1-p_i}\right) &= \log\left(\frac{p_G f(\mathbf{x}|G)}{p_B f(\mathbf{x}|B)}\right)\\ &= \mathbf{x}\cdot\Sigma^{-1}2(\mu_B-\mu_G)^{\mathbf{T}} + (\mu_G\cdot\Sigma^{-1}\mu_G^{\mathbf{T}} + \mu_B\cdot\Sigma^{-1}\mu_B^{\mathbf{T}}) + \log\left(\frac{p_G}{p_B}\right).\end{aligned} \tag{4.34}$$

Since this is a linear combination of the x_i, it satisfies the logistic regression assumption. However, other classes of distribution also satisfy the logistic assumption, including ones that do not lead to linear discriminant functions if the Bayes loss approach of section 4.2 is applied. Consider, for example, the case where the characteristics are all binary and independent of each other. This means that

$$\begin{aligned}&\text{Prob}(X_i = 1|G) = p_G(i); \qquad \text{Prob}(X_i = 0|G) = 1 - p_G(i);\\ &\text{Prob}(X_i = 1|B) = p_B(i); \qquad \text{Prob}(X_i = 0|B) = 1 - p_B(i).\end{aligned}$$

Hence if p_G, p_B are the prior probabilities of goods and bads in the population

$$\text{Prob}(G|\mathbf{x}) = \frac{\text{Prob}(\mathbf{x}|G)p_G}{\text{Prob}(\mathbf{x})} = \frac{\prod_i p_G(i)^{x_i}(1-p_G(i))^{1-x_i}p_G}{\text{Prob}(\mathbf{x})}, \tag{4.35}$$

then

$$\begin{aligned}\log\left(\frac{\text{Prob}(G|\mathbf{x})}{\text{Prob}(B|\mathbf{x})}\right) &= \Sigma_i x_i(\log(p_G(i)) - \log(p_B(i)))\\ &\quad + \Sigma_i(1-x_i)(\log(1-p_G(i)) - \log(1-p_B(i))) + \log\left(\frac{p_G}{p_B}\right)\\ &= \Sigma_i x_i\left(\log\left(\frac{p_G(i)(1-p_B(i))}{p_B(i)(1-p_G(i))}\right)\right) + \Sigma_i\log\left(\frac{1-p_G(i)}{1-p_B(i)}\right) + \log\left(\frac{p_G}{p_B}\right).\end{aligned} \tag{4.36}$$

This is again of the form of (4.31) and hence satisfies the logistic regression assumption.

The only difficulty with logistic regression compared with ordinary regression is that it is not possible to use the ordinary least-squares approach to calculate the coefficients **w**. Instead one has to use the maximum likelihood approach to get estimates for these coefficents. This leads to an iterative Newton–Raphson method to solve the equations that arise. With the power of modern computers this is no problem, even for the large samples that are often available when building credit scorecards.

One of the surprising results is that although theoretically logistic regression is optimal for a much wider class of distributions than linear regression for classification purposes, when comparisons are made on the scorecards developed using the two different methods on the same data set, there is very little difference in their classification. The difference is that linear regression is trying to fit the probability p of defaulting by a linear combination of the attributes while logistic regression is trying to fit $\log(\frac{p}{1-p})$ by a linear combination of the attributes. As Figure 4.3 shows, if one maps a linear shift of p and $\log(\frac{p}{1-p})$, then they are very similar until either p becomes close to 0 or close to 1. In the credit-scoring context, this means that the scores being produced by the two methods are very similar except for those where the probability of default is very low or very high. These are the applicants for whom it should be easy to predict whether they will default. For the more difficult regions of default prediction—around $p = 0.5$—the two curves are very similar. This may explain why there is less variation in the methods than one would have expected.

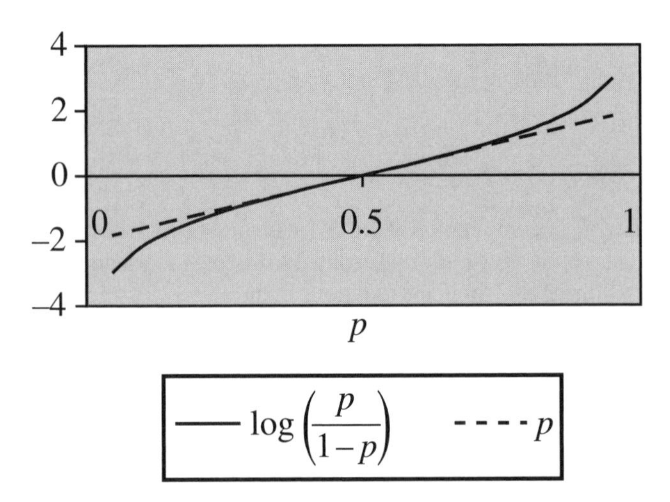

Figure 4.3. *Graph of* $\log(\frac{p}{1-p})$ *and* $ap + b$.

4.6 Other nonlinear regression approaches

Two other nonlinear functions have been used in credit scoring. The first of these is the probit function, which was used in credit scoring by Grablowsky and Talley (1981). In probit analysis, if $N(x)$ is the cumulative normal distribution function so that

$$N(x) = \frac{1}{\sqrt{2\pi}} \int_{-\infty}^{x} e^{-\frac{y^2}{2}} dy,$$

then the aim is to estimate $N^{-1}(p_i)$ as a linear function of the characteristics of the applicant, so

$$N^{-1}(p_i) = \mathbf{w} \cdot \mathbf{x}_i^{\mathbf{T}} = w_0 + w_1 x_{1i} + w_2 x_{2i} + \cdots + w_p x_{pi}. \tag{4.37}$$

Again, although p_i takes only values between 0 and 1, $N^{-1}(p_i)$ takes values between $-\infty$ and $+\infty$ and so allows the linear function to vary over this full range. One way to think about the probit approach is to say that

$$W = \mathbf{w} \cdot \mathbf{x}_i^{\mathbf{T}} = w_1 X_1 + w_2 X_2 + \cdots + w_p X_p$$

is a measure of the goodness of an applicant, but whether the applicant is actually bad or good depends on whether the value of W is greater or less than a cutoff level C. If we assumed that C is not fixed but is a variable with standard normal distribution, then this gives the same probability of applicants being good as the probit equation (4.37). One can never test this assumption since the W is not a variable with any real meaning. However, one can always fit this model to any sample of past applicants. As in the case of logistic regression, one uses maximum likelihood estimation to obtain the values $\mathbf{w}$ and again one has to use iterative techniques to solve for these values. There are situations where an iteration procedure may not converge, and then alternative iterative procedures must be tried.

The other approach used in credit scoring and that is very common in economic models is tobit analysis. The tobit transformation assumes that one can estimate p_i by

$$p_i = \max\{\mathbf{w} \cdot \mathbf{x}_i^{\mathbf{T}}, 0\} = \max\{w_0 + w_1 x_{1i} + w_2 x_{2i} + \cdots + w_p x_{pi}, 0\}. \tag{4.38}$$

In this case, one is dealing with the mismatch in limiting values between the two sides of the regression approach to discriminant analysis (4.22) by limiting the right-hand side to be positive. Since there are many economic situations where the variable in which one is interested shows itself only when it is positive, the statistical analysis of tobit regressions is fully worked out and many of the standard statistical packages will produce estimates of the parameters. An example of where tobit analysis may be appropriate is if one assumes that for consumers debt = income − expenditure and then one decides to regress debt on income and expenditure. One has to use tobit analysis since negative debt does not show itself as debt.

In the credit-scoring context, there is something unsatisfactory about the asymmetry of the tobit transformation. It has dealt with the difficulty of estimating negative probabilities but not of estimating probabilities greater than 1. A more symmetrical model would be

$$p_i = \mathrm{Min}\{1, \max\{\mathbf{w} \cdot \mathbf{x}_i^{\mathbf{T}}, 0\}\}, \tag{4.39}$$

but unfortunately the analysis and parameter estimation for that transformation is not available in standard packages.

Probit and tobit are not approaches that find much favor with credit-scoring practitioners since they are more concerned with getting the decision correct for the less clear-cut cases than with the fact that the probability for a clear bad is 0.05 or −0.05. However, it has to be remembered that the statistical techniques do not know this, and their objective is to minimize the total sum of the errors. Thus an outlandish case might affect the parameters considerably and hence change the classifications in the difficult region. This suggests one might try a two-stage approach. In the first stage, one tries to estimate the likely cutoff region, while in the second stage, one concentrates on getting the classification in that region to be as accurate as possible.

4.7 Classification trees (recursive partitioning approach)

A completely different statistical approach to classification and discrimination is the idea of classification trees, sometimes called recursive partitioning algorithms (RPA). The idea

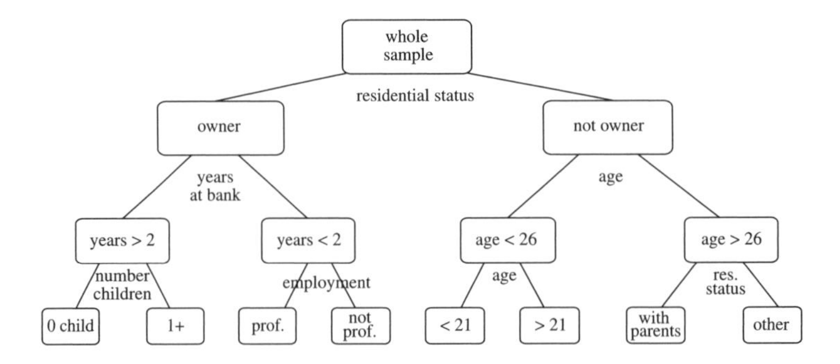

Figure 4.4. *Classification tree.*

is to split the set of application answers into different sets and then identify each of these sets as good or bad depending on what the majority in that set is. The idea was developed for general classification problems independently by Breiman and Friedman in 1973, who described a number of statistical applications (but not credit scoring) in their book (Breiman et al. 1984). Its use in credit scoring quickly followed (Makowski 1985, Coffman 1986). The idea was also picked up in the artificial intelligence literature. Similar ideas were used in classification problems, and useful computer software was developed for its implementation. Although the software has different names—CHAID and C5, for example—the basic steps are the same in both literatures (Safavian and Landgrebe 1991).

The set of application data A is first split into two subsets so that looking at the sample of previous applicants, these two new subsets of application attributes are far more homogeneous in the default risk of the applicants than the original set. Each of these sets is then again split into two to produce even more homogeneous subsets, and the process is repeated. This is why the approach is called recursive partitioning. The process stops when the subsets meet the requirements to be terminal nodes of the tree. Each terminal node is then classified as a member of A_G or A_B and the whole procedure can be presented graphically as a tree, as in Figure 4.4.

Three decisions make up the classification tree procedure:

- what rule to use to split the sets into two—the splitting rule;
- how to decide that a set is a terminal node—the stopping rule;
- how to assign terminal nodes into good and bad categories.

The good-bad assignment decision is the easiest to make. Normally, one would assign the node as good if the majority of sample cases in that node are good. An alternative is to minimize the cost of misclassification. If D is the debt incurred by misclassifying a bad as a good and L is the lost profit caused by misclassifying a good as a bad, then one minimizes the cost if one classifies the node as good when the ratio of goods to bads in that node in the sample exceeds $\frac{D}{L}$.

The simplest splitting rules are those that look just one step ahead at the result of the proposed split. They do this by finding the best split for each characteristic in turn by having some measure of how good a split is. Then they decide which characteristic's split is best under this measure. For any continuous characteristic X_i, one looks at the splits $\{x_i < s\}$, $\{x_i \geq s\}$ for all values of s and finds the value of s where the measure is best. If X_i is a categorical variable, then one looks at all possible splits of the categories into two and checks the measure under these different splits. Usually one can order the categories in increasing good:bad ratios and know that the best split will divide this ordering into two groups. So what measure is used? The most common is the Kolmogorov–Smirnov statistic, but there are at least four others—the basic impurity index, the Gini index, the entropy index, and the half-sum of squares. We look at each in turn and use them to calculate the best split in the following simple example.

Example 4.1. Residential status has three attributes with the numbers of goods and bads in each attribute in a sample of a previous customer, shown in Table 4.1. If one wants to split the tree on this characteristic, what should the split be?

Table 4.1.

Residential status	Owner	Tenant	With Parents
Number of goods	1000	400	80
Number of bads	200	200	120
Good:bad odds	5:1	2:1	0.67:1

4.7.1 Kolmogorov–Smirnov statistic

For a continuous characteristic X_i, let $F(s|G)$ be the cumulative distribution function of X_i among the goods and $F(s|B)$ be the cumulative distribution among the bads. Assuming bads have a greater propensity for low values of X_i than the goods and that the definitions of the costs D and L are as in the earlier paragraph, the myopic rule would be to split on the value s which minimizes

$$LF(s|G)p_G + D(1 - F(s|B))p_B. \tag{4.40}$$

If $Lp_G = Dp_B$, this is the same as choosing the Kolmogorov–Smirnov distance between the two distributions, as Figure 4.5 shows; i.e., one wants to minimize $F(s|G) - F(s|B)$ or more obviously maximize $F(s|B) - F(s|G)$.

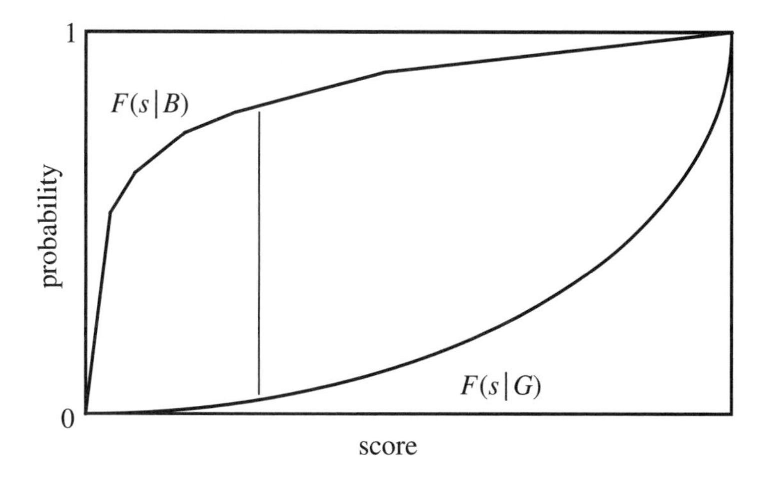

Figure 4.5. *Kolmogorov–Smirnov distance.*

If one thinks of the two subgroups as the left (l) and the right (r) group, then this is like maximizing the difference between $p(l|B)$, the probability a bad appears in the left group (i.e., $F(s|B)$ in the continuous case), and $p(l|G)$, the probability a good appears in the left group (i.e., $F(s|G)$ in the continuous case). One can rewrite $p(l|B)$ as $\frac{p(B|l)p(l)}{p(B)}$ using Bayes's theorem, and this is easier to calculate. Thus for categorical and continuous variables, the Kolmogorov–Smirnov (KS) criterion becomes the following: Find the left-right split that maximizes

$$\text{KS} = |p(l|B) - p(l|G)| = \left| \frac{p(B|l)}{p(B)} - \frac{p(G|l)}{p(G)} \right| \cdot p(l). \tag{4.41}$$

For Example 4.1, it is obvious from the good:bad odds that the best split must either be with parent in one group and owner + tenant in the other, or tenant and with parent in one group and owner in the other:

$$\begin{aligned}
&l = \text{parent}, \quad r = \text{owner + tenant:}\\
&p(l|B) = \frac{120}{520} = 0.231, \quad p(l|G) = \frac{80}{1480} = 0.054, \quad \text{KS} = 0.177,\\
&l = \text{parent + tenant}, \quad r = \text{owner:}\\
&p(l|B) = \frac{320}{520} = 0.615, \quad p(l|G) = \frac{480}{1480} = 0.324, \quad \text{KS} = 0.291.
\end{aligned}$$

Thus the best split is parent + tenant in one group and owner in the other.

4.7.2 Basic impurity index $i(v)$

There is a whole class of impurity index measures that try to assess how impure each node v of the tree is, where purity corresponds to that node being all of one class. If one then splits the node into a left node l and a right node r with the proportion going into l being $p(l)$ and the proportion going into r being $p(r)$, one can measure the change in impurity that the split has made by

$$I = i(v) - p(l)i(l) - p(r)i(r). \tag{4.42}$$

The greater this difference, the greater the change in impurity, which means the new nodes are much more pure. This is what we want so we choose the split that maximizes this expression. This is equivalent to minimizing $p(l)i(l) + p(r)i(r)$. Obviously, if there was no split with a positive difference, then one should not split the node at all.

The most basic of these impurity indices is to take $i(v)$ to be the proportion of the smaller group in that node so that

$$\begin{aligned}
i(v) = p(G|v) \quad &\text{if } p(G|v) \le 0.5,\\
i(v) = p(B|v) \quad &\text{if } p(B|v) < 0.5.
\end{aligned} \tag{4.43}$$

For Example 4.1, this gives the following calculation of the indices to see which is the better split, where v is the whole set before any split:

$$\begin{aligned}
&l = \text{parent}, \quad r = \text{owner + tenant:}\\
&i(v) = \frac{520}{2000} = 0.26, \quad p(l) = \frac{200}{2000} = 0.1, \quad i(l) = \frac{80}{200} = 0.4,\\
&p(r) = \frac{1800}{2000} = 0.9, \quad i(r) = \frac{400}{1800} = 0.22,\\
&I = 0.26 - 0.1(0.4) - 0.9(0.22) = 0.02;
\end{aligned}$$

$$
\begin{aligned}
&l = \text{parent} + \text{tenant}, \quad r = \text{owner}: \\
&i(v) = \frac{520}{2000} = 0.26, \quad p(l) = \frac{800}{2000} = 0.4, \quad i(l) = \frac{320}{800} = 0.4, \\
&p(r) = \frac{1200}{2000} = 0.6, \quad i(r) = \frac{200}{1200} = 0.167, \\
&I = 0.26 - 0.4(0.4) - 0.6(0.167) = 0.
\end{aligned}
$$

This suggests the best split is parent in one group and owner + tenant in the other.

Although this looks useful, appearances can be deceptive. I in the second split is 0 because the bads are the minority in all three nodes v, l, and r. This will always happen if the same group is in the minority in all three nodes, and for many credit situations this is the case. Since all splits give the same difference—0—this criterion is useless to help decide which is the best. Moreover, Breiman et al. (1984) gave an example in which there were 400 goods and 400 bads. One partition split this into one class of 300 goods and 100 bads and the other of 100 goods and 300 bads, while another partition split it into one class of just 200 goods and the other of 200 goods and 400 bads. Both these splits have the same index difference $i(v)$, but most people would think the second partition that has been able to identify a complete group of goods is a better split. What is needed is an index that rewards the purer nodes more than this one does.

4.7.3 Gini index

Instead of being linear in the proportion of the impurity probability, the Gini index is quadratic and so puts relatively more weight on purer nodes. It is defined by

$$
\begin{aligned}
&i(v) = p(G|v)p(B|v), \\
\text{so} \quad &G = p(G|v)p(B|v) - p(l)(G|l)p(B|l) - p(r)(G|r)p(B|r).
\end{aligned}
\tag{4.44}
$$

In the example we are interested in, this gives

$$
\begin{aligned}
&l = \text{parent}, \quad r = \text{owner} + \text{tenant}: \\
&i(v) = \left(\frac{1480}{2000}\right)\left(\frac{520}{2000}\right) = 0.1924, \\
&p(l) = \frac{200}{2000} = 0.1, \quad i(l) = \left(\frac{80}{200}\right)\left(\frac{120}{200}\right) = 0.24, \\
&p(r) = \frac{1800}{2000} = 0.9, \quad i(r) = \left(\frac{400}{1800}\right)\left(\frac{1400}{1800}\right) = 0.1728, \\
&I = 0.1924 - 0.1(0.24) - 0.9(0.1728) = 0.01288;
\end{aligned}
$$

$$
\begin{aligned}
&l = \text{parent} + \text{tenant}, \quad r = \text{owner}: \\
&i(v) = \left(\frac{520}{2000}\right)\left(\frac{1480}{2000}\right) = 0.1924, \\
&p(l) = \frac{800}{2000} = 0.4, \quad i(l) = \left(\frac{320}{800}\right)\left(\frac{480}{800}\right) = 0.24, \\
&p(r) = \frac{1200}{2000} = 0.6, \quad i(r) = \left(\frac{200}{1200}\right)\left(\frac{1000}{1200}\right) = 0.1389, \\
&I = 0.1924 - 0.4(0.24) - 0.6(0.1389) = 0.01306.
\end{aligned}
$$

This index suggests that the best split is parent + tenant in one node and owner in the other node.

4.7.4 Entropy index

Another nonlinear index is the entropy one, where

$$i(v) = -p(G|v)\ln(p(G|v)) - p(B|v)\ln(p(B|v)). \tag{4.45}$$

As its name suggests, this is related to the entropy or the amount of information in the split between the goods and bads in the node. It is a measure of how many different ways one could end up with the actual split of goods to bads in the node, and it is related to the information statistic used to in coarse classifying measurements (see section 8.7 for more details).

Using this measure the splits in Example 4.1 are measured as follows:

l = parent, r = owner + tenant:

$$i(v) = -\left(\frac{520}{2000}\right)\ln\left(\frac{520}{2000}\right) - \left(\frac{1480}{2000}\right)\ln\left(\frac{1480}{2000}\right) = 0.573,$$

$$p(l) = \frac{200}{2000} = 0.1, \quad i(l) = -\left(\frac{80}{200}\right)\ln\left(\frac{80}{200}\right) - \left(\frac{120}{200}\right)\ln\left(\frac{120}{200}\right) = 0.673,$$

$$p(r) = \frac{1800}{2000} = 0.9, \quad i(r) = -\left(\frac{400}{1800}\right)\ln\left(\frac{400}{1800}\right) - \left(\frac{1400}{1800}\right)\ln\left(\frac{1400}{1800}\right) = 0.530,$$

$$I = 0.573 - 0.1(0.673) - 0.9(0.530) = 0.0287;$$

l = parent + tenant, r = owner:

$$i(v) = -\left(\frac{520}{2000}\right)\ln\left(\frac{520}{2000}\right) - \left(\frac{1480}{2000}\right)\ln\left(\frac{1480}{2000}\right) = 0.573,$$

$$p(l) = \frac{800}{2000} = 0.4, \quad i(l) = -\left(\frac{320}{800}\right)\ln\left(\frac{320}{800}\right) - \left(\frac{480}{800}\right)\ln\left(\frac{480}{800}\right) = 0.673,$$

$$p(r) = \frac{1200}{2000} = 0.6, \quad i(r) = -\left(\frac{200}{1200}\right)\ln\left(\frac{200}{1200}\right) - \left(\frac{1000}{1200}\right)\ln\left(\frac{1000}{1200}\right) = 0.451,$$

$$I = 0.573 - 0.4(0.673) - 0.6(0.451) = 0.0332.$$

Hence this index also suggests the best split is with parent + tenant in one node and owner in the other node.

4.7.5 Maximize half-sum of squares

The last measure we look at is not an index but comes instead from the χ^2 test, which would check if the proportion of goods is the same in the two daughter nodes we split into. If this χ^2 statistic Chi is large, we usually say the hypothesis is not true; i.e., the two proportions are not the same. The larger the value, the more unlikely they are the same, but the latter could be reinterpreted to say the greater the difference between them there is. That is what we want from a split, so we are led to the following test.

If $n(l)$ and $n(r)$ are the total numbers in the left and right modes, then maximize

$$\text{Chi} = n(l)n(r) - \frac{(p(G|l) - p(G|r))^2}{n(l) + n(r)}.$$

Applying this test to the data in Example 4.1 gives

$$l = \text{parent}, \quad r = \text{owner} + \text{tenant}:$$

$$n(l) = 200, \qquad p(G|l) = \frac{80}{200} = 0.4,$$

$$n(r) = 1800, \quad p(G|r) = \frac{1400}{1800} = 0.777, \qquad \text{so Chi} = 25.69;$$

$$l = \text{parent} + \text{tenant}, \quad r = \text{owner}:$$

$$n(l) = 800, \qquad p(G|l) = \frac{480}{800} = 0.6,$$

$$n(r) = 1200, \quad p(G|r) = \frac{1000}{1200} = 0.833, \qquad \text{so Chi} = 26.13.$$

Hence this measure is maximized when the "with parents" class and the tenants class are put together. One can construct examples where the tests suggest different splits are best.

There are other methods of splitting. Breiman et al. (1984) suggested that a better criterion than just looking at the next split would be to consider what the situation is after r more generations of splits. This takes into account not just the immediate improvement caused by the current split but also the long-run strategic importance of that split. It is usually the case that one is splitting using different characteristics at different levels of the tree, and also different characteristics come to the fore for different subsets at the same level of the tree, as Figure 4.4 illustrates. In this way, the tree picks up the nonlinear relationships between the application characteristics.

Although we talk about when we stop the tree and recognize a node as a terminal node, it is perhaps more honest to talk of a stopping and pruning rule. This emphasizes that we expect the initial tree to be too large and it will need to be cut back to get a tree that is robust. If one had a tree where every terminal node had only one case from the training sample in it, then the tree would be a perfect discriminator on the training sample but it would be a very poor classifier on any other set. Thus, one makes a node a terminal node for one of two reasons. The first reason is the number of cases in the node is so small that it makes no sense to divide it further. This is usually when there are fewer than 10 cases in the node. The second reason is the split measurement value if one makes the best split into two daughter nodes is hardly any different from the measurement value if one keeps the node as is. One has to make "hardly any difference" precise in this context, and it could be that the difference in the measure is below some prescribed level β.

Having obtained such an overlarge tree, one can cut back by removing some of the splits. The best way to do this is to use a holdout sample, which was not used in building the tree. This sample is used to estimate empirically the expected losses for different possible prunings of the tree. Using this holdout sample and a classification tree T, define T_G (T_B) to be the set of nodes that are classified as good (bad). Let $r(t, B)$ be the proportion of the holdout sample that is in node t and that are classified as bad and $r(t, G)$ be the proportion of the sample that is in t and is good. Then an estimate of the expected loss is

$$r(T) = \sum_{t \in T_G} Dr(t, B) + \sum_{t \in T_B} Lr(t, G). \tag{4.46}$$

If $n(T)$ is the number of nodes in tree T, define $c(T) = r(T) + dn(T)$ and prune a tree T^* by looking at all subtrees of T^* and choosing the tree T that minimizes $c(T)$. If $d = 0$, we end up with the original unpruned tree, while as d becomes large, the tree will consist of only one node. Thus the choice of d gives a view of how large a tree is wanted.

4.8 Nearest-neighbor approach

The nearest-neighbor method is a standard nonparametric approach to the classification problem first suggested by Fix and Hodges (1952). It was applied in the credit-scoring contect first by Chatterjee and Barcun (1970) and later by Henley and Hand (1996). The idea is to choose a metric on the space of application data to measure how far apart any two applicants are. Then with a sample of past applicants as a representative standard, a new applicant is classified as good or bad depending on the proportions of goods and bads among the k nearest applicants from the representative sample—the new applicant's nearest neighbors.

The three parameters needed to run this approach are the metric, how many applicants k constitute the set of nearest neighbors, and what proportion of these should be good for the applicant to be classified as good. Normally, the answer to this last question is that if a majority of the neighbors are good, the applicant is classified as good; otherwise, the applicant is classified as bad. However, if, as in section 4.2, the average default loss is D and the average lost profit in rejecting a good is L, then one could classify a new applicant as good only if at least $\frac{D}{D+L}$ of the nearest neighbors are good. This criterion would minimize the expected loss if the likelihood of a new applicant being good is the proportion of neighbors who are good.

The choice of metric is clearly crucial. Fukanaga and Flick (1984) introduced a general metric of the form

$$d(\mathbf{x}_1, \mathbf{x}_2) = (\mathbf{x}_1 - \mathbf{x}_2)\mathbf{A}(\mathbf{x}_1)((\mathbf{x}_1 - \mathbf{x}_2)^{\mathbf{T}})^{\frac{1}{2}}, \tag{4.47}$$

where $\mathbf{A}(x)$ is a $p \times p$ symmetric positive definite matrix. $\mathbf{A}(\mathbf{x})$ is called a local metric if it depends on $\mathbf{x}$ and is called a global metric if it is independent of $\mathbf{x}$. The difficulty with the local metric is that it picks up features of the training set that are not appropriate in general, and so most authors concentrate on global metrics. The most detailed examination of nearest-neighbors approach in the credit-scoring context was by Henley and Hand (1996), who concentrated on metrics that were mixtures of Euclidean distance and the distance in the direction that best separated the goods and the bads. One gets this direction from Fisher's linear discriminant function of section 4.3. Thus if $\mathbf{w}$ is the p-dimensional vector defining that direction, given in (4.21), Henley and Hand suggest a metric of the form

$$d(\mathbf{x}_1, \mathbf{x}_2) = \{(\mathbf{x}_1 - \mathbf{x}_2)^{\mathbf{T}}(\mathbf{I} + D\mathbf{w} \cdot \mathbf{w}^{\mathbf{T}})(\mathbf{x}_1 - \mathbf{x}_2)\}^{\frac{1}{2}}, \tag{4.48}$$

where I is the identity matrix. They perform a large number of experiments to identify what might be a suitable choice of D. Similarly, they choose k, the ideal number of nearest neighbors, by experimenting with many choices of k. Although there are not large variations in the results, the best choice of D was in the range 1.4 to 1.8. The choice of k clearly depends on the size of the training sample, and in some cases changing k by one makes a noticeable difference. However, as Figure 4.6 suggests, in the bigger picture there is not much difference in the rate of misclassifying bads for a fixed acceptance rate as k varies over quite a range from 100 to 1000 (with a training sample of 3000). To avoid picking a locally poor value of k, one could smooth k by choosing a distribution for k. Thus for each point, there may be a different number of nearest neighbors. However, satisfactory results can be obtained without resorting to this level of sophistication.

Nearest-neighbor methods, although they are far less widely in credit scoring than the linear and logistic regression approaches, have some potentially attractive features for actual implementation. It would be easy to dynamically update the training sample by adding new

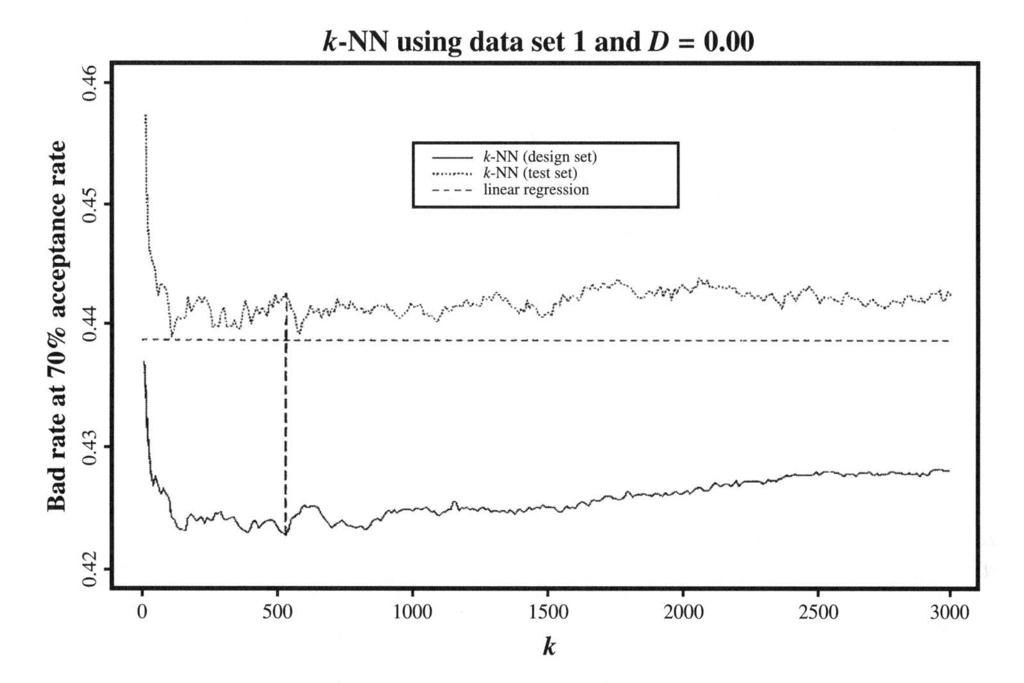

Figure 4.6. *Default (bad) rate in nearest-neighbor systems as k—size of neighbourhood—varies.*

cases to the training sample when it is known if they are good or bad and dropping the cases that have been in the sample longest. This would partially overcome the need to update the scoring system regularly because of the changes in the population, though a metric d like (4.48) would need to be updated as well to take account of the population shifts and this could not be done dynamically. The concern that there is a lot of calculation needed to work out whether a new case is good or bad—finding out which are its k nearest neighbors in the training sample—is misplaced as modern computers can do such calculations in a few seconds. However, in many ways, finding a good metric in the first place is almost the equivalent of the regression approach to scorecard building. Thus many users are content to stop at that point and use a traditional scorecard. As is the case with the decision tree approach, the fact that the nearest-neighbors technique is not able to give a score for each particular applicant's characteristics deprives users of an anchor that helps them feel that they understand what the system is really doing.

It is also the case that monitoring the performance of the system is nigh on impossible. How does one know when the original metric is no longer appropriate? In section 8.12, we discuss the various ways one can do this for regression base scorecards, but those methods will not work for the nearest-neighbor approach.

4.9 Multiple-type discrimination

There may be circumstances in credit scoring where the applicants need to be divided into more than two groups. As was argued earlier in this chapter, one has to be careful about this because the groupings formed may be due to the way the organization goes about its decision making. However, some organizations may split past applicants into the goods whom they want as customers, those whom they do not want because they default, and those whom they do not want because they do not use the loan facility enough for the organization to make a profit. Similarly, in the U.S., consumers can default or they can seek the protection of personal bankruptcy against their debts. In that case, the sample of past applicants can be split into goods, defaulters, and bankrupts since it is considered that the characteristics of the latter two groups are different. A detailed analysis of bankruptcy is found in Chapter 13.

If one wants to divide the population into g groups, most of the previous methods will continue to apply if suitable modifications are made. Thus one can change the decision-making approach to discriminant analysis as follows.

Suppose that $c(i, j)$ is the cost of misassigning an applicant of group j to group i, and let p_j be the proportion of the population in group j. Let $p(\mathbf{x}|j)$ be the probability that applicants in group j have application attributes $\mathbf{x}$ (similar to $p(\mathbf{x}|G)$ in section 4.2). Then the probability that an applicant with attributes $\mathbf{x}$ is in group j, which we define as $P(j|\mathbf{x})$, satisfies

$$P(j|\mathbf{x}) = \frac{p_j p(\mathbf{x}|j)}{\Sigma_i p_i p(x|i)}. \tag{4.49}$$

If one wants to minimize the expected loss, one would assign applicant with an attribute $\mathbf{x}$ to group i if

$$\Sigma_j c(i, j) p_j p(\mathbf{x}|j) < \Sigma_j c(k, j) p_j p(\mathbf{x}|j) \quad \text{for all } k,\ k \neq i. \tag{4.50}$$

In a similar way, one can extend the logistic regression approach to classifying three or more groups by taking (4.32) as the probability the customer is in one class with similar forms for estimates of probabilities of being in the other classes. Although the statistical methods of multiple group classification are well developed, there has been little application of these ideas in credit scoring.

Chapter 5

Nonstatistical Methods for Scorecard Development

5.1 Introduction

The original idea in the development of credit scorecards was to use statistical analysis of a sample of past customers to help decide which existing or new customers were likely to be satisfactory. One can also look at nonstatistical approaches to the same problem. This has happened to all such classification problems in the last 25 years. Until the 1980s, the only approaches were statistical ones, but then it was realized (Freed and Glover 1981a, 1981b) that finding the linear function of the characteristics that best discriminates between groups can be modeled as a linear programming problem. The linear programming approach measures goodness of fit by taking the sum of the absolute errors involved or the maximum error involved. This is discussed in section 5.2. If one wants to take the number of cases where the discrimination is incorrect as a measure of goodness of fit, then one has to introduce integer variables into the linear program, and this leads to the integer programming models of section 5.3.

The 1970s saw tremendous research in the area of artificial intelligence, whereby researchers tried to program computers to perform natural human skills. One of the most successful attempts was expert systems, in which computers were given a database of information on a certain field obtained from experts in that field and a mechanism for generating rules from the information. The computer programs then used this combination to analyze new situations and come up with ways to deal with the new situation, which would be as good as experts would suggest. Successful pilot systems were developed for medical diagnosis, and since this is essentially a classification problem, researchers considered using expert systems in credit scoring. So far this has had limited success, as outlined in section 5.6.

In the 1980s another variant on the artificial intelligence–based approach to classification problems, neural networks, suddenly came to the fore and has continued to be a very active area of research. Neural networks are ways to model the decision process in a problem in the way the cells in the brain use neurons to trigger one another and hence set up learning mechanisms. A system of processing units is connected together, each of which gives an output signal when it receives input signals. Some of the processing units can receive external input signals and one can give an output signal. The system is given a set of data where each example is a set of input signals and a specific output signal. It tries to learn from this data how to reproduce the relationship between the input and output signals by adjusting the way each processing unit relates its output signal to the corresponding input signal. If the input signals are taken to be the characteristics of a customer and the output signal is whether their credit performance is good or bad, one can see how this approach can be used in credit scoring. Section 5.4 looks at the application of neural networks in credit scoring.

One can also think of the development of a scorecard as a type of combinatorial optimization problem. One has a number of parameters—the possible scores given to the various attributes—and a way to measure how good each set of parameters is (say, the misclassification error when such a scorecard is applied to a sample of past customers). There have been a number of generic approaches to solving such problems in the last decade—simulated annealing, tabu search, and genetic algorithms. The last of these has found some applications in pilot projects in credit scoring. These ideas are described in section 5.5.

This chapter outlines these approaches and their relative advantages and disadvantages and concludes by describing the results of comparisons of their classification accuracy and those of statistically based methods in credit and behavioral scoring applications.

5.2 Linear programming

Mangasarian (1965) was the first to recognize that linear programming could be used in classification problems where there are two groups and there is a separating hyperplane, i.e., a linear discriminant function, which can separate the two groups exactly. Freed and Glover (1981a) and Hand (1981) recognized that linear programming could also be used to discriminate when the two groups are not necessarily linearly separable, using objectives such as minimization of the sum of absolute errors (MSAE) or minimizing the maximum error (MME).

Recall that the decision that has to be made by the lending organization is to split A, the set of all combinations of values of the application variables $\mathbf{X} = (X_1, X_2, \ldots, X_p)$ can take, into two sets: A_G, corresponding to the answers given by the goods, and A_B, the set of answers given by the bads. Suppose one has a sample of n previous applicants. If n_G of the sample are goods, then for ease of notation we assume these are the first n_G in the sample. The remaining n_B of the sample $i = n_G + 1, \ldots, n_G + n_B$ are bad. Assume applicant i has characteristics $(x_{i1}, x_{i2}, \ldots, x_{ip})$ in response to the application variables $\mathbf{X} = (X_1, X_2, \ldots, X_p)$. We want to choose weights or scores $(w_1, w_2, \ldots, w_p)$ so that the weighted sum of the answers $w_1X_1 + w_2X_2 + \cdots + w_pX_p$ is above some cutoff value c for the good applicants and below the cutoff value for the bads. If the application variables have been transformed into binary variables, then one can think of the weights w_i as the scores for the different answers given.

Usually, one cannot hope to get a perfect division of the goods from the bads, so one introduces variables a_i, all of which are positive or zero, which allow for possible errors. Thus if applicant i in the sample is a good, we require $w_1x_{i1} + w_2x_{i2} + \cdots + w_px_{ip} \geq c - a_i$, while if applicant j is a bad, we require $w_1x_{j1} + w_2x_{j2} + \cdots + w_px_{jp} \leq c + a_j$. To find the weights $(w_1, w_2, \ldots, w_p)$ that minimize the sum of the absolute values of these deviations (MSD), one has to solve the following linear program:

$$\begin{aligned} \text{Minimize} \quad & a_1 + a_2 + \cdots + a_{n_G+n_B} \\ \text{subject to} \quad & w_1x_{i1} + w_2x_{i2} + \cdots + w_px_{ip} \geq c - a_i, && 1 \leq i \leq n_G, \\ & w_1x_{i1} + w_2x_{i2} + \cdots + w_px_{ip} \leq c + a_i, && n_G + 1 \leq i \leq n_G + n_B, \\ & a_i \geq 0, && 1 \leq i \leq n_G + n_B. \end{aligned} \tag{5.1}$$

If instead one wished to minimize the maximum deviation (MMD), one simplifies to the same error term in each constraint, namely,

$$
\begin{array}{lll}
\text{Minimize} & a & \\
\text{subject to} & w_1x_{i1} + w_2x_{i2} + \cdots + w_px_{ip} \geq c - a, & 1 \leq i \leq n_G, \\
& w_1x_{i1} + w_2x_{i2} + \cdots + w_px_{ip} \leq c + a, & n_G + 1 \leq i \leq n_G + n_B, \\
& a \geq 0. &
\end{array}
\tag{5.2}
$$

In credit scoring, an advantage of linear programming over statistical methods is that if one wants a scorecard with a particular bias, linear programming can easily include the bias into the scorecard development. Suppose, for example, that X_1 is the binary variable of being under 25 or not and X_2 is the binary variable of being over 65, and one wants the score for under 25s to be higher than for retired people. All that is necessary is to add the constraint $w_1 \geq w_2$ to the constraints in (5.1) or (5.2) to obtain such a scorecard. Similarly, one could ensure that the weighting on the application form variables exceeds those on the credit bureau variables, where the former are the variable X_1 to X_s and the latter are X_{s+1} to X_p, by adding to (5.1) or (5.2) constraints for each applicant that

$$w_1x_{i1} + w_2x_{i2} + \cdots + w_sx_{is} \geq w_{s+1}x_{is+1} + w_{s+2}x_{is+2} + \cdots + w_px_{ip} \quad \text{for all } i. \tag{5.3}$$

However, one has to be slightly careful because in the linear programming formulation one has made the requirements that the goods are to have scores higher than or the same as the cutoff score and bads are to have scores lower that or the same as the cutoff score since linear programming cannot deal with strict inequalities. This means that if one were able to choose the cutoff score as well as the weights, there is always a trivial solution where one puts the cutoff at zero and all the weights at zero also, i.e., $w_i = 0$ for all i. Thus everyone has a zero total score and sits exactly on the cutoff. One way out of this would seem to be to set the cutoff score to be some nonzero value, say, 1. However, Freed and Glover (1986a, 1986b) pointed out this is not quite correct and one will need to solve the model twice—once with the cutoff set to a positive value and once with it set to a negative value.

To see this, consider the following two examples with one variable X_1 and three applicants.

Example 5.1. In Figure 5.1(a), one has the two goods having values 1 and 2, respectively, and the bad has a value 0. It is easy to see that if the cutoff is 1, one can choose $w_1 = 1$ and there are no errors. For Figure 5.1(b), however, the linear program that minimizes MSD is

$$
\begin{array}{ll}
\text{Minimize} & a_1 + a_2 + a_3 \\
\text{subject to} & 2w \leq 1 + a_1, \quad w \geq 1 - a_2, \quad 0 \geq 1 - a_3.
\end{array}
$$

This is solved by taking $w = \frac{1}{2}$ with $a_1 = 0$, $a_2 = 0.5$, and $a_3 = 1$, so the total error is 1.5. Yet if we allow the cutoff point to be -1, then minimizing MSD becomes

$$
\begin{array}{ll}
\text{Minimize} & a_1 + a_2 + a_3 \\
\text{subject to} & 2w \leq -1 + a_1, \quad w \geq -1 - a_2, \quad 0 \geq -1 - a_3.
\end{array}
$$

This is solved exactly by letting $w = -\frac{1}{2}$ with 0 total error.

However, if we try to solve Figure 5.1(a) with a cutoff of -1, then the best we can do is $w = -\frac{1}{2}$, with a total error of 1.

The point is that if the goods generally have higher characteristic values than the bads, then one wants the characteristics to have positive weights and hence a positive cutoff score. If the goods tend to have lower values in each characteristic than the bads, then to get the total score for a good to be higher than the total score for a bad, the weights w_i need to

B	G	G
0	1	2

(a)

G	G	B
0	1	2

(b)

Figure 5.1.

be negative. This will make the total scores all negative, and so the cutoff value will be a negative score.

Fixing the cutoff value causes another problem, namely, those cases where the ideal cutoff score would have been zero. Formally, this means that the solutions obtained by linear programming are not invariant under linear transformations of the data. One would hope that adding a constant to all the values of one variable would not affect the weights the scorecard chose nor how good its classification was. Unfortunately, that is not the case, as the following example shows.

Example 5.2. There are two characteristic variables (X_1, X_2), and let there be three good cases with values $(1, 1)$, $(1, -1)$, and $(-1, 1)$ and three bad cases with values $(0, 0)$, $(-1, -1)$, and $(0.5, 0.5)$. If one is trying to minimize the sum of the deviations using a cutoff boundary of $w_1X_1 + w_2X_2 = 1$, then by symmetry one can assume the cutoff goes through the point $(1, -1)$ and so is of the form $(c + 1)X_1 + cX_2 = 1$ (see Figure 5.2). With $0.5 < c$, the deviation for $(-1, 1)$ is 2 and for $(0.5, 0.5)$ is $c - 0.5$, and so the deviation error is minimized at 2 by the scorecard $1.5X_1 + 0.5X_2$.

If we add 1 to the values of both variables so that the goods have values $(2, 2)$, $(2, 0)$, and $(0, 2)$ and the bads have values $(1, 1)$, $(0, 0)$, and $(1.5, 1.5)$, then the cutoff that minimizes MSD is $0.5X_1 + 0.5X_2 = 1$ (see Figure 5.2) with a deviation of 0.5, which is quite a different scorecard.

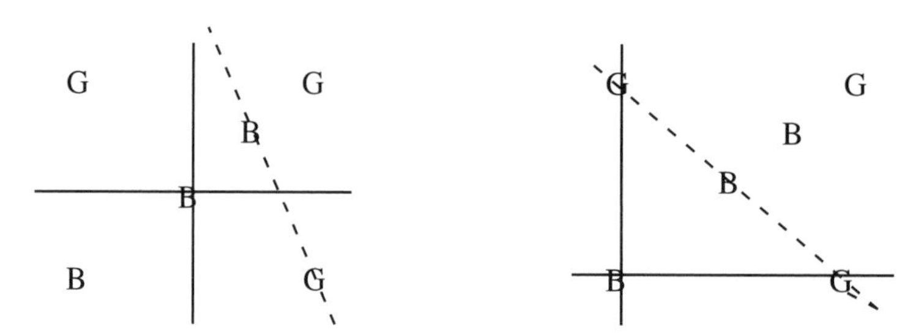

Figure 5.2. *Points in Example* 5.2.

Various modifications of the formulation have been suggested to overcome these problems such as requiring the one constraint in (5.1) to become

$$w_1x_{i1} + w_2x_{i2} + \cdots + w_px_{ip} \leq c - e + a_i, \quad n_G + 1 \leq i \leq n_G + n_B \tag{5.4}$$

so there is a gap between the two regions. However, then one has to develop techniques to decide into which set the points in this gap should be put. These difficulties have been overcome in the case when the mean values of the characteristics of the goods and the bads are different by using the normalization proposed by Glover (1990), namely, by adding the constraint

$$\sum_{j=1}^{p}\left(n_B\sum_{i=1}^{n_G}x_{ij}-n_G\sum_{i=n_G+1}^{n_G+n_B}x_{ij}\right)w_j=1. \tag{5.5}$$

One of the most general of the linear programming formulations was proposed by Freed and Glover (1986b). As well as the errors or external deviations a_i, by which a constraint is not satisfied, it also looks at the internal deviations e_i, which are how far from the boundary are the scores of the applicants who are correctly classified. Freed and Glover proposed an objective that was a weighted combination of the maximum internal and external deviations and the sum of the absolute values of the internal and external deviations. This is solved by the following linear program:

$$\begin{aligned}
\text{Minimize}\quad & k_0a_0-l_0e_0+\sum_{i=1}^{n_G+n_B}k_ia_i+\sum_{i=1}^{n_G+n_B}l_ie_i\\
\text{subject to}\quad & w_1x_{i1}+w_2x_{i2}+\cdots+w_px_{ip}\geq c-a_0-a_i+e_0+e_i, \quad 1\leq i\leq n_G,\\
& w_1x_{i1}+w_2x_{i2}+\cdots+w_px_{ip}\leq c+a_0+a_i-e_0-e_i,\\
& \qquad\qquad n_G+1\leq i\leq n_G+n_B,\\
& \sum_{j=1}^{p}\left(n_B\sum_{i=1}^{n_G}x_{ij}-n_G\sum_{i=n_G+1}^{n_G+n_B}x_{ij}\right)w_j=1,\\
& a_i,e_i\geq 0, \quad 0\leq i\leq n_G+n_B.
\end{aligned} \tag{5.6}$$

This formulation of the problem always gives a nontrivial solution and is invariant under linear transformation of the data.

The practical complaint about the linear programming approach is that since it has no statistical underpinning, one cannot evaluate whether the estimated parameters are statistically significant. Ziari et al. (1997) suggested a way to estimate the parameters by using jackknife or bootstrap resampling techniques. Both procedures assume that the sample used in the linear program represents the whole population of clients. By resampling a subset from the original sample, each new subsample will generate different parameter estimates. The distribution of these estimates reflects some information about the distribution of the parameters obtained from the full sample. In the jackknife procedure, a subset of t clients is left out of the sample and the linear program is solved on the remaining ones (usually $t = 1$). This is repeated using different subsets of t to get estimates for the parameters. In the bootstrap approach, a sample of $n = n_G + n_B$ is taken with replacement from the original data set of the same size and the linear programming parameters are calculated. This is repeated a number of times, and the mean and standard deviation of these bootstrap estimates of the parameters give approximations for the standard deviation in the original parameter estimates. Further details of bootstrapping and jackknifing are found in section 7.4.

Another advantage claimed for regression over linear programming is that in the former you can introduce variables one at a time into the scorecard starting with the most powerful—the forward approach of many statistical packages. This means you can develop lean and mean models wherein you decide in advance the number m of characteristics you want and regression then finds the m that are most discriminating. Nath and Jones (1988) showed how the jackknife approach applied to linear programming can be used to select the m most powerful characteristics. It does, however, involve solving the linear program a large number of times.

The review papers by Erenguc and Koehler (1990) and Nath, Jackson, and Jones (1992) compared the linear programming and regression approaches to classification on several data sets (but none of the sets was credit data). Their results suggest that the linear programming approach, while competitive, does not classify quite as well as the statistical methods.

5.3 Integer programming

Linear programming models minimize the sum of the deviation in the credit score of those who are misclassified. However, a more practical criterion would be to minimize the number of misclassifications, or the total cost of misclassifying if it is thought that D, the cost of misclassifying a bad as a good is very different from L, the cost of misclassifying a good as a bad. Building scorecards under these criteria can again be done by linear programming, but as some of the variables will have to be integer (0, 1, etc.), this technique is called integer programming. Such a model was given by Koehler and Erenguc (1990) as follows:

$$\begin{array}{lll} \text{Minimize} & L(d_1 + \cdots + d_{n_G}) + D(d_{n_{G+1}} + \cdots + d_{n_{G+B}}) & \\ \text{subject to} & w_1 x_{i1} + \cdots + w_p x_{ip} \geq c - M d_i, & 1 \leq i \leq n_G, \\ & w_1 x_{i1} + \cdots + w_p x_{ip} \leq c + M d_i, & n_G + 1 \leq i \leq n_G + n_B, \\ & 0 \leq d_i \leq 1, & d_i \text{ integer.} \end{array} \tag{5.7}$$

Thus d_i is a variable that is 1 if customer i in the sample is misclassified and 0 if not. Again, as it stands (5.7) is minimized by $c = 0$, $w_i = 0$, $i = 1, \ldots, p$, so one has to add normalization conditions, for example,

$$\begin{array}{l} \sum_{j=1}^{p} (s_j^+ + s_j^-) = 1, \\ \qquad 0 \leq s_j^+, \quad s_j^- \leq 1, \quad \text{and} \quad s_j^+, s_j^- \text{ integer}, \quad j = 1, \ldots, p, \\ -1 + 2s_j \leq w_j \leq 1 - 2s_j, \quad j = 1, \ldots, p. \end{array} \tag{5.8}$$

This set of constraints requires one of the s_i^+, s_i^- to be 1 and the corresponding w_i to be either greater than 1 or less than -1 (equivalent to forcing c to be positive or negative). Thus it is similar to requiring that

$$\sum_{j=1}^{p} w_i = 1 \quad \text{and} \quad c = +1 \text{ or } -1.$$

Joachimsthaler and Stam (1990) and Koehler and Erengac (1990) found that the integer model (5.7) is a better classification model than the linear programming models. However, it has two major disadvantages. First, it takes much longer than linear programming to solve and hence can deal with only very small sample sets with number of cases in the hundreds. Second, there are often a number of optimal solutions with the same number of misclassifications on the training set but with quite different performances on holdout samples.

The computational complexity of the solution algorithms for integer programming is a real difficulty, which probably rules it out for commercial credit-scoring applications. Several authors suggested ways to exploit the structure of the classification problem to increase the speed of the branch-and-bound approach to solving integer programming in this case (Rubin 1997, Duarte Silva and Stam 1997). However, this still makes the approach feasible only when the sample set is 500 or smaller.

Various authors added extra conditions and secondary objectives to (5.7) to ensure a classification that is robust in the holdout sample, as well as minimizing the number of misclassifications in the training sample. Pavar, Wanarat, and Loucopoulus (1997) suggested models that also maximize the difference between the mean discriminant scores; Bajgier and Hill (1982) sought to minimize the sum of the exterior deviations (the difference between the score of the misclassified and the cutoff score) as well as the number of misclassifications. Rubin (1990) took a secondary goal of maximizing the minimum interior deviation—the difference between the score of the correctly classified and the cutoff score.

Thus far, integer programming has been used to solve the classification problem with the minimum number of misclassifications criterion, but it can also be used to overcome two of the difficulties that arise in the linear programming approach to minimizing the absolute sum of errors. The first is the problem about removing trivial solutions and ensuring the weights are invariant if the origin shifts in the data. This was overcome in the minimum misclassification formulation (5.7) by using the normalization (5.8). A similar normalization can be applied to the MSD or MMD problem. This would lead to the following integer program to minimize the sum of the absolute value of the deviations (MSD) (see Glen (1999)):

$$\begin{aligned}
\text{Minimize} \quad & a_1 + a_2 + \cdots + a_{n_G+n_B} \\
\text{subject to} \quad & (w_1^+ - w_1^-)x_{i1} + (w_2^+ - w_2^-)x_{i2} + \cdots + (w_p^+ - w_p^-)x_{ip} \geq c - a_i, \\
& \qquad 1 \leq i \leq n_G, \\
& (w_1^+ - w_1^-)x_{i1} + (w_2^+ - w_2^-)x_{i2} + \cdots + (w_p^+ - w_p^-)x_{ip} \leq c + a_i, \\
& \qquad n_{G+1} \leq i \leq n_G + n_B, \\
& (w_1^+ + w_1^-) + (w_2^+ + w_2^-) + \cdots + (w_p^+ + w_p^-) = 1, \\
& cs_j^+ \leq w_i^+ \leq s_j^+, \quad cs_j^- \leq w_i^- \leq s_j^-, \quad c = 1, \ldots, p, \\
& s_j^+ + s_j^- \leq 1, \quad j = 1, \ldots, p,
\end{aligned} \tag{5.9}$$

where $w_i^+, w_i^- \geq 0$, $0 \leq s_j^+, s_j^- \leq 1$, s_j^+, s_j^- integer, $j = 1, \ldots, p$, and c is any positive constant less than 1.

These conditions guarantee that at most one of w_i^+, w_i^- is positive, but some of the $w_i^{\pm}$ must be nonzero. The result is the same as the normalization in (5.8). With these extra $2p$ integer variables, one can ensure there are no trivial solutions and the results are invariant under change of origin.

Another criticism of the linear programming approach was that it was hard to choose the best scorecard using only m characteristics. Jackknifing methods were suggested to overcome this problem, but one can also use the integer variable formulation of (5.9) to find the best scorecard using only m characteristics. All one needs to do is to add another constraint to (5.9), namely,

$$\sum_{j=1}^{p} (s_j^+ + s_j^-) = m. \tag{5.10}$$

This ensures that only m of the weights w_i^+, w_i^- are positive and so only m of the characteristics are nonzero. One could solve this problem for different values of m to get a view of which characteristics should be entered when m variables are already included. One technical difficulty is that several different scorecards may be optimal using m characteristics, so going to $m + 1$ characteristics might mean that several characteristics leave the scorecard and several new ones enter. One can overcome this by finding all optimal m characteristic scorecards as follows. If solving the problem with (5.9) and (5.10) gives a solution where the nonzero characteristics are the set $C = \{i_1, i_2, \ldots, i_m\}$, add the constraint

$$\sum_{j \in C} (s_j^+ + s_j^-) \leq m - 1$$

and solve again. This will give a different solution and one can check if it gives the same objective function. If it does not, the solution is unique, but if it does, then repeat the process until all the optimal solutions are found.

5.4 Neural networks

Neural networks were originally developed from attempts to model the communication and processing information in the human brain. In the brain, large numbers of dendrites carry electrical signals to a neuron, which converts the signals to a pulse of electricity sent along an axon to a number of synapses, which relate information to the dendrites of other neurons. The human brain has an estimated 10 billion neurons (Shepherd and Koch, 1990). Analogous to the brain, a neural network consists of a number of inputs (variables), each of which is multiplied by a weight, which is analogous to a dendrite. The products are summed and transformed in a "neuron" and the result becomes an input value for another neuron.

5.4.1 Single-layer neural networks

A single-layer neural network consists of the components just described where instead of the transformed value becoming the input for another neuron, it is the value we seek. Thus it may predict whether a case is to be accepted or rejected. A single-layer neural network may be represented diagrammatically as in Figure 5.3.

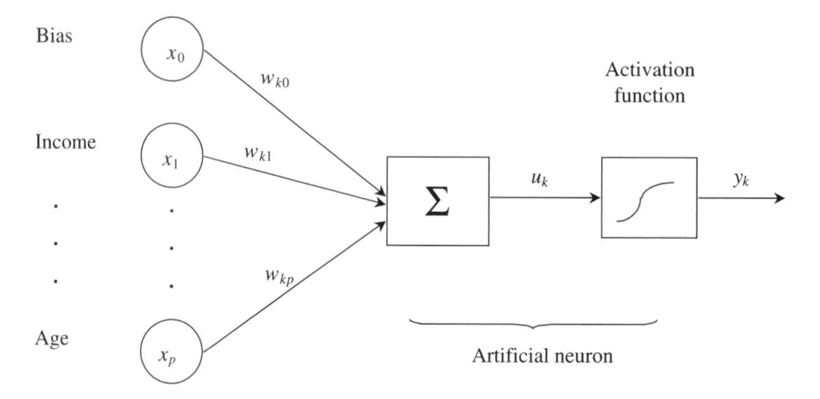

Figure 5.3. *A single-layer neural network.*

We can represent a single-layer neural network algebraically as

$$u_k = w_{k0}x_0 + w_{k1}x_1 + \cdots + w_{kp}x_p = \sum_{q=0}^{p} w_{kq}x_q, \tag{5.11}$$

$$y_k = F(u_k). \tag{5.12}$$

Each of $x_1, \ldots, x_p$ is a variable, such as a characteristic of a credit card applicant. Each takes on a value known as a signal. The weights, often known as synaptic weights, if positive are known as excitory because they would increase the corresponding variable, and if negative are called inhibitory because they would reduce the value of u_k for positive variables. Notice that the subscripts on each weight are written in the order (k, p), where k indicates the neuron to which the weight applies and p indicates the variable. In a single-layer neuron, $k = 1$

because there is only one neuron. Notice also that variable x_0 is ascribed the value $+1$ so that the $w_{k0}x_0$ term in (5.11) is just w_{k0}, often known as bias. This has the function of increasing or decreasing u_k by a constant amount.

The u_k value is then transformed using an activation (or transfer or squashing) function. In early networks, this function was linear, which severely limited the class of problems that such networks could deal with. Various alternative transfer functions are used, and they include the following:

- Threshold function:

$$F(u) = 1 \quad \text{if } u \geq 0, \\ = 0 \quad \text{if } u < 0, \tag{5.13}$$

  which implies that if u is 0, or greater the neuron outputs a value of 1; otherwise, it outputs a 0.

- Logistic function:

$$F(u) = \frac{1}{1 + e^{-au}}. \tag{5.14}$$

Both are represented in Figure 5.4. The value of a in the logistic function determines the slope of the curve. Both functions constrain the output of the network to be within the range (0, 1). Sometimes we wish the output to be in the range $(-1, +1)$ and the tanh function, $F(u) = \tanh(h)$, is used.

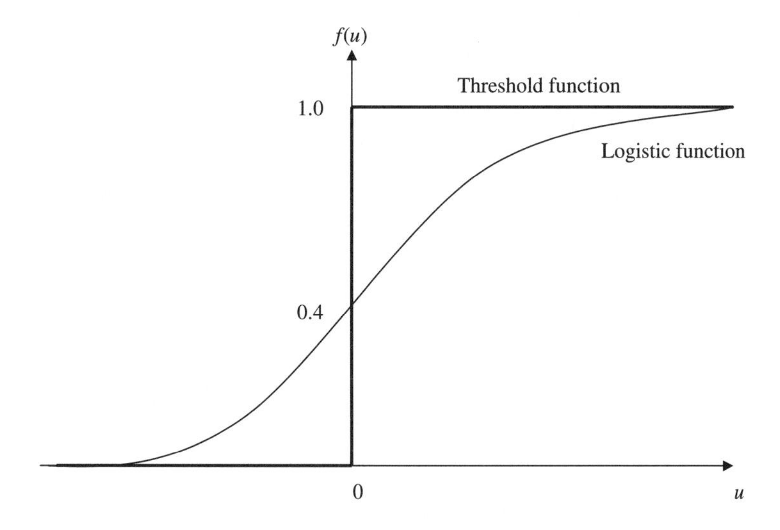

Figure 5.4. *Threshold and logistic activation functions.*

Given the values for the weights and the transfer function, we can predict whether an application for a credit card is to be accepted or rejected by substituting the applicant's characteristics into (5.11), calculating the value of y_k from (5.12), and comparing this value with a cutoff value.

A model consisting of a single neuron and a threshold activation function is known as a perceptron. Rosenblatt (1958, 1960) showed that if the cases to be classified were linearly separable, that is, they fall either side of a straight line if there are two input signals (and of a hyperplane when there are p signals), then an algorithm that he developed would converge to establish appropriate weights. However, Minsky and Papert (1969) showed that the perceptron could not classify cases that were not linearly separable.

In 1986, Rumelhart, Hinton, and Williams (1986a, 1986b) showed that neural networks could be developed to classify nonlinearly separable cases using multiple-layer networks with nonlinear transfer functions. At approximately the same time, methods for estimating the weights in such models using back-propagation were described in Rumelhart and McClelland (1986), Parker (1982), and LeCun (1985). Since this is the most commonly used method, we describe it below.

5.4.2 Multilayer perceptrons

A multilayer perceptron consist of an input layer of signals, an output layer of output signals (different y_v values), and a number of layers of neurons in between, called hidden layers. Each neuron in a hidden layer has a set of weights applied to its inputs which may differ from those applied to the same inputs going to a different neuron in the hidden layer. The outputs from each neuron in a hidden layer have weights applied and become inputs for neurons in the next hidden layer, if there is one; otherwise, they become inputs to the output layer. The output layer gives the values for each of its member neurons whose values are compared with cutoffs to classify each case. A three-layer network is shown in Figure 5.5.

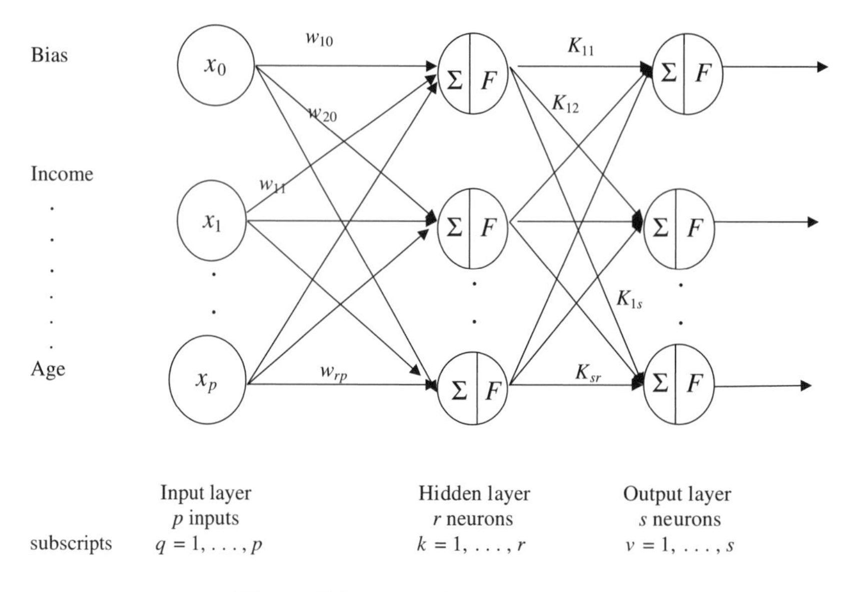

Figure 5.5. *A multilayer perceptron.*

We can represent a multilayer perceptron algebraically as follows. We do this for Figure 5.5. From (5.11) and (5.12), we have

$$y_k = F_1\left(\sum_{q=0}^{p} w_{kq}x_q\right), \tag{5.15}$$

where the subscript 1 on F indicates it is the first layer after the input layer. The y_k, $k = 1, \ldots, r$, are the outputs from the first hidden layer. Since the output of one layer is an input to the following layer, we can write

$$z_v = F_2\left(\sum_{k=1}^{r} K_{vk} y_k\right) = F_2\left(\sum_{k=1}^{r} K_{vk}\left(F_1\left(\sum_{q=0}^{p} w_{kq} x_q\right)\right)\right), \tag{5.16}$$

where z_v is the output of neuron v in the output layer, $v = 1, \ldots, s$, F_2 is the activation function in the output layer, and K_{vk} is the weight applied to the y_k layer that joins neuron k in the hidden layer and neuron v in the output layer.

The calculation of the vector of weights is known as training. There are many such methods, but the most frequently used method is the back-propagation algorithm. Training pairs, consisting of a value for each of the input variables for a case and the known classification of the case, are repeatedly presented to the network and the weights are adjusted to minimize some error function.

5.4.3 Back-propagation algorithm

Initially all weights are set equal to some randomly chosen numbers. A training pair is selected and the x_p values used to calculate the z_v values using (5.16). The difference between the z_v values and the known values, o_v, are calculated. This is known as a forward pass. The backward pass consists of distributing the error back through the network in proportion to the contribution made to it by each weight and adjusting the weights to reduce this portion of the error. Then a second training pair is selected and the forward and backward passes are repeated. This is repeated for all cases, called an epoch. The whole procedure is repeated many times until a stopping criterion is reached.

The change in the weights effected when a new case is presented is proportional to the first derivative of the error term with respect to each weight. This can be understood as follows. Define the error when the training case t is presented, $e_v(t)$, as

$$e_v(t) = o_v(t) - y_v(t), \tag{5.17}$$

where $o_v(t)$ is the actual observed outcome for case t in neuron v and $y_v(t)$ is the predicted outcome. The aim is to choose a vector of weights that minimizes the average value over all training cases of

$$E(t) = 0.5\sum_{v=1}^{s} e_v^2(t), \tag{5.18}$$

where s is the number of neurons in the output layer. This average value is

$$E_{\text{mean}}(t) = \frac{1}{N}\sum_{t=1}^{N} E(t), \tag{5.19}$$

where N is the number of training cases.

For any neuron v in any layer c, we can write

$$u_v^{[c]} = \sum_{k=0}^{r} w_{vk} y_k^{[c-1]}, \tag{5.20}$$

$$y_v^{[c]} = F(u_v^{[c]}), \tag{5.21}$$

which are simply (5.11) and (5.12) rewritten to apply to any neuron in any layer rather than just to the input layer and first neuron.

Hence the partial derivative of $E(t)$ with respect to weight w_{vk} can, according to the chain rule, be written as

$$\frac{\partial E(t)}{\partial w_{vk}(t)} = \frac{\partial E(t)}{\partial e_v(t)} \cdot \frac{\partial e_v(t)}{\partial y_v(t)} \cdot \frac{\partial y_v(t)}{\partial u_v(t)} \cdot \frac{\partial u_v(t)}{\partial w_{vk}(t)}. \tag{5.22}$$

From (5.18),

$$\frac{\partial E(t)}{\partial e_v(t)} = e_v(t). \tag{5.23}$$

From (5.17),

$$\frac{\partial e_v(t)}{\partial y_v(t)} = -1. \tag{5.24}$$

From (5.21),

$$\frac{\partial y_v(t)}{\partial u_v(t)} = F'(u_v(t)). \tag{5.25}$$

From (5.20),

$$\frac{\partial u_v(t)}{\partial w_{vk}(t)} = y_k(t). \tag{5.26}$$

By substitution,

$$\frac{\partial E(t)}{\partial w_{vk}(t)} = -e_v(t) \cdot F'(u_v(t)) \cdot y_k(t). \tag{5.27}$$

The change in weights between the forward pass and the backward pass is therefore

$$\Delta w_{vk}(t) = -\eta \frac{\partial E(t)}{\partial w_{vk}(t)} = \eta \delta_v(t) y_k(t), \tag{5.28}$$

where $\delta_v(t) = e_v(t)F'(u_v(t))$. The constant η, called a training rate coefficient, is included to alter the change in w so as to make the changes smaller or larger. Smaller values improve accuracy but extend the training time. Equation (5.28) is known as the delta rule (or the Widrow–Hoff rule).

The implementation of this rule depends on whether neuron v is in the output layer or a hidden layer. If neuron v is in the output layer, the value of e_v is directly observable because we know both the observed outcome o_v and the predicted outcome y_v. If neuron v is in a hidden layer, one component in e_v, o_v, is not observable. In this case, we still use (5.28), except that we must calculate $\delta_v(t)$ in a different way. The method used is to calculate the value of δ for each neuron in the output layer, multiply each δ by the weight that connects the neuron for which it was calculated and a neuron in the previous layer, and then sum these products for each neuron in the previous layer. Figure 5.6 illustrates this for two neurons in the output layer.

In general,

$$\delta_k^{[c-1]} = F_k^{[c-1]'} \sum_{v=1}^{s} \delta_v^{[c]} w_{vk}^{[c]}. \tag{5.29}$$

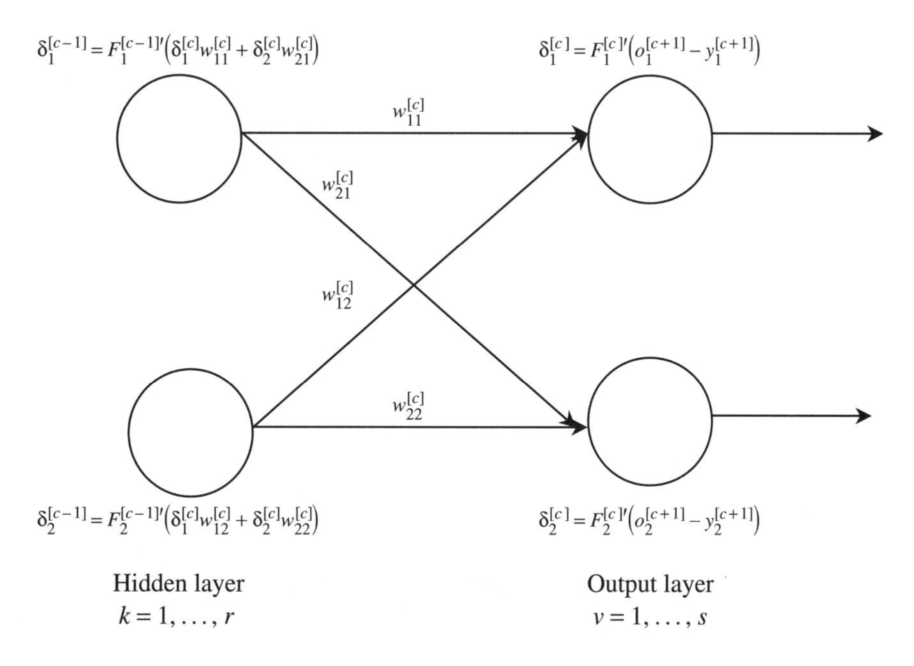

Figure 5.6. *Back-propagation.*

The δ value for any neuron in one layer is distributed recursively, according to the connecting weights, to neurons in the previous layer, and then, for each previous layer neuron, the weighted δ's connecting to it are summed. This gives a value of δ for the previous layer. The procedure is repeated back, layer by layer, to the original set of weights from the input layer.

The change in weights connecting any layer, $[c-1]$, where variables are subscripted k, to the next layer, $[c]$, where variables are subscripted v, is therefore, from (5.28) and (5.29),

$$\Delta w_{vk} = \eta \delta_v^{[c]} y_k^{[c-1]}. \tag{5.30}$$

This rule is often amended by the inclusion of a momentum term, $\alpha \cdot \Delta w_{vk}(t-1)$:

$$\Delta w_{vk}(t) = \alpha \Delta w_{vk}(t-1) + \eta \delta_v^{[c]} y_k^{[c-1]}. \tag{5.31}$$

This is now called the generalized delta rule because (5.30) is a special case of (5.31), where $\alpha = 0$. The momentum may be added to speed up the rate at which the w values converge while reducing the effect of the calculated weights oscillating—that is, on successive iterations alternately increasing and decreasing because the error surface in weight space has a local minimum. Haykin (1999) showed that for the weights to converge, α must be set within the range ± 1. In practice it is normally set to $0 < \alpha < +1$.

The back-propagation algorithm is a gradient descent method. Intuitively, we can think of a diagrammatic representation if we consider only two inputs, as in Figure 5.7. The very idealized surface represented by $ABCD$ represents the size of the error E as a function of the vector of weights. The aim is to find the weight vector $\mathbf{w}^*$ that minimizes the error, as at point E. Our initial choice of weight vector $\mathbf{w}_a$ gives an error value indicated by point F for training case 1 from a feed-forward pass. During a backward pass the weight vector is changed to be $\mathbf{w}_b$ to give an error indicated by point G. The move from point F to point G given by the algorithm is an approximation to the change which would be the

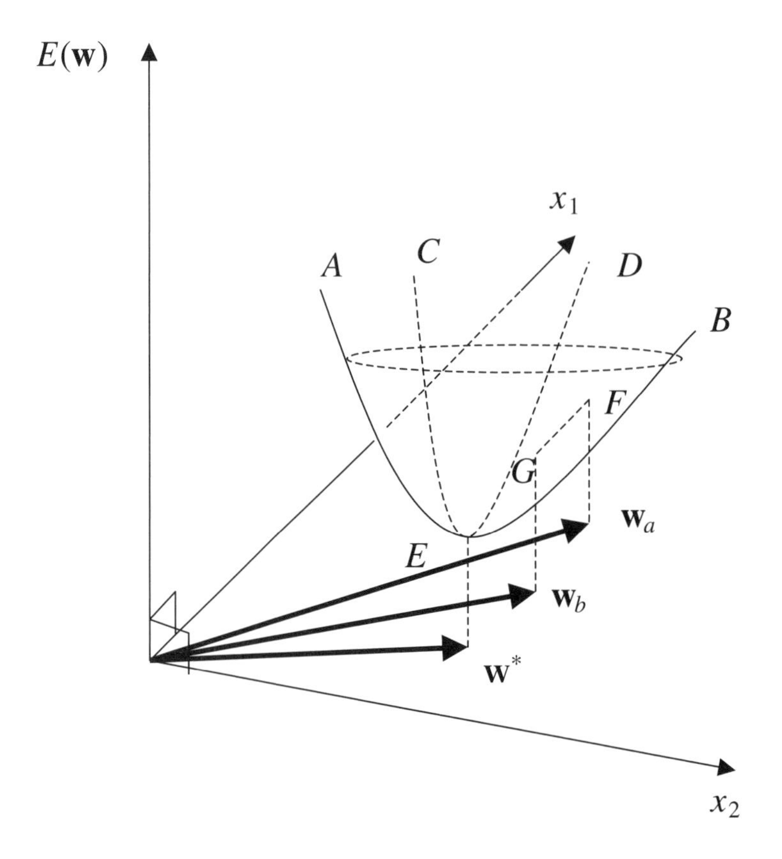

Figure 5.7. *Back-propagation and an error surface.*

greatest descent toward E. In practice, the surface $ABCD$ will contain very many bumps and wrinkles giving many local maxima and minima. To identify a minimum value of the error, we can require the first partial derivative of the error surface with respect to the weight vector $\frac{\partial E}{\partial w_1}, \frac{\partial E}{\partial w_2}, \ldots$ to be zero. The algorithm would then be stopped when this occurred. However, this is extremely time-consuming and requires knowledge of these derivatives. An alternative method used in practice is to stop the algorithm when the absolute change in $E_{\text{mean}}(\mathbf{w})$ is sufficiently small—for example, less than 0.01 in each epoch.

5.4.4 Network architecture

A number of aspects of a neural network have to be chosen by the experimenter. These include the number of hidden layers, the number of neurons in the hidden layers, and the error function to use (we have, so far, considered only one as represented by (5.17) and (5.18)). Here we concentrate on the number of hidden layers.

First, consider why we have hidden layers. If we have a single layer, we can classify only groups that are linearly separable. When we include hidden layers together with a nonlinear activation function, the final network can correctly classify cases into classes that are not linearly separable. The hidden layers model the complicated nonlinear relationships between the variables and transform the input variables into hidden space, which is further transformed. As such, these hidden layers extract features in the data. In particular, introducing a single nonlinear hidden layer results in a network that will give a value above or

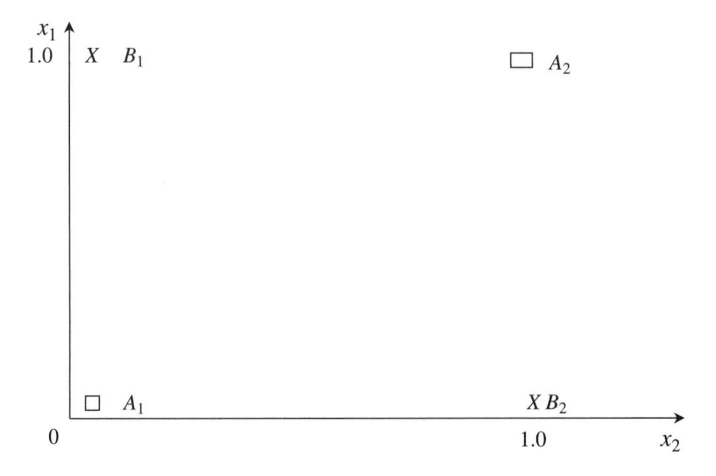

Figure 5.8. *Data for the XOR problem.*

below a cutoff value in a convex region of the input variables. Having a second hidden layer will allow these convex regions to be combined, which can give nonconvex or even totally separate regions. Hand (1997) argued that for this reason, "in principle two hidden layers are sufficient for any problem."

The effect of introducing a hidden layer and nonlinear activation function can be shown using the XOR problem. Consider just two inputs (perhaps characteristics of credit card applicants) which have values of 0 or 1. Cases in class A have either $x_1 = 0$ and $x_2 = 0$ or $x_1 = 1$ and $x_2 = 1$. Cases in class B have either $x_1 = 1$ and $x_2 = 0$ or $x_1 = 0$ and $x_2 = 1$. In Figure 5.8, we have plotted these four possibilities in a graph with axes x_1 and x_2. No straight line can separate the two groups. However, if the values of x_1 and x_2 are weighted and subject to a nonlinear transformation to give y_1 and y_2 values, these y_1 and y_2 values may then be used to separate the groups by using a straight line. A hidden layer consisting of a nonlinear transformation function of the sum of the weighted inputs has been included. If weights with values $+1$ in all four cases, together with biases of $-\frac{3}{2}$ and $-\frac{1}{2}$, are applied, we have

$$u_1 = x_1 + x_2 - \frac{3}{2}, \qquad u_2 = x_1 + x_2 - \frac{1}{2}.$$

If we then use the activation function

$$\begin{aligned} u_1 < 0 &\Rightarrow y_1 = 0, \\ u_1 \geq 0 &\Rightarrow y_1 = 1, \\ u_2 < 0 &\Rightarrow y_2 = 0, \\ u_2 \geq 0 &\Rightarrow y_2 = 1, \end{aligned}$$

then the four cases have values as plotted in Figure 5.9. Class A cases, which in terms of (x_1, x_2) values were $(0, 0)$ or $(1, 1)$, now become, in terms of (y_1, y_2) values, $(0, 0)$ or $(1, 1)$. Class B cases, which in x space were $(0, 1)$ or $(1, 0)$, now become, in y space, $(0, 1)$ and $(0, 1)$. Now a straight line can separate the two groups. This straight line represents the output layer. However, for practical reasons more than two layers may occasionally be used.

In addition, a practitioner needs to decide how many neurons should be included in each hidden layer. The optimal number is not known before training begins. However, a number of heuristics are available for guidance (Garson 1998).

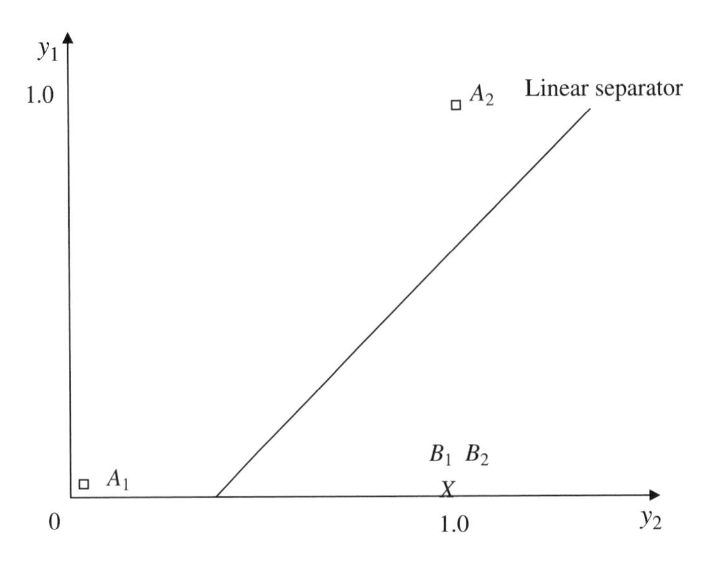

Figure 5.9. *Transformed data.*

5.4.5 Classification and error functions

In credit scoring, we are often interested in classifying individuals into discrete groups, for example, good payers and poor payers or good, poor, and bad payers (Desai et al. 1996, 1997). In principle, when we wish to classify cases into Z groups, we need Z outputs from a multilayer perceptron. Using results by White (1989) and Richard and Lipman (1991), it can be shown that a multilayer perceptron, trained by back-propagation to minimize the value of (5.19) using a finite number of independent cases and identically distributed values of inputs, leads asymptotically to an approximation of the posterior probabilities of class membership (see Haykin 1999, Bishop 1995). Therefore, provided the training sample fulfils these conditions, we can adopt the decision rule that we allocate a case to a group C_g, $g = 1, \ldots, Z$, if the output value of the output layer on which that group has been trained, $F_g(\mathbf{x})$, is greater than the output value of the output layer on which any other group has been trained, $F_h(\mathbf{x})$—that is, if

$$F_g(\mathbf{x}) > F_h(\mathbf{x}), \quad g \neq h. \tag{5.32}$$

However, the error function of (5.19) is not necessarily the best error function for classification problems. It is derived from maximizing the likelihood that the overall network weights and functions generate the set of output values, o_v, which are themselves deterministically determined with additional normally distributed noise. But in classification problems, the output values are binary variables: a case either is a member of a group or is not a member.

Instead of (5.19), we may derive an alternative error function. Suppose we have a network with one output per group g, y_{vg}. The probability of observing the output values, given the input vector $x(t)$, is y_g and the distribution of these probabilities is

$$P(\mathbf{o}(t)|\mathbf{x}(t)) = \prod_{g=1}^{Z} (y_g^t)^{o_g^t}. \tag{5.33}$$

Forming the likelihood function, taking its log, and multiplying by -1, we gain

$$E_2 = -\sum_t \sum_{g=1}^{Z} O_g^t \ln y_g^t. \tag{5.34}$$

This is the relative entropy criterion.

Since the y_v values are to be interpreted as probabilities, they need to have the following properties: $0 \leq y_{vg} \leq 1$ and $\sum_{g=1}^{Z} y_{vg} = 1$. To achieve this, the softmax activation function is often used:

$$y_g = \frac{e^{u_g}}{\sum_{g=1}^{Z} e^{u_g}}. \tag{5.35}$$

Precisely this error and activation function were used by Desai et al. (1997) in their comparison of the classificatory performance of logistic regression, linear discriminant analysis, and neural networks.

5.5 Genetic algorithms

Very simply, a genetic algorithm (GA) is a procedure for systematically searching through a population of potential solutions to a problem so that candidate solutions that come closer to solving the problem have a greater chance of being retained in the candidate solution than others. GAs were first proposed by Holland (1975) and have analogies with the principle of evolutionary natural selection proposed by Darwin (1859).

5.5.1 Basic principles

Suppose that we wish to calculate the parameters $a_1, a_2, \ldots, a_p, b_1, b_2, \ldots, b_p$, and c in the following credit-scoring equation to classify applicants for a loan:

$$f(x_i) = a_1 x_{i1}^{b_1} + a_2 x_{i2}^{b_2} + \cdots + a_p x_{ip}^{b_p} + c, \tag{5.36}$$

where $x_{i1}, \ldots, x_{ip}$ are characteristic values for applicant i.

Once the parameters are estimated, an applicant may be classified as good or bad according to whether $f(x_i)$ is greater than or less than 0.

The procedures followed in the GA method of calculation are shown in Figure 5.10. First, the population of candidate values for the a's, b's, and c's is selected. For example, the range of possible a_1 values may be -1000 to $+1000$ and so on for each a, the range for b_1 may be 0 to 6, and so on. For the purposes of the algorithm, each number in the solution is represented in its binary form. A solution to the calculation problem is a complete set of $\{0, 1\}$ values for $a_1, \ldots, a_p, b_1, \ldots, b_p$, and c.

At this stage, we introduce some terminology. A collection of 0s and 1s is known as a string or chromosome. Within a string are particular features or genes, each of which takes on particular values or alleles. Thus a solution to the credit-scoring function problem consists of sets of genes arranged in a row, each gene having a value of 0 or 1 and each set relating to $a_1, x_1, n_1, a_2, x_2, n_2$, and so on. The whole row is a chromosome or string.

At the second stage, a number of solutions are selected for inclusion in the intermediate population. These may be chosen randomly unless the analyst has prior knowledge of more appropriate values. To select members of the intermediate population, the performance of

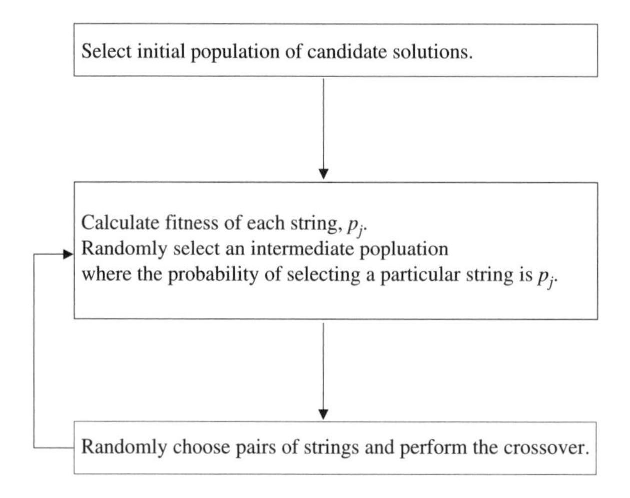

Figure 5.10. *Procedural stages of a genetic algorithm.*

each solution in the initial population is calculated. The performance is often called fitness. In the credit-scoring problem, the fitness might be calculated as the percentage of cases correctly classified. Let j indicate a particular solution. We wish to compare the fitness of each solution, f_j, but the value of f_j depends on the fitness function used. To avoid this, we calculate the normalized fitness function for each candidate solution, p_j, as

$$p_j = \frac{f_j}{\sum_{j=1}^{n_{\text{pop}}} f_j}, \tag{5.37}$$

where n_{pop} is the number of candidate solutions in the population. The intermediate population is selected by randomly selecting strings from the initial population, where p_j is the probability that a string is selected on any occasion. This may be implemented by spinning a roulette wheel where the portion of the circumference that corresponds to each string is equal to p_j of the whole. The wheel is spun n_{pop} times.

At this second stage, an intermediate population with members of the original population only is created. No new strings are created. In the third stage, new strings are created. A given number of pairs of solutions from the intermediate population is chosen and genetic operators are applied. A genetic operator is a procedure for changing the values within certain alleles in one or a pair of strings. Either of two operators may be followed: crossover and mutation. Each chromosome has the same chance of selection for crossover, p_c, which is determined by the analyst. This may be implemented by repeatedly generating a random number r, and if the value of r for the kth number is less than p_c, the kth chromosome is selected.

Crossover is implemented when the first or last n bits of one string (indexed from, say, the left) are exchanged with the first or last n bits of another string. The value of n would be chosen randomly. For example, in the following two parent strings, the last five bits have been exchanged to create two children:

$$0110|11100 \rightarrow 011010110,$$
$$1100|10110 \rightarrow 110011100.$$

The children replace their parents within the population.

Mutation is implemented when an element within a string is randomly selected and its value flipped, that is, changed from 0 to 1 or vice versa. The analyst selects the probability p_m that any element will be mutated.

The selected chromosome, including the children resulting from crossover and after mutation, form the new population. The second and third stages are then repeated a chosen number of times.

The parameters chosen by the analyst are the number of candidate solutions in the population, the probabilities of crossover and of mutation (and hence the expected number of crossovers and mutations), and the number of generations. Michalewicz (1996) suggested some heuristic rules and that the population size of 50–100, with p_c between 0.65 and 1.00 and p_m of 0.001 to 0.01, is often used. Thus Albright (1994) in his application of GAs to the construction of credit union score cards used a population of 100 solutions and up to 450 generations with p_c around 0.75 and p_m around 0.05.

Equation (5.36) is one example of a scoring function that has been used in the scoring literature. Usually the literature does not reveal such detail. A paper that does give such detail (Yobas, Crook, and Ross 2000) has a different function. This can be explained by the chromosome of a solution in Figure 5.11.

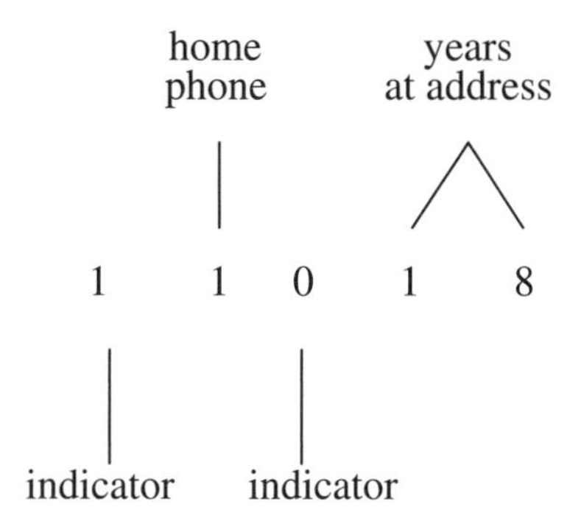

Figure 5.11.

Here there are five genes arranged in groups consisting of two parts. Each group relates to a characteristic (e.g., home phone). Within each group, the first gene, a single bit, indicates whether the variable enters the classification rule (value = 1 if it does, 0 if it does not). The second gene indicates the value of the variable (in the case of binary variables) or the second and third genes indicate a range of values (in the case of continuous variables). If a case fulfils the condition represented by the chromosome, it is predicted to be a good payer; otherwise, it is a bad payer. The condition shown in the chromosome in Figure 5.11 is that the applicant has a home phone. Years at address does not enter the classification rule because the indicator for this variable has the value 0. If the indicator for years at address had been 1, this would also have been relevant to the prediction. In that case the condition would read, "if the applicant has a home phoneand his/her years at address is in the range 1 to 8 inclusive, then (s)he is a good payer."

5.5.2 Schemata

In this section we consider how similarity of values in particular portions within strings can speed up the search for the optimal solution, and so we try to give an understanding of how and why GAs work. To do this, it is helpful to consider a particular problem. Suppose that we wish to use a GA to find the value of x that maximizes $y = -0.5x^2 + 4x + 100$. By simple differentiation, we know that the answer is 4.

Suppose we set the population to be integer values between 0 and 15. Then we randomly choose four candidate solutions, which are shown in Table 5.1. Examination of Table 5.1 suggests that strings in which the value furthest to the left is 0 have a greater relative fitness p_j than the others.

Table 5.1. *Candidate solutions for a genetic algorithm.*

String no.	String	x (decimal value)	y (fitness)	p_j
1	1010	10	90.0	0.2412
2	0111	7	103.5	0.2775
3	1100	12	72.0	0.1930
4	0011	2	107.5	0.2882

A schema (a term first used by Holland (1968)) is a template that states the values in particular positions within a string. If we use the symbol $*$ to mean "don't care" what the value is, we can see that the schema $0**1$ is possessed by strings 2 and 4 but not by 1 and not by 3. The similarity of strings is used to show why a GA is likely to progress toward an optimum solution.

Notice that each string matches 2^l schemata. (In each position, there is either the particular 0 or 1 value or a $*$ and there are l positions.) If there are n strings in a population, between 2^l and $n2^l$ different schemata match the population, depending on how many strings have the same schema.

Now consider the effects of the creation of a new population on the expected number of schemata within it. We show that as new generations are created, the expected number of strings in a given-size population that match fitter schemata will increase. That is, fitter schemata will come to dominate the population of solutions.

Suppose the population is set at n_{pop} strings. The expected number of strings that match schema s that will exist in the new generation at time $(t + 1)$ equals

$$\begin{pmatrix} \text{number of strings} \\ \text{which match} \\ \text{schema } s\text{, time } t \end{pmatrix} \times \begin{pmatrix} \text{probability that strings} \\ \text{which match } s \text{ are selected} \end{pmatrix} \times (\text{number of selections}). \tag{5.38}$$

Let $m(s, t)$ denote the number of strings that match s at time t. This is the first term. Remember from (5.37) that in the selection process, the probability that a string j is selected, p_j, is the fitness of the string j divided by the total fitness of all strings in the population. Thus the probability that a string that matches schema s is selected is the average fitness of the strings that match s divided by the total fitness of all strings in the population. That is,

$$\frac{f(s,t)}{\sum_{j=1}^{n_{\text{pop}}} f_j}, \tag{5.39}$$

where $f(s, t)$ is the average fitness of all strings that match s at time t. Finally, the number of selections is just the population size. Substituting into (5.38) gives

$$m(s,t+1) = m(s,t) \cdot \frac{f(s,t)}{\sum_{j=1}^{n_{pop}} f_j} \cdot n_{pop}, \tag{5.40}$$

where $m(s,t+1)$ is the expected number of strings that match s at time $t+1$. Writing

$$\frac{\sum_{j=1}^{n_{pop}} f_j}{n_{pop}}$$

as $\bar{f}$, the average fitness of all strings in the population, we can rewrite (5.40) as

$$m(s,t+1) = m(s,t) \cdot \frac{f(s,t)}{\bar{f}}. \tag{5.41}$$

Equation (5.41) shows that as each generation is created, if the average fitness of schema s is greater than that in a population, then the expected number of strings in the population with this schema will increase. Correspondingly, the expected number of strings within the population with a schema with a relatively low fitness will decrease. In fact, it can be shown that if the relative fitness of a schema exceeds the average, the number of strings with this schema increases exponentially as further generations are created. To show this, suppose that $f(s,t) = (1+k)\bar{f}$, that is, the average fitness of a schema is a constant proportion of $\bar{f}$ larger than $\bar{f}$. Then from (5.41), $m(s,1) = m(s,0)(1+k)$, $m(s,2) = m(s,1)(1+k) = m(s,o)(1+k)^2$, and in general $m(s,t) = m(s,0)(1+k)^t$.

We now amend (5.41) to include crossover and mutation. To understand this, we need to define the order and the defining length of a schema. The order of a schema H, labeled $O(H)$ is the number of numbers (0 and 1) in fixed positions in the schema. The defining length of a schema (denoted δ) is the difference in positions between the first and last fixed numbers in the schema. For example, in the schema

$$10 * * * * 1 *,$$

the order is 3 and the defining length is $7 - 1 = 6$.

Now consider crossover. Consider two schemata:

$$\begin{aligned} S_1&: \quad 11 * * * * * 01, \\ S_2&: \quad * * * 11 * * * *. \end{aligned}$$

Since the crossover point is chosen randomly and each point to be chosen is equally likely to be chosen, and supposing that the crossover operator is a swapping of the last four elements between two strings, crossover is more likely to destroy S_1 than S_2. The last four elements of S_2 are "don't care," and so, if replaced by any string, the new child would still be an example of schema S_2. On the other hand, it is much less likely that a string with 01 as the last two elements would be selected to replace the last four elements of S_1. The relevant difference between S_1 and S_2 is the defining length, which is 8 for S_1 $(9-1)$ and 1 for S_2 $(5-4)$.

Thus the greater the defining length, the greater the probability that the schema will be destroyed. In particular, if the crossover location is chosen from $(L-1)$ sites, the probability that a schema will be destroyed is $\frac{\delta}{(L-1)}$ and the probability that it will survive is

$$1 - \left(\frac{\delta}{L-1}\right). \tag{5.42}$$

Given that the probability that a string is selected for crossover is p_c, the probability of survival can be amended to $1 - p_c(\frac{\delta}{L-1}.)$. However, it is possible that because of the particular positions of the particular numbers (0, 1s) in two strings that are to be crossed over, after crossover one of the schema matching the two strings will still match the child. For example, if the two strings to be mated began with 11 and ended with 01—and thus both are an example of S_1 above—then crossover at a point between the second 1 and before the 0 would give a child that still matched S_1. Therefore, the probability that a schema survives is at least $1 - p_c(\frac{\delta}{L-1}\cdot)$.

We can now amend (5.41) to allow for crossover. $m(s, t+1)$ is the expected number of strings that match s after the creation of a new generation. Equation (5.41) allowed for selection only. However, when we also allow for crossover, the expected number of strings that match s in the new generation equals the number in the reproduced population multiplied by the chance that the string survives. Remembering the "at least" expression of the last paragraph, we can amend (5.41) to be

$$m(s, t+1) \geq m(s, t) \cdot \frac{f(s, t)}{\bar{f}} \cdot \left(1 - \left(\frac{p_c \delta}{L-1}\right)\right). \tag{5.43}$$

For a given string length $(L-1)$, given defining length (δ), and given probability of selection for crossover (p_c), our original interpretation of (5.41) applies. However, in addition, we can say that the expected number of strings that match schema s will increase between generations if the probability of crossover decreases and/or if the defining length of the strings in the solution decreases.

Equation (5.43) can be further amended to allow for mutation. (See a text on GAs for this amendment, for example, (Goldberg 1989).) Intuitively, it can be appreciated that as the order (i.e., number of fixed bit values) increases, given the probability that a bit is flipped, the lower the expected number of strings that match a schema in the following generation will be. There is a greater chance that one of the fixed bit values will be flipped and that the child fails to match schema s.

Equation (5.43), as amended by the preceding paragraph, allows us to conclude that the number of strings matching schemata with above-average fitness, which have a short defining length and low order, will increase exponentially as new generations of a given size are created. This is known as the schemata theorem.

5.6 Expert systems

In general terms, an expert system is a collection of processes that emulate the decision-making behavior of an expert. An expert system consists of a number of parts. First, there is a knowledge base, which usually consists of rules. Second is a series of facts that, in the inference engine, are matched with the rules to provide an agenda of recommended actions. This inference engine may provide an explanation for the recommendations and the system may have a facility for updating knowledge. The rules, often called production rules, may be of the IF ..., THEN ... form. For example, a rule might be

IF annual payments exceed 50% of annual income, THEN the loan will not be repaid.

The construction of the knowledge base is one of the earlier stages in the construction of an expert system. This is done by a knowledge engineer who repeatedly discusses with an expert the behavior and decisions made by the expert whose behavior is to be emulated. The knowledge engineer may represent the heuristics followed by an expert in symbolic logic. The rules may be extracted by the knowledge engineer directly or with the use of specialized software. An example of the derivation of such rules is the training of a neural network so that when new facts—that is, application characteristics—are fed into it, a decision is given (Davis, Edelman, and Gammerman 1992). The rules created from a neural network are uninterpretable, and yet one aim of an expert system is to provide explanations for the actions recommended. Therefore, recent research in the area of knowledge acquisition, or machine learning, is in symbolic empirical learning (SEL), which is concerned with creating symbolic descriptions where the structure is not known in advance. An example of a SEL program is Quinlan's C5 (a development from his ID3), which constructs decision trees from input data (see section 4.7 above).

The expert system may consist of literally thousands of rules—for example, XCON, DEC's expert system for configuring computers, contained 7000 rules. In an expert system, the rules need not be sequential, and the matching of facts to rules is complex. Such matching is carried out by algorithms such as the Rete algorithm, devised by Forgy (1982).

From the foregoing, it can be seen that an expert system is particularly applicable when the decision maker makes decisions that are multiple and sequential or parallel and where the problem is ill-defined because of the multiplicity of decisions that can be made—in short, where it is not efficient to derive an algorithmic solution.

Relatively few examples of an expert system used for credit scoring have been published, and because the details of such systems are usually proprietary, none that have been published give exact details. However, an example by Talebzadeh, Mandutianu, and Winner (1994) describes the construction of an expert system, called CLUES, for deciding whether to underwrite a mortgage loan application, which was received by the Countrywide Funding Corporation.

They considered using expert systems, neural nets, mentoring systems, and case-based reasoning. They chose expert systems rather than neural networks, for example, because neural networks require large numbers of cases for retraining and so were less amenable to updating than an expert system into whose knowledge base additional rules could be inserted. In addition, unlike neural nets, expert systems could provide reasons for a recommendation. Loan applications to Countrywide contained 1500 data items, on the basis of which an underwriter decides whether to grant a loan by considering the strengths and weaknesses of the applicant. Knowledge engineers interviewed and observed underwriters as they decided whether to grant a loan.

Talebzadeh, Mandutianu, and Winner (1994) identified three types of analysis that each loan officer undertook: an examination of the ability of the borrowers to make the payments, an examination of the past repayment performance of the application, and consideration of the appraisal report. Eventually, the expert system had around 1000 rules, which evaluated each item within each of the three analyses, for example, assets and income.

Extensive comparisons between the decisions made by the system and those made by underwriters showed that virtually all the decisions made by CLUES were confirmed by the underwriters.

In another example of an expert system, Jamber et al. (1991) described the construction of a system by Dun and Bradstreet to produce credit ratings of businesses. After fine-tuning the rules, the system agreed with the experts in 98.5% of cases in a test sample. In another example, Davis, Edelman, and Gammerman (1992) achieved agreement between their system and that of loan officers for consumer loans in 64.5% of test cases, but their "expert system" was really a neural network applied to application data rather than the construction of a knowledge base from the behavior of experts. In a related context, Leonard (1993a, 1993b) constructed an expert system to predict cases of fraud among credit card users. His rule extraction technique was an algorithm by Biggs, De Ville, and Suen (1991), which constructs decision trees from raw financial data.

5.7 Comparison of approaches

In the last two chapters, a number of methods for developing credit-scoring systems were outlined, but which method is best? This is difficult to say because commercial consultancies have a predilection to identify the method they use as best. On the other hand, comparisons by academics cannot reflect exactly what happens in the industry since some of the significant data, like the credit bureau data, are too sensitive or too expensive to be passed on to them by the users. Thus their results are more indicative than definitive. Several comparisons of classification techniques have been carried out using medical or scientific data but fewer by using credit-scoring data.

Table 5.2 shows the results of five comparisons using credit-scoring data. The numbers should be compared across the rows but not between the rows because they involve different populations and different definitions of good and bad.

Table 5.2. *Comparison of classification accuracy for different scoring approaches.*

Authors	Linear reg.	Logistic reg.	Classification trees	LP	Neural nets	GA
Srinivisan (1987b)	87.5	89.3	93.2	86.1	—	—
Boyle (1992)	77.5	—	75.0	74.7	—	—
Henley (1995)	43.4	43.3	43.8	—	—	—
Yobas (1997)	68.4	—	62.3	—	62.0	64.5
Desai (1997)	66.5	67.3	—	—	66.4	—

The entries in the table are the percent correctly classified if the acceptance rate. However, there is no difference made between goods classified as bads and vice versa. In Srinivasan's comparison (1987b) and Henley's (1995) work, classification trees is just the winner; Boyle et al.'s (1992) and Yobas, Crook, and Ross's (1997) linear regression classifies best, while the paper by Desai et al. (1997) suggests logistic regression is best. However, in almost all cases, the difference is not significant.

The reason for the similarity may be the flat maximum effect. Lovie and Lovie (1986) suggested that a large number of scorecards will be almost as good as each other as far as classification is concerned. This means there can be significant changes in the weights around the optimal scorecard with little effect on its classification performance, and it perhaps explains the relative similarity of the classification methods.

This relative stability of classification accuracy to choice of method used has prompted experts to wonder if a scoring system is also relatively stable to the choice of customer sample on which the system is built. Can one build a scorecard on one group of customers and use it to score a different group of customers; i.e., are there generic scorecards? Platts and Howe (1997) experimented by trying to build a scorecard that could be used in a number of European countries. Overstreet, Bradley, and Kemp (1992) tried to build a generic scorecard for U.S. credit unions by putting together a sample of customers from a number of unions because each union did not have enough customers to build a stable scoring system. In both cases, the results were better than not using a scorecard at all, but the classification accuracy was well below what is usual for credit scorecards. Thus it does seem the systems are very sensitive to differences in the populations that make up the scorecard and suggest that segmenting the population into more homogeneous subgroups is the sensible thing to do. Further details of these generic scorecards are found in section 12.2.

If classification accuracy is not the way to differentiate between the approaches, what should be used? An obvious answer is the special features the different methods bring to the scoring process. The regression approaches, both linear and logistic, have all the underpinning of statistical theory. Thus one can perform statistical tests to see whether the coefficients (the scores) of an attribute are significant, and hence whether that attribute should really be in the scorecard. Correlation analysis shows how strongly related are the effects of different characteristics and so whether one really needs both in the scorecard. The regressions can be performed using all the characteristics or by introducing them one by one, each time introducing the one that increases the existing scorecard's discrimination the most. Thus these tests allow one to identify how important the characteristics are to the discrimination and whether two characteristics are essentially measuring the same effect. This allows one to drop unimportant characteristics and arrive at lean, mean, and robust scorecards. It is also useful in identifying which characteristics should be included and which can be dropped when application forms are being updated or information warehousing systems are being developed.

The linear programming approach has to be extended to integer programming, which is impractical for commercial systems, to give similar tools on the relative importance of variables, as outlined in section 5.3. However, linear programming deals very easily with constraints that lenders might impose on the scorecard. For example, they may want to bias the product toward younger customers and so require the score for being under 25 to be greater than the score for being over 65. This is easily done in any of the linear programming formulations by adding the constraint $w(\text{under } 25) \geq w(\text{over } 65)$. One could require other constraints; for example, the weighting on residential states is no more than 10% of the total score, or the weighting on income must be monotone in income. Regression approaches find it almost impossible to incorporate such requirements. One of the other advantages of linear programming is that it is easy to score problems with hundreds of thousands of characteristics, and so splitting characteristics into many binary attributes causes no computational difficulties, whereas the statistical analysis of such data sets with large number of variables can cause computational problems.

The methods that form groups like classification trees and neural networks have the advantage that they automatically deal with interactions between the characteristics, whereas for the linear methods, these interactions have to be identified beforehand and appropriate complex characteristics defined. For example, Table 5.3 describes data on the percentage goods in different categories of residential status and phone ownership.

Table 5.3. *Percent of goods in residential status and phone ownership.*

	Own phone	No phone	
Owner	95%	50%	90%
Tenant	75%	65%	70%
	91%	60%	

Since the percentage of goods is greater among owners than tenants (90% to 70%) and among those with a phone than those without a phone (91% to 60%), the highest score in a linear system will be of owners who have a phone and the lowest of tenants who do not have a phone. In fact, owners who do not have a phone are significantly worse, and this would not be picked up by a linear system if it had not been identified beforehand. Classification trees and neural networks do tend to pick up such interactions. Some scorecard developers will use classification trees to identify major interactions and then segment on the characteristic involved in several such interactions (see section 8.6), i.e., in this example, they build separate linear scorecards for owners and for tenants. Age is one of the most common variables on which to segment.

Nearest-neighbor methods and genetic algorithms have been insufficiently used in practice to detail their extra features. However, it is clear that nearest-neighbor methods could be used to develop continuously updated scorecards. When the history on past customers is sufficiently long to make an accurate judgment on whether they are good or bad, then it can be added to the sample, while very old sample points could be deleted. Thus the system is being continuously updated to reflect the trends in the population. Genetic algorithms allow a wide number of scorecards to be considered in the construction of a fit population.

One final point to make in any comparison of methods regards how complex the lender wants the system to be. To be successfully operated, it may be necessary to sacrifice some classification accuracy for simplicity of use. One extreme of this is the scorecard developed for Bell Telephone and described by Showers and Chakrin (1981). The scorecard consisted of 10 yes-no questions and the cutoff was at least 8 yeses. Thus a scorecard with a very small number of binary attributes may be successfully implemented by nonexperts, while more complex scorecards will fail to be implemented properly and hence lose the classification accuracy built into them (Kolesar and Showers 1985).

Chapter 6

Behavioral Scoring Models of Repayment and Usage Behavior

6.1 Introduction

The statistical and nonstatistical classification methods described in the last two chapters can be used to decide whether to grant credit to new applicants and to decide which of the existing customers are in danger of defaulting in the near or medium-term future. This latter use is an example of behavioral scoring—modeling the repayment and usage behavior of consumers. These models are used by lenders to adjust credit limits and decide on the marketing and operational policy to be applied to each customer. One can also model consumers' repayment and usage behavior using Markov chain probability models. Such models make more assumptions about the way consumers behave, and thus they can be used to forecast more than just the default probability of the consumer.

In section 6.2, we explain how the classification methods are used to obtain behavioral scores. Section 6.3 describes the uses and variants of this form of behavioral score. Section 6.4 describes the Markov chain models used in behavioral scoring, where the parameters are estimated using data from a sample of previous customers. Section 6.5 explains how this idea can be combined with dynamic programming to develop Markov decision process models that optimize the credit limit policy. Section 6.6 considers some of the variants of these models, especially ones where the population is segmented and different models are built for each segment. It also looks at the way the parameters for the model can be estimated and at the tests that are used to check that the assumptions of the models are valid. Section 6.7 outlines the Markov chain models where the parameter values for an individual customer are estimated and updated using that individual's performance. This Bayesian approach to modeling contrasts with the orthodox approach of the models in sections 6.4 and 6.5.

6.2 Behavioral scoring: Classification approaches

The main difference between the classification approaches of behavioral scoring and credit scoring is that more variables are available in the former. As well as all the application-form characteristics and the data from the credit bureau (whose values are regularly updated into the scoring system), there are characteristics that describe the repayment and usage behavior of the customer. These are obtained from a sample of histories of customers as follows. A particular point of time is chosen as the observation point. A period preceding this point, say the previous 12 to 18 months, is designated the performance period, and the characteristics of the performances in this period are added to the credit bureau and application information. A time—say, 12 months after the observation point—is taken as the outcome point, and the customer is classified as good or bad depending on their status at that time. Figure 6.1 shows the relationship between the observation and outcome points and the performance period.

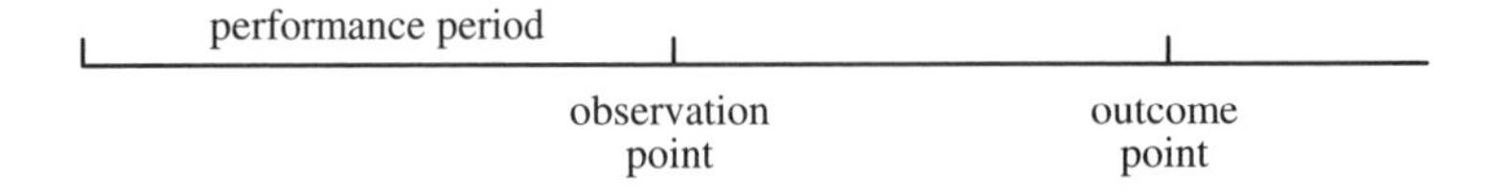

Figure 6.1. *Behavioral scoring time line.*

The extra performance variables in behavioral scoring systems include the current balance owed by the account and various averages of this balance. There will be similar records of the amount repaid in the last month, six months, etc., as well as the amount of new credit extended and the usage made of the account over similar periods. Other variables refer to the status of the account, such as the number of times it had exceeded its limits, how many warning letters had been sent, and how long since any repayment had been made. Thus there can be a large number of very similar performance variables with strong correlations. Usually, one chooses only a few of these similar variables to be in the scoring system. Thus a common first step is to apply a linear regression of the status of the account at the outcome point against these similar variables and to leave in only those that have the greatest impact. These can be either the ones with the most significant coefficients if all variables are in the regression or those that enter first if an iterative regression procedure is applied, with one new variable entering at each stage. For more details, see section 8.8, on choosing the characteristics.

The definition of the status of the account at the outcome point is usually compatible with that used in the application scoring system. Thus common definitions of bad are accounts that have three, possibly consecutive, months of missed payments during the outcome period.

6.3 Variants and uses of classification approach-based behavioral scoring systems

A very common variant of the classification-based behavioral scoring system outlined in the previous section is to segment the population and apply different scoring systems to the different segments. The population can be segmented on characteristics like age that have strong interactions with other characteristics. If one uses a classification tree, then the highest-level splits are good indications of what the appropriate segments of the population might be to deal with such interaction effects.

Another more obvious segmentation is between new and established customers. The former may have insufficient history for the performance variables to have values. Someone who opened an account within the last six months cannot have an average balance for the last year or the last half-year. A separate behavior scorecard is built for those with so little history. Not surprising, these scorecards put greater weight on the application-form data than do scorecards for more mature accounts. It is then necessary to calibrate the scorecards so that there are no discontinuities in score and hence in good:bad odds at the time when customers move from one scorecard to another. Ways to ensure consistency in calibrating the scorecard are outlined in section 8.12, and this consistency is required for all portfolios of segmented scorecards.

Another variant on the standard behavioral scoring methodology deals with the length of the outcome period. Since one is anxious to identify the behavior of all defaulters and not just these who are well on the way to defaulting, one wants to make this a reasonable period of time; 12 or 18 months is normal. However, with a performance period of 12 months, one is building a scoring system on a sample that is two to three years old. Population drift means that the typical characteristics of the population as well as the characteristics that are indicative of bad outcomes may start to change over such a period. To overcome this, some organizations take a two-stage approach, using an outcome point only six months after the observation point. First, they identify what behavior in a six-month period is most indicative of the customer going bad in the next 12 months, using the normal definition of bad. This is usually done using the classification tree approach on a sample with a suitably long history. This bad behavior in the six-month period is then taken as the definition of bad for a behavioral score using more recent data with an outcome period of just the last six months. Thus one can use data that are only 12 or 18 months old depending on the performance period used.

One of the problems with behavioral scoring is how the score should be used. Using the typical lengths of outcome period and definition of bad, the score is a measure of the risk that the customer will default in the next 12 months, assuming the operating policy is the same as that being used in the sample period. It can therefore be argued that using this score to change the operating policy invalidates the effectiveness of the score. Yet this is what is done when one uses the behavioral score to set the credit limits. The assumption is that those with higher behavioral scores, and hence less risk of defaulting at the current credit limit policy, should be given higher credit limits. This is analogous to saying that only those who have no or few accidents when they drive a car at 30 mph in the towns should be allowed to drive at 70 mph on the motorways. It does not accept that there might be other skills needed to drive faster. Similarly, there may be other characteristics that are needed to manage accounts with large credit limits, which are not required in managing accounts with limited available credit. If one accepts that this is a leap of faith, then it is a reasonable one to make, and one that many lenders are willing to accept. What is commonly done is to cut the behavioral score into a number of bands and to give a different credit limit to each band, increasing the limit with the behavioral score. Other organizations think of the behavioral score as a measure of the risk involved and take another measure, such as credit turnover, of the return involved. They then produce a matrix of behavioral score bands and credit turnover bands with different credit limits in each; see Table 6.1 as an example.

Table 6.1. *Example of a credit limit matrix.*

Monthly credit turnover	**<£50**	**£50–£150**	**£150–£300**	**£300+**
Behavioral score				
<200	£0	£0	£500	£500
200–300	£0	£500	£1000	£2000
300–500	£2000	£3000	£3000	£5000
500+	£5000	£5000	£5000	£5000

This then raises the question of how these limits are chosen. Ideally, one would like to have a profit model, where one could put in the behavioral score as a measure of the riskiness of a customer and the variables that measure the profit of the customer and then find the optimal credit limit. Lenders do not usually have such models and so are reduced to setting credit limits using hunches and experience. A more scientific approach advocated by Hopper and Lewis (1992) is the champion-versus-challenger approach, where they suggest trying out new credit policies on subsets of the customers and comparing the performance of these subsets with the group using the existing policy. This approach emphasizes the point that it takes time to recognize whether a new policy is better or worse than the old one. Increasing credit limits gives an immediate boost to the usage or sales and so apparently to the profit, but it takes another 6 to 12 months for the consequent increase in bad debt to become apparent. Thus such competitions should be kept going for at least 12 months.

The classification-based behavioral scores have other uses, such as estimating whether a customer will continue to use the credit line, whether they will close the account and move to another lender, and whether they can be persuaded to buy other products from the lender. Chapter 10 looks at some of these other applications in detail, but it is sufficient to say that in some cases one needs to know further information like the age of the account or when certain events occurred. Customers in the U.K. have, until the recent changes in car registration lettering, tended to change their cars about the same time of year every two, three, or four years. Thus knowing when the last car loan was taken out is likely to be a significant factor in whether a customer is thinking about getting another car loan.

6.4 Behavioral scoring: Orthodox Markov chain approach

The idea of building a model of the repayment and usage behavior of a consumer, as opposed to the black box approach of classification-based behavioral scoring, was first suggested in the early 1960s. The idea is to identify the different states that a borrower's account can be in, and then to estimate the chance of the account moving from one state at one billing period to another at the next. The states depend mainly on information concerning the current position of the account and its recent history, but they can also depend on the initial application information. Thus the current balance, the number of time periods overdue, and the number of reminder letters in the last six months might be typical information used. The object is to define states in which the probability of moving to any particular state at the next billing period is dependent only on the current state the account is in and not on its previous history. This is the definition of a Markov chain. A more formal definition follows.

Definition 6.1. *Let* $\{X_0, X_1, X_2, X_3, \dots\}$ *be a collection of random variables that take values in one of* M *states. The process is said to be a finite-valued Markov chain if*

$$\text{Prob}\{X_{t+1} = j | X_0 = k_0,\ X_1 = k_1, \dots, X_{t-1} = k_{t-1},\ X_t = i\} = P\{X_{t+1} = j | X_t = i\} \tag{6.1}$$

for all t *and* i, j, *where* $1 \le i, j \le M$. *The conditional probabilities* $P\{X_{t+1} = j | X_t = i\}$ *are called transition probabilities and represented* $p_t(i, j)$, *and the probability properties mean that one requires that* $p_t(i, j) \ge 0$ *and* $\sum_j p_t(i, j) = 1$.

The matrix of these probabilities is denoted P_t, so $(P_t)(i, j) = p_t(i, j)$. The Markov property (6.1) means that one can obtain the distribution of the X_t given the value of X_0 by multiplying the matrices $P_0, P_1, \dots, P_{t+1}$ together since

$$P\{X_{t+1} = j | X_0 = i\} = \sum_{k(1)\cdots k(t-1)} P\{X_{t+1} = j | X_t = k(t)\} P\{X_t = k(t) | X_{t+1} = k(t-1)\}$$
$$\cdots P\{X_2 = k(2) | X_1 = k(1)\} P\{X_1 = k(1) | X_0 = i\}$$

$$= \sum_{k(1)\cdots k(t-1)} p_0(i, k(1))p_1(k(1), k(2)) \cdots p_t(k(t), j)$$

$$= (P_0 \cdot P_1 \cdot \cdots \cdot P_t)(i, j). \tag{6.2}$$

If the $p_t(i, j) = p(i, j)$ for all t, i, and j, the process is a stationary Markov chain. In this case, the k-stage transition probabilities are obtained by multiplying P by itself k times, so

$$P\{X_{t+1} = j | X_0 = i\} = \sum_{k(i)\cdots k(t-1)} p_0(i, k(1))p_1(k(1), k(2)) \cdots p_t(k(t), j)$$
$$= (P \cdot P \cdot \cdots \cdot P)(i, j) = P^{t+1}(i, j). \tag{6.3}$$

If π_{t+1} is the distribution of X_{t+1}, so $\pi_{t+1}(i) = \text{Prob}(X_{t+1} = i)$, then (6.3) corresponds to

$$\pi_{t+1} = P\pi_t = P^{t+1}\pi_0 \tag{6.4}$$

in vector notation. For aperiodic Markov chains (ones where there is no $k > 2$, so the $p^n(i, j) \neq 0$ for some i, j only if k divides n exactly), in the long run π_t converges to a long-run distribution π^*. Replacing π_t and π_{t+1} by π^* in (6.4) shows that π^* must satisfy

$$\pi^* = P\pi^*. \tag{6.5}$$

The states of the Markov chain divide into persistent and transient ones. Persistent states i are ones where the chain is certain to return to and correspond to states where $\pi_i^* > 0$; i.e., there is a positive probability of being in them in the long run. Transient states i are ones where the chain has a probability of less than 1 of ever returning to and corresponding to states where $\pi_i^* = 0$.

Consider the following Markov chain model.

Example 6.1. One of the simplest examples of this type of model is to take the status of a credit account as one of the following states $\{NC, 0, 1, 2, \ldots, M\}$, where NC is no-credit status, where the account has no balance; 0 is where the account has a credit balance but the payments are up to date; 1 is where the account is one payment overdue; i is where the account is i payments overdue; and one assumes M payments overdue is classified as default. The transition matrix of the Markov chain would be

From/To	***NC***	**0**	**1**	**2**	**...**	***M***
NC	$p(NC, NC)$	$p(NC, 0)$	0	0	...	0
0	$p(0, NC)$	$p(0, 0)$	$p(0, 1)$	0	...	0
1	$p(1, NC)$	$p(1, 0)$	$p(1, 1)$	$p(1, 2)$	...	0
2	$p(2, NC)$	0	$p(2, 1)$	$p(2, 2)$	...	0
...	...	...	...	...	...	...
M	$p(M, NC)$	$p(M, 0)$	0	0	...	$p(M, M)$

(6.6)

This is the model described by Kallberg and Saunders (1983), in which the accounts jump from state to state. The data in their example led to that stationary transition matrix

From/To	***NC***	**0**	**1**	**2**	**3**
NC	0.79	0.21	0	0	0
0	0.09	0.73	0.18	0	0
1	0.09	0.51	0	0.40	0
2	0.09	0.38	0	0	0.55
3	0.06	0.32	0	0	0.62

(6.7)

Thus if one starts with all the accounts having no credit $\pi_0 = (1, 0, 0, 0, 0)$, after one period the distribution of accounts is $\pi_1 = (0.79, 0.21, 0, 0, 0)$. After subsequent periods, it becomes

$$\begin{aligned} \pi_2 &= (0.64, 0.32, 0.04, 0, 0), \\ \pi_3 &= (0.540, 0.387, 0.058, 0.015, 0), \\ \pi_4 &= (0.468, 0.431, 0.070, 0.023, 0.008), \\ \pi_5 &= (0.417, 0.460, 0.077, 0.028, 0.018), \\ \pi_{10} &= (0.315, 0.512, 0.091, 0.036, 0.046). \end{aligned} \tag{6.8}$$

This proves a useful way for estimating the amount of bad debt (state 3) that will appear in future periods. After 10 periods, it estimates that 4.6% of the accounts will be bad. This approach seems to be more robust to variations in sales volumes than more standard forecasting techniques, which use lagged values of sales volumes in their forecasts.

One of the first Markov chain models of consumer repayment behavior was a credit card model suggested by Cyert, Davidson, and Thompson (1962), in which each dollar owed jumped from state to state. The Cyert model had difficulties, however, with accounting conventions. Suppose an account was £20 overdue—£10 was three months overdue and £10 was one month overdue. If the standard payment is £10, and £10 is paid this month (so the account is still £20 overdue), is £10 of this four months overdue and £10 two months overdue or is £10 two months overdue and £10 one month overdue? This problem was addressed by van Kueler, Spronk, and Corcoran (1981), who used a partial payment of each amount compared with paying off the oldest debt suggested by Cyert, Davidson, and Thompson (1962). Other variants of the model were suggested by Corcoran (1978), who pointed out that the system would be even more stable if different transition matrices were used for accounts of different-size loans. Two similar models were investigated by Kallberg and Saunders (1983), in which the state space depends on the amount of the opening balance and whether payments made were (1) less than the total outstanding balance but greater than the required amount, (2) within a small interval of the required amount, (3) some payment but significantly less than the required amount, or (4) no payment at all.

One can define much more sophisticated Markov chain models in which each state s in the state space has three components, $s = (b, n, i)$, where b is the balance outstanding, n is the number of current consecutive periods of nonpayment, and i represents the other characteristics of importance. If one defines the expected one-period reward the lender makes from a customer in each state, one can calculate the expected total profit from the customer under any credit limit policy. In fact, one can go further and calculate the credit limit policy that maximizes the rewards for a given level of bad debt or minimizes the bad debt for a given level of reward. To see this, consider the following examples in the mail-order context.

Example 6.2. Let the states of consumer repayments be $s = (b, n, i)$, where b is the balance outstanding, n is the number of periods of nonpayment, and i are the other characteristics. From the sample of customer histories, we estimate the following:

- $t(s, a)$ the probability an account in state s repays a next period;
- $w(s, i')$ the probability an account in state s changes its other characteristics to i' next period;
- $r(s)$ the expected value of orders placed per period by those in state s.

If no more orders were allowed, we can use dynamic programming to calculate the chance of defaulting, $D(s)$, given the account is in state s. We assume that default corresponds to

N successive periods of nonpayment. The means that

$$\begin{aligned} D(b, N, i) &= 1 \quad \text{for all } b, i, j, \\ D(0, n, i) &= 0 \quad \text{for all } i, n < 1, \end{aligned} \tag{6.9}$$

whereas

$$D(b, n, i) = \sum_{i', a \neq 0} t(s, a) w(s, i') D(b - a, 0, i') + \sum_{i'} t(s, 0) w(s, i') D(b, n + 1, a). \tag{6.10}$$

This follows since if the system is in state (b, n, i) and there is a payment of a and if i changes to i', which occurs with probability $t(s, a)w(s, i)$, the state of the system becomes $(b - a, 0, i)$. With probability $t(s, 0)w(s, i')$, there is no payment and the state moves to $(b, n + 1, i')$. One can use (6.8) and (6.10) to calculate $D(b, n, i)$ for each state since, as Figure 6.2 shows, the jumps in state move the state either closer to the bottom of the grid or closer to the right-hand side, and the value $D(b, n, i)$ is known on these line edges.

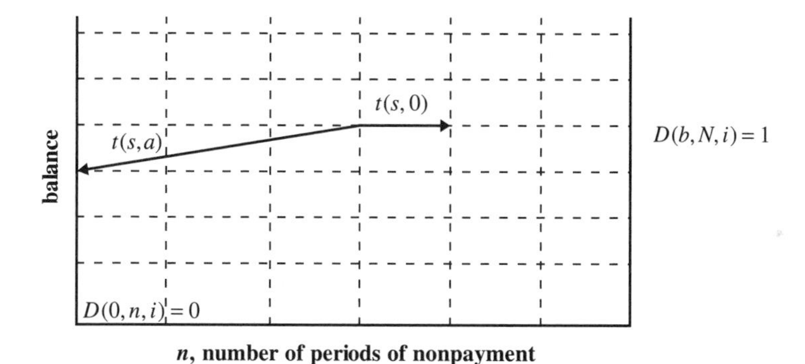

Figure 6.2. *Calculation of D.*

It is obvious that the chance of default $D(b, n, i)$ increases as b increases—if n and i are kept fixed—since there have to be more repayments before the balance is paid off.

Using past data, we are able to calculate the default probability for each state using the above calculations and also the expected value of the orders in each state. The states can then be ordered in increasing probability of default, say, for example, Table 6.2, which is a hypothetical nine-state example, where the total expected value of orders is £1,000. The lenders then have to decide what default level and value of orders lost is acceptable. If they say a default level of .1 is acceptable, then they will accept orders in the states up to s_8 and reject those in s_4, s_6, and s_5 and so lose $\frac{60+40+50}{1000} = 15\%$ of the orders. If they say that a level of .025 is acceptable, then they will reject orders from those in states s_9, s_8, s_4, s_6, and s_5 and so lose $\frac{120+80+60+40+50}{1000} = 25\%$ of the orders. Thus management has to choose which default level D^* is acceptable. Since $D(b, n, i)$ is increasing in b, one can solve

$$D(L(n, i), n, i) = D^* \tag{6.11}$$

to find the credit limit $L(n, i)$ for each state (n, i).

Table 6.2. *Value and default probability at each state.*

State	s_7	s_1	s_3	s_2	s_9	s_8	s_4	s_6	s_5
Default probability	.001	.004	.008	.02	.06	.09	.12	.15	.2
Value of orders (£)	50	150	250	200	120	80	60	40	50

6.5 Markov decision process approach to behavioral scoring

The last Markov chain model of the previous section had two elements—states, describing the repayment performance of the consumers with a transition matrix describing the dynamics of the repayment behavior, and a reward function, describing the value to the organization of the consumer being in that state. If one adds a third element—decisions the lender can make that impact on the rewards and the transitions between the states—one has a Markov decision process that is an example of stochastic dynamic programming.

Formally, a Markov decision process is a stochastic process X_t, $t = 1, 2, \dots,$ taking values in a state space S. For each state $i \in S$, there is a set of actions $k \in K_i$, and one of these has to be chosen each time the process enters a state. The result of choosing action k in state i is that there is an immediate reward $r^k(i)$, and with probability $p^k(i, j)$ the system moves into state j at the next period. The objective is to choose the actions $k \in K_i$ so as to maximize some function of the rewards. This could be the reward over a finite number of periods or the reward over an infinite number of periods. In the latter, one criterion is the expected discounted total reward. In this, one takes the net present value of the future rewards, as developed in (3.1) and (3.2), and writes the discounting term as β, where $\beta = \frac{1}{1+i}$ if i is the interest rate given. Thus the rewards in period two are discounted by a factor β, those in period three by a factor β^2, and so on. The second possible criterion in the infinite horizon case is the expected average reward. This is the limit as n goes to infinity of the average reward over the first n periods. Whatever criterion is used, under very mild conditions, which always hold if the state space is finite and the number of actions per state is finite (see Puterman 1994 for details), the optimal value and the optimal actions satisfy what is known as the Bellman optimality equation. For example, if $v_n(i)$ is the optimal total reward over n periods with the system starting in state i, then

$$v_n(i) = \max_{k \in ki} \left\{ r^k(i) + \sum_j p^k(i, j) v_{n-1}(j) \right\} \quad \text{for all } i \in S \text{ and all } n = 1, 2, \dots. \quad (6.12)$$

Similarly, if β is the discount factor and $v(i)$ is the optimal expected discounted reward with the system starting in state i, then

$$v(i) = \max_{k \in ki} \left\{ r^k(i) \sum_j p^k(i, j) \beta v(j) \right\} \quad \text{for all } i \in S. \quad (6.13)$$

Standard analysis (see Puterman 1994) shows that the actions that maximize the right-hand side of (6.13) make up the optimal policy for the discounted infinite horizon problem, while the maximizing actions of the right-hand side of (6.12) form the optimal policy for maximizing the total reward over a finite horizon. Thus solving the optimality equations (6.12) and (6.13) will give both the optimal values and the optimal actions. The finite horizon optimality equations (6.12) can be solved by first deciding what the boundary conditions of

$v_0(.)$ should be (normally $v_0(.) = 0$). Then one can solve for $v_1(i)$. Having solved for $v_1(.)$, this allows the equation for $v_2(.)$ to be solved and this process can be repeated to obtain $v_n(.)$ for all n. This process of value iteration can also be used for solving the infinite horizon discounted reward problems with optimality equation (6.13). One solves the following set of equations iteratively:

$$v_n(i) = \max_{k \in K} \left\{ r^k(i) + \sum_{j \in S} p^k(i, j)\beta v_{n-1}(j) \right\} \quad \text{for all } i \in S, \quad n = 1, 2, \ldots. \tag{6.14}$$

It can be shown that the solutions $v_n(0)$ converge to $v(0)$, the solution of (6.13). Similarly, as Puterman (1994) proved, after some value n^*, all the iterations $n \geq n^*$ have maximizing actions that also maximize the infinite horizon problem.

Returning to the models of the previous section, we can now extend them to become Markov decision processes. Consider the following extension of Example 6.2.

Example 6.3. The states of consumer repayment behavior are $s = (b, n, i)$, where b is the balance outstanding, n is the number of periods since the last payment, and i is any other information. The actions are the credit limit L to set in each state. The Markov decision process behavioral scoring system is well determined once we have defined $p^L(s, s')$ and $r^L(s)$. Define $p^L(s, s')$ by estimating the following from past data:

$t^L(s, a)$ the probability account in state s with credit limit L repays a next period;

$q^L(s, e)$ the probability account in state s with credit limit L orders e next period;

$w^L(s, i')$ the probability account in state s with credit limit L changes to information state i'.

Then

$$\begin{aligned}
&p^L(b, n, i; b + e - a, 0, i') = t^L(s, a)q^L(s, e)w^L(s, i'), \quad b + e - a \leq L, \quad a > 0,\\
&\quad p^L(b, n, i; b - a, 0, i') = t^L(s, a)\left(q^L(s, 0) + \sum_{e \geq L-b+a} q^L(s, e)\right) w^L(s, i'), \quad a > 0,\\
&p^L(b, n, i; b + e, n + 1, i') = t^L(s, 0)q^L(s, e)w^L(s, i'), \quad b + e \leq L,\\
&\quad p^L(b, n, i; b, n + 1, i') = t^L(s, 0)\left(q^L(s, 0) + \sum_{e \geq L-b} q^L(s, e)\right) w^L(s, i').
\end{aligned} \tag{6.15}$$

The reward to the firm is the proportion f of the purchase amount that is profit for the lender, less the amount lost when a customer defaults. If we assume that N consecutive periods of nonpayment is what the company considers as default, then

$$r^L(b, n, i) = f \sum_e eq^L(s, e) - bt^L(s, 0)\delta(n - (N - 1)). \tag{6.16}$$

If one assumes profits are discounted by β each period, then let $v_n(s)$ be the optimal total discounted expected profit over an n-period horizon starting in state s. The optimality equation that $v_n(s)$ must satisfy is

$$v_n(s) = \max_{L} \left\{ r^L(s) + \sum_{s' \in S} p^k(s, s')\beta v_{n-1}(s') \right\} \quad \text{for all } s \in S, \quad n = 1, 2 \ldots. \tag{6.17}$$

This would give a credit limit policy where the limit L is a function of the state s and the number of periods n until the end of the time horizon. To make the policy independent of the period, one needs to consider the infinite horizon expected discounted profit criterion, where the optimal value $v(s)$ starting in state s satisfies

$$v(s) = \max_L \left\{ r^L(s) + \sum_{s' \in S} p^k(s, s')\beta v(s') \right\} \quad \text{for all } s \in S, \quad n = 1, 2, \ldots . \tag{6.18}$$

The actions that maximize the right-hand side of the (6.18) gives the optimal credit limits that maximize profit.

6.6 Validation and variations of Markov chain models

The Markov chain and Markov decision process models of the last two sections make more assumptions about the form of consumer repayment behavior than do the statistical classification models. One still needs, however, to use the data on previous applicants to estimate the parameters of the probability model chosen as well as check that the data support the assumptions of the model. Following are the estimators used to estimate the parameters and the tests needed to check the assumptions of the model.

6.6.1 Estimating parameters of a stationary Markov chain model

Suppose one takes the simplest model where the state space of the Markov chain is S which has m values and the transition probabilities from one state to another are the same for all time periods; i.e., the chain is stationary. Thus one needs to estimate $p(i, j), i, j \in S$, from a sample of histories of R previous customers. The rth customer's history is $s_0^r, s_1^r, s_2^r, \ldots, s_T^r$.

Let $n^r(i)$ be the number of times state i appears in this sequence from times 0 to $T-1$, and let $n^r(i, j)$ be the number of times the sequence i followed by j appears in times 1 to T. Let $n(i) = \sum_{r \in R} n^r(i)$ and $n(i, j) = \sum_{r \in R} n^r(i, j)$ be the total number of times state i and sequence i, j appeared in times 1 to T of all histories. Bartlett (1951) and Hoel (1954) showed that the maximum likelihood estimate of $p(i, j)$ is

$$\hat{p}(i, j) = \frac{n(i, j)}{n(i)}. \tag{6.19}$$

This is the obvious estimate.

6.6.2 Estimating parameters of nonstationary Markov chains

If one believes that the process is a Markov chain but the transition probabilities vary depending on the period under question, one has a nonstationary Markov chain. In this case, $p_t(i, j)$ is the probability the chain moves from state i to state j in period t. If one has a sample of histories of previous customers, let $n_t(i), i \in S$, be the number who are in state i at period t, whereas let $n_t(i, j)$ be the number who move from i at period t to state j at the next state. The maximum likelihood estimator of $p_t(i, j)$ is then

$$\hat{p}_t(i, j) = \frac{n_t(i, j)}{n_t(i)}. \tag{6.20}$$

Models in which the transition matrix changes every period need lots of data to estimate the parameters and are not very useful for future predictions. It is more common to assume the chain has a seasonal effect, so t is indexed by the month or the quarter of the year. Hence we would estimate $p_{\text{Jan}}(i, j)$ by using (6.20) over all cases when t = January. Other models imply that the trend is more important than the season, so one assumes the transition matrix varies from year to year but stays constant within the year. Hence one assumes that there is a $p_{1999}(i, j)$, $p_{2000}(i, j)$, etc., and uses (6.20), where t is all the 1999 periods, to estimate $p_{1999}(i, j)$.

6.6.3 Testing if $p(i, j)$ have specific values $p^0(i, j)$

If one wants to test if $p(i, j)$ have specific values $p^0(i, j)$ for all $j \in S$ (recall that S has m values), one can use the result (Anderson and Goodman 1957) that

$$\sum_{j \in S} \frac{n(i)(\hat{p}(i, j) - p^0(i, j))^2}{p^0(i, j)} \tag{6.21}$$

has a χ^2 asymptotic distribution with $m - 1$ degrees of freedom. Thus one would accept this hypothesis at a significance level α (say, 95%) if (6.21) has a value that is less than the α-significant part of the χ^2 distribution with $m - 1$ degrees of freedom.

Since the variables $n(i)(\hat{p}(i, j) - p^0(i, j))^2$ for different i are asymptotically independent, the test quantities (6.21) can be added together and the result is still a χ^2 variable with $m(m - 1)$ degrees of freedom. Thus if one wants to check whether $\hat{p}(i, j) = p^0(i, j)$ for all i, j, one calculates

$$\sum_{i \in S} \sum_{j \in S} \frac{n(i)(\hat{p}(i, j) - p^0(i, j))^2}{p^0(i, j)} \tag{6.22}$$

and tests it using the χ^2 variable tests with $m(m - 1)$ degrees of freedom.

6.6.4 Testing if $p_t(i, j)$ are stationary

Checking whether rows of transition matrices at different periods are essentially the same is similar to checking on the homogeneity of contingency tables, and thus the χ^2 tests of the latter work in the former case (Anderson and Goodman 1957).

Thus for a given state i, we are interested in whether the hypothesis that the ith row of the t-period transition matrix $p_t(i, j)$, $j = 1, \dots, m$, is the same for all periods t; i.e., $p_t(i, j) = p(i, j)$, $j = 1, \dots, m$. Let $\hat{p}_t(i, j)$, $\hat{p}(i, j)$, $j = 1, 2, \dots, m$, be the estimates for these probabilities obtained by (6.19) and (6.20). Then to test the hypothesis, calculate

$$\sum_{t=1}^{T-1} \sum_{j \in S} \frac{n_t(i)[\hat{p}_t(i, j) - \hat{p}(i, j)]^2}{\hat{p}(i, j)}. \tag{6.23}$$

If the null hypothesis is true and there are $T + 1$ periods in total $(0, 1, 2, \dots, T)$, this has the χ^2 distribution with $(m - 1)(T - 1)$ degrees of freedom.

The previous test looked at the time independence of transitions from state i only, but normally one would like to check that the whole transition matrix is time independent. This corresponds to the null hypothesis that $p_t(i, j) = p(i, j)$ for all $i = 1, \dots, m$, $j = 1, \dots, m$, and $t = 0, \ldots, T-1$. Since $p_t(i, j)$ and $p(i, j)$ for different values of i are asymptotically independent random variables, one can also show that

$$\chi^2 = \sum_{i \in S} \chi_i^2 = \sum_{i \in S} \sum_{t=1}^{T-1} \sum_{j \in S} \frac{n_t(i)[\hat{p}_t(i, j) - \hat{p}(i, j)]^2}{\hat{p}(i, j)} \tag{6.24}$$

has a χ^2 distribution with $m(m-1)(T-1)$ degrees of freedom, and one can apply the appropriate χ^2 test.

An alternative approach is to recognize that

$$L_i = -2 \log \prod_{t=0}^{T-1} \prod_{j=1}^{m} \left(\frac{\hat{p}(i, j)}{\hat{p}_t(i, j)} \right)^{n_t(i,j)} \tag{6.25}$$

is χ^2 with $(m-1)(T-1)$ degrees of freedom as a test for the homogeneity of row i. $\sum_{i \in S} L_i$ has a χ^2 distribution with $m(m-1)(T-1)$ degrees of freedom and so can be used to test if the whole matrix is stationary. Notice that if the matrix is nonstationary, then we cannot estimate the transition matrix for the current year until several months into the year. In addition, forecasting future years can be done only if it is assumed they are like certain previous years.

Thus far, we have assumed that the process is Markov so that all we need to estimate the transition probabilities $p(i, j)$ for this period is the current state of the system i. If this were not the case, we would need to know the history of the process and in particular what previous state k it was in. This is a way to check on the Markovity of the chain.

6.6.5 Testing that the chain is Markov

Take a sample of histories of previous customers and define $n_t(i)$, $n_t(i, j)$ to be the number of times that a customer was in state i at time t and the number of times that a customer was in state i at time t followed by moving to j at $t+1$. Similarly define $n_t(i, j, k)$ to be the number of times that a customer was in state i at time t followed by being in j at time $t+1$ and k at time $t+2$. For a stationary chain, estimate that

$$\hat{p}(i, j, k) = \frac{\sum_{t=0}^{T-2} n_t(i, j, k)}{\sum_{t=0}^{T-2} n_t(i, j)}. \tag{6.26}$$

These are estimators of $p(i, j, k) = p_{ijk}$, the probability of moving from i to j to k. The Markovity of the chain corresponds to the hypothesis that $p_{1jk} = p_{2jk} = \cdots = p_{mjk} = p_{jk}$ for $j, k = 1, 2, \ldots, m$. One can again make parallels with contingency tables, saying that for a fixed j, etc., the rows $p_{1jk}, p_{2jk}, \ldots, p_{mjk}$ are homogeneous. Thus a χ^2 test again works. Let

$$\chi_j^2 = \sum_{i \in S} \sum_{K \in S} \frac{n^*(i, j)[\hat{p}(i, j, k) - \hat{p}(j, k)]^2}{\hat{p}(j, k)}, \tag{6.27}$$

where

$$\hat{p}(j, k) = \frac{\sum_{t=1}^{T-1} n_t(j, k)}{\sum_{t=1}^{T-1} n_t(j)}$$

and

$$n^*(i, j) = \sum_{t=1}^{T-1} n_t(i, j).$$

Then it has a χ^2 distribution with $(m - 1)^2$ degrees of freedom.

This test checks on the non-Markovity or otherwise of state j. To check that the whole chain is Markov, i.e., $p_{ijk} = p_{jk}$ for all $i, j, k = 1, 2, \ldots, m$, one computes

$$\chi^2 = \sum_j \chi_j^2 = \sum_{i,j,k} \frac{n^*(i, j)[\hat{p}(i, j, k) - \hat{p}(j, k)]^2}{\hat{p}(j, k)} \tag{6.28}$$

and checks whether this value falls inside or outside the significant value for a χ^2 test with $m(m - 1)^2$ degrees of freedom.

If it is found the chain is not Markov, one can recover a useful model of consumer behavior in one of two ways. One can either complicate the state space by including information about previous states visited as part of the current state description or segment the population into separate groups and build a Markov chain model for each group. In the first approach, if the consumer is thought of as being in one of the states i in S at each period, then the state of the system at period t would be extended to be of the form $(i(t), i(t-1), \ldots, i(t+1-r))$, where $i(s)$ is the state the consumer was in at period s. Thus the state includes the consumer's current repayment behavior and their repayment behavior in the previous $r-1$ periods. One can then apply the estimates (6.19) and (6.20) and the tests (6.27) and (6.28) to check if the chain is Markov on this extended state space. This really corresponds to estimating the probabilities $p(i_1, i_2, \ldots, i_r)$ that the process goes through the sequence $i_1, i_2, \ldots, i_r$ of states and checking whether these probabilities are independent of the first components (see Anderson and Goodman 1957). If one requires the state at time t to include $(i(t), i(t - 1), \ldots, i(t + 1 - r))$, the Markov chain is said to be of order r. (A normal Markov chain is of order 1.)

One can segment the population into a number of groups in different ways, for example, by splitting on characteristics like age or amount of data available, and thus have different chains for new customers and established customers. Edelman (1999) developed different chains for different products and different months of the year. He then looked to see whether it was better to segment by product or by period. Another segmentation that has received some attention because it has proved very successful in labor mobility and in consumer brand preferences is the mover-stayer model.

6.6.6 Mover-stayer Markov chain models

In the mover-stayer model, the population is segmented into two groups—stayers, who never leave their initial states, which is usually up-to-date repayment, and movers, who make transitions according to a Markov chain. The assumption is that there is a whole group of people who will never leave the "nonrisky" safe states, while others are more volatile in their behavior. Assume that the movers move between states according to a Markov chain with transition matrix P, whose entries are $p(i, j)$, $i, j \in S$. Assume that $s(i)$ is the percentage of the population who are stayers in state i. Hence let $s = \sum_{i \in S} s(i)$ be the proportion of stayers. Then if D is the diagonal matrix with entries $(s(1), s(2), \ldots, s(m))$, the one-period transitions are given by the matrix

$$D + (1 - s)P. \tag{6.29}$$

However, the two-period transitions (changes in state two periods apart) are given by the matrix $D + (1-s)P^2$ and the t-period transitions by $D + (1-s)P^t$.

If one wishes to fit such a model to data on previous customers, one needs to estimate $s(i), i \in S$, and $p(i,j), i,j \in S$. These estimations were calculated in two papers (Frydman, Kallberg, and Kao 1985 and Frydman 1984). Suppose one has the credit history for customers over $T+1$ periods, $t = 0, 1, 2, \ldots, T$, and $n_t(i), n(i) = \sum_{t=0}^{T-1} n_t(i), n_t(i,j)$, and $n(i,j) = \sum_{t=0}^{T-1} n_t(i,j)$ are defined, as before, as the number of customers in state i at time t, the total number of visits by customers to state i in periods 0 to $T-1$, the number of i,j transitions made at time t, and the total number of i,j transitions made, respectively. Let n be the total number of customers in the sample and let $s(i)$ be the proportion of customers who are in state i for all $T+1$ periods of the credit history. The estimators of the $s(i)$ and $p(i,j)$ are not straightforward because there is a probability $(p(i,i))^T$ that a mover who starts in state i in period 1 will stay there for the rest of the time and so has the same behavior as a stayer. If $\hat{p}(i,i)^T$ is the estimator for that behavior, it would then seem reasonable that the maximum likelihood estimator $\hat{s}(i)$ for $s(i)$ would satisfy

$$\hat{s}(i) = \begin{cases} \left(\dfrac{ns(i) - n_0(i)(\hat{p}(i,i))^{T-1}}{n_0(i)(1-\hat{p}(i,i)^{T-1})}\right) & \text{if } ns(i) - n_0(i)\hat{p}(i,i)^T > 0, \\ 0 & \text{otherwise.} \end{cases} \tag{6.30}$$

$\hat{p}(i,i)$ satisfies

$$\begin{aligned} &\hat{p}(i,i)(n(i) - Tns(i)) \\ &\quad = (n(i,i) - Tns(i)) + (\hat{p}(i,i))^T((n(i) - Tn_0(i))\hat{p}(i,i) - n(i,i) + Tn_0(i)) \end{aligned} \tag{6.31}$$

(see Frydman et al. 1985). If T is very large and thus $\hat{p}(i,i)^T$ goes to zero, the second term on the right-hand side disappears, and then one can think of $n(i) - Tns(i)$ as the number of times that movers are in state i and $(n(i,i) - Tns(i))$ as the number of times that movers move from i to i. This extra term is a compensation because with finite T, one may be misclassifying some movers as stayers. Having solved (6.31) to obtain $\hat{p}(i,i)$, we can compute $\hat{p}(i,k)$ iteratively beginning with $k = 1$ by

$$\hat{p}(i,k) = \frac{n(i,k)\left(1 - \hat{p}(i,i) - \sum_{r=1, r\neq i}^{k-1} \hat{p}(i,r)\right)}{\sum_{r=k, r\neq i}^{m} n(i,r)}. \tag{6.32}$$

Having obtained these estimates—no matter how unpleasant they are—one can now test whether the new model is an improvement on the standard Markov chain and use likelihood ratio tests; see (Frydman, Kallberg, and Kao 1985), for example.

6.7 Behavioral scoring: Bayesian Markov chain approach

Section 6.4 looked at behavioral scoring systems based on Markov chain models, where the parameters were estimated from samples of previous customers. This is very much the orthodox statistical approach, where one assumes that such probabilities are fixed and are the same for groups of similar customers. The alternative approach, Bayesian statistics, is to assume that such probabilities are subjective and one's view on their value is likely to change as new information becomes available. In behavioral scoring systems, this implies that the probability of a consumer repaying an amount in the next period depends very much on that consumer's history and will change from period to period depending on whether repayment was made. The parameters of the model are automatically updated in light of payment or nonpayment each period, and in the long run the parameters of an individual customer will depend totally on the credit history of that customer.

Example 6.4 (fixed-amount repayment loan). Consider the simple case in which there is a fixed amount a of a loan that has to be paid each period. Thus the only unknown is whether there will be a repayment next period. Hence one has a Bernoulli random variable, $X = 1$ if repayment and 0 if no repayment, and one has to estimate $p = \text{Prob}(X = 1)$. At any point, the state of the system must reflect one's belief of the distribution of the values of p. In Bayesian statistics, one chooses beliefs that are part of a family of distributions so that when new information appears, the updated belief distribution will still be part of that family. These families are the conjugate distributions for the variable that one is trying to estimate. In the case of a Bernoulli variable, the beta distributions are the family of conjugate distributions.

Using a slightly different parameterization from normal, $B(r, n)$, a beta distribution with parameters (r, n), where $0 \leq r \leq n$, has density function

$$f_{r,m}(p) = \frac{(m-1)!}{(r-1)!(m-r-1)!} p^{r-1}(1-p)^{m-r-1}, \quad 0 \leq p \leq 1; \quad 0 \text{ outside this range.} \tag{6.33}$$

Suppose that the initial (prior) belief f_0 about the value of p is $B(r, m)$ and then a repayment is made. Then the belief at the start of the next period f_1 should satisfy

$$f_1(p) = \frac{P(X_1 = 1|p) f_0(p)}{P(X_1 = 1)} = \frac{p\left(\frac{(m-1)!}{(r-1)!(m-r-1)!} p^r (1-p)^{m-r}\right)}{P(X_1 = 1)}. \tag{6.34}$$

Since

$$P(X_1 = 1) = \frac{(m-1)!}{(r-1)!(m-r-1)!} \int_0^1 p \cdot p^{r-1}(1-p)^{m-r-1} dp = \frac{r}{m},$$

then

$$f_1(p) = \frac{m!}{r!(m-r-1)!} p^{r+1}(1-p)^{m-r};$$

i.e., the new distribution is $B(r + 1, m + 1)$. Similarly, if there is no payment,

$$f_1(p) = \frac{P(X_1 = 0|p) f_0(p)}{P(X_1 = 0)} = \frac{(1-p)\left(\frac{(m-1)!}{(r-1)!(m-r-1)!} p^r (1-p)^{m-r}\right)}{P(X_1 = 0)}. \tag{6.35}$$

Since $P(X_1 = 0) = \frac{m-r}{m}$, then

$$f_1(p) = \frac{m!}{(r-1)!(m-r)!} p^r (1-p)^{m+1-r}.$$

This is the density function for $B(r, m + 1)$. With this parameterization, one can think of m as the number of payment periods for which there is history and r as the number of payments in that time.

Thus in the repayment loan case, the state of the customer is given as (b, n, r, m), where b is the balance still left to pay, n is the number of periods since the last payment, and the belief about the probability of repayment next period is a beta distribution, $B(r, m)$. Let $D(b, n, r, m)$ be the probability the customer will default on the loan if the repayment amount is a. Then the transitions in the Markov chain mean that this default probability satisfies

$$D(b, n, r, m) = \frac{r}{m} D(b - a, 0, r + 1, m + 1) + \left(1 - \frac{r}{m}\right) D(b, n + 1, r, m + 1)$$
$$\text{for } n = 0, 1, 2, \ldots, N, \quad 0 \leq r \leq m, \tag{6.36}$$
$$\text{with } D(0, n, r, m) = 0 \quad \text{and} \quad D(b, N, r, m) = 1.$$

The boundary conditions follow if one takes default to be N consecutive periods of nonpayment. In a sense, we have a model very similar to that in Example 6.3; see Figure 6.2. The difference, as Figure 6.3 implies, is that the probability of the jumps varies from period to period according to the history of the process.

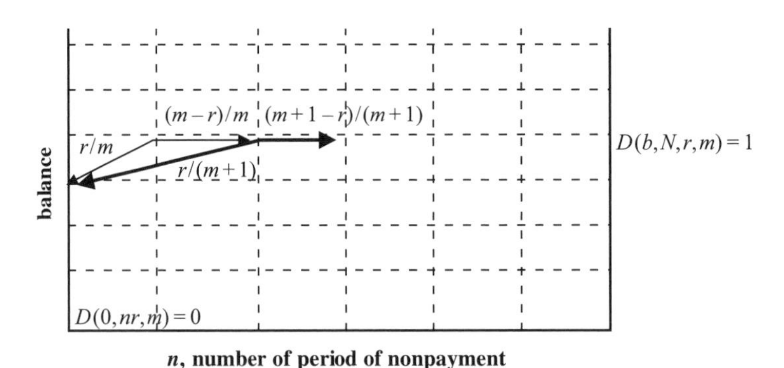

Figure 6.3. *Calculation of* $D(b, n, r, m)$.

To implement this model, one has to decide what is the prior distribution on the repayment probability at the beginning of the loan—say, (r_0, n_0). One could take it to be an uninformative prior $r_0 = 1, n_0 = 2$, which says all probabilities are equally likely. Better would be to perform some credit scoring and relate (r_0, n_0) to the credit score with $\frac{r_0}{n_0}$ increasing as the credit score increases. Whatever is chosen, after m periods in which there have been r repayments, the belief about the repayment probability is $B(r + r_0, m + m_0)$, where the actual repayment history of the individual (r, m) dominates the initial condition (r_0, m_0).

Example 6.5 (Bierman–Hausman charge card model). The first Bayesian model of repayment behavior was suggested by Bierman and Hausman (1970) in the context of a charge card, for which the amount borrowed has to be completely repaid each period before more can be borrowed the next month. They modeled this as a Markov decision process, where the state space is (r, m), the parameter of the beta distribution describing the belief of the repayment probability. Let P be the profit obtained this period when credit is given and collected and let D $(D < 0)$ be the loss (in current terms) when credit is given but not collected. If costs and profits are discounted by β each period (Srinivasan and Kim (1987a) describe a more realistic discounting), then $v(r, m)$, the maximum expected total profit that can be obtained from a customer in state (r, m), satisfies

$$v(r,m) = \max\left\{\frac{r}{m}\left(P + \beta v(r+1, m+1) + \left(1 - \frac{r}{m}\right)\right)(D + \beta v(r, m+1)); 0\right\}. \quad (6.37)$$

The two actions in this model are to allow credit or not, and once credit is refused, it will never be allowed to the customer again. The model can be solved using value iteration as outlined in section 6.5.

Bierman and Hausman sought to extend the model by allowing the decision to be the amount of credit that can be extended, rather than the all-or-nothing decision of (6.37). They assumed that the amount of credit offered is some fraction Y of a standard amount. They allowed for the repayment probability to depend on the credit allowed by saying Prob$\{\text{repayment}|Y = y\} = p_y$, where p was the probability of repaying the standard amount. This extension shows the limitations of this Bayesian approach. What one wants is that if the initial belief p on the repayment probability of the standard amount, $f_0(p)$, is a beta distribution, then the beliefs after the first repayment period are still in the beta family. These are the equivalent calculations to (6.34) and (6.35). Let $X_1 = y$ if there is full repayment of the credit, which amounts to y of the standard amount, and $X_1 = (0, y)$ if there is no repayment of a fraction y. Calculations akin to (6.34) show that if $f_0(p)$ is $B(r, m)$ and $X_1 = y$,

$$f_1(p) = \frac{P(X_1 = y|p)f_0(p)}{P(X_1 = y)} = \frac{(m+y)!}{(r+y-1)!(m-r-1)!}p^{r+y-1}(1-p)^{m-r-1}. \quad (6.38)$$

This is still a beta distribution. However, if there is no repayment ($X_1 = (0, y)$),

$$f_1(p) = \frac{P(X_1 = (0,y)|p)f_0(p)}{P(X_1 = (0,y))} \propto (1 - p^y)p^{r-1}(1-p)^{m-r-1}. \quad (6.39)$$

This is not a beta distribution. Bierman and Hausman overcome this by assuming that after every nonpayment, the expected discounted payoff will be zero; i.e., no more profit will be made from that customer. This leads to the optimality equation for $v(r, m)$ of

$$v(r,m) = \max_y\left\{\left(\frac{r}{m}\right)^y\left(Py + \beta v(r+y, m+y) + \left(1 - \left(\frac{r}{m}\right)^y\right)\right)(-Dy); 0\right\}, \quad (6.40)$$

where P and D are the profit and loss from a standard amount of credit repaid or defaulted on. The first terms correspond to lending y of the standard amount and the last term to lending nothing.

If one feels that this assumption of no further profit after one nonpayment is too severe, then a different set of problems arises. As the posterior beliefs of the repayment parameter p are no longer beta distributions, the state of the customer can no longer be described by (r, m). Instead, the state of the customer must involve the whole history of payments and nonpayments, i.e., $((y_1), (0, y_2), (y_3), \ldots, (y_n))$ would be a description of someone who paid y_1 in period 1, missed paying in period 2, paid y_3 in period 3, etc. Thus the size of the state space blows up exponentially and makes calculations nearly impossible. Dirickx and Wakeman (1976) investigated this approach and showed that the posterior distributions can be expressed as a linear combination of beta distributions and that if the optimal decision in any period was to advance no credit ($y = 0$), then this would remain optimal at all subsequent periods. However, it did not seem possible to use the model for practical calculations.

Example 6.6 (credit card repayment model). Modeling credit card or mail-order repayments is more complicated since the required repayment level R is usually some percentage of the balance outstanding and so can vary from month to month as new items are purchased. Thomas (1992) looked at a model in which there are two parameters, described by Bayesian beliefs. The first is again the probability p that the customer pays something next period. The second is that there is a maximum affordable repayment amount M that the customer can repay, so if $M < R$, the customer will pay M, and if $M \geq R$, the customer will pay between R and M. As before, assume that the belief about p is given by a beta distribution with parameters (r, m). To deal with the belief about M, split the repayment amount into k possible levels, $1, 2, 3, \ldots, k$, etc. (Think of it as £1, £2, £3, ... for convenience.) Instead of estimating M, one defines Bernoulli random variables M_1, M_2, M_3, where

$$p_i = P\{M_i = 1\} = P\{M \geq i | M \geq i - 1\} \quad \text{for } i = 1, 2, \ldots, K. \tag{6.41}$$

Thus $M_i = 1$ means that if the affordable repayment amount is at least £$i-1$, it will be at least £i. Given these marginal conditional distributions M_i, one can reconstruct the distribution of M, at least as a discrete variable with values £1, £2, etc. Since the M_i are all Bernoulli random variables and independent of each other, one can describe the belief about each probability p_i by a beta distribution with parameters (r_i, m_i). Concentrating on the repayment aspect of the model, the state of a consumer is given by $(b, n, r, m, r_i, m_i, r_2, m_2, \ldots, r_k, m_k)$, where b is the balance outstanding, n is the number of periods since a repayment, r and m are the parameters of the beta distribution describing the repayment parameter p, and r_i and m_i are the parameters of the beta distribution describing the marginal probabilities of the affordable amount, p_i. The transitions are as follows:

$$\begin{aligned}
\text{if no payment,} \quad & (b, n, r, m, r_1, m_1, \ldots, r_K, m_K) \\
& \to (b, n+1, r, m+1, r_1, m_1, \ldots, r_K, m_K); \\
\text{if payment } a < R, \quad & (b, n, r, m, r_1, m_1, \ldots, r_K, m_K) \\
& \to (b-a, 0, r+1, m+1, r_1+1, m_1+1, \ldots, r_{a+1}, m_a+1, \\
& \qquad r_{a+1}, m_{a+1}+1, \ldots, r_K, m_K); \\
\text{if payment } a \geq R, \quad & (b, n, r, m, r_1, m_1, \ldots, r_K, m_K) \\
& \to (b-a, 0, r+1, m+1, r_1+1, m_1+1, \ldots, \\
& \qquad r_{a+1}, m_a+1, r_{a+1}, m_{a+1}, \ldots, r_K, m_K).
\end{aligned} \tag{6.42}$$

The idea here is that if there is no repayment, then the mean of the belief goes down from $\frac{r}{m}$ to $\frac{r}{m+1}$. If payment is made for an amount a greater than R, then the belief the customer can pay at all levels up to a goes up from $\frac{r_i}{m_i}$ to $\frac{r_{i+1}}{m_{i+1}}$. The beliefs about paying above a do not change. If, however, the payment is $a < R$, then the belief the customer can pay at all levels up to a goes up again, but the belief the consumer can pay at level $a + 1$ goes down from $\frac{r_{a+1}}{m_{a+1}}$ to $\frac{r_{a+1}}{m_{a+1}+1}$. The probability that a repayment will be made next period is $\frac{r}{m}$, and the expected affordable repayment amount is

$$E(M) = \frac{r_1}{m_1}\left(1 + \frac{r_2}{m_2}\left(1 + \frac{r_3}{m_3}\left(1 + \cdots\left(1 + \frac{r_K}{m_K}\right)\right)\right)\right). \tag{6.43}$$

Although the state space is much larger than in the simple model of Example 6.4 and (6.37), it does not suffer from the exponential expansion that occurred in the extension of Example 6.5. One can add on the purchase side by saying there is a probability $q(e)$ that the customer will take e of new credit next period as in (6.15) to develop this into a Markov chain or Markov decision process model.

Chapter 7

Measuring Scorecard Performance

7.1 Introduction

Having built a credit or behavioral scorecard, the obvious question is, "How good is it?" This begs the question of what we mean by good. The obvious answer is in distinguishing the goods from the bads because we want to treat these groups in different ways in credit-scoring systems—for example, accepting the former for credit and rejecting the latter. Behavioral scoring systems are used in a more subtle way, but even if we stick to the idea that we will measure the scoring system by how well it classifies, there are still problems in measuring its performance. This is because there are different ways to define the misclassification rate, mainly due to the sample that we use to check this rate. If we test how good the system is on the sample of customers we used to build the system, the results will be much better than if we did the test on another sample. This must follow because built into the classification system are some of the nuances of that data set that do not appear in other data sets. Thus section 7.2 looks at how to test the classification rate using a sample, called the holdout sample, separate from the one used to build the scoring system. This is a very common thing to do in the credit-scoring industry because of the availability of very large samples of past customers, but it is wasteful of data in that one does not use all the information available to help build the best scoring system. There are times, however, when the amount of data is limited, for example, when one is building a system for a completely new group of customers or products. In that case, one can test the validity of the system using methods that build and test on essentially the same data set but without causing the misclassification errors to be optimistically biased. Section 7.3 looks at the cross-validation methods of leave-one-out and bootstrapping, which do this, while section 7.4 shows how jackknifing can also be used for this purpose.

There are a number of standard ways to describe how different two populations are in their characteristics. These can be used in the case of scorecard-based discrimination methods to see how different the scores are for the two groups of goods and bads. These measure how well the score separates the two groups, and in section 7.5, we look at two such measures of separation—the Mahalanobis distance and Kolmogorov–Smirnov statistics. These give a measure of what is the best separation of the groups that the scorecard can make. One can extend the Kolmogorov–Smirnov approach to get a feel for how good the separation of the two groups is over the whole range of the scores. One way of doing this will generate a curve—the receiver operating characteristic (ROC) curve—which is a useful way of comparing the classification properties of different scorecards. This curve and the related Gini coefficient are considered in section 7.6. Finally, in section 7.7, we look at some ways that the performance of the scorecard can be checked and, if necessary, corrected without having to completely rebuild the whole system.

7.2 Error rates using holdout samples and 2 x 2 tables

In section 4.2, we defined the decision making in a credit-scoring system as follows: $\mathbf{X} = (X_1, X_2, \ldots, X_p)$ are the application variables, and each applicant gives a set of answers $\mathbf{x}$ to these variables and thereafter is found to be either good (G) or bad (B). Assuming that the application characteristics are continuous (a similar analysis works for the discrete case), let $f(\mathbf{x})$ be the distribution of application characteristics, $f(G|\mathbf{x})$ be the probability of being a good if the application characteristics were $\mathbf{x}$, and $f(B|\mathbf{x}) = 1 - f(G|\mathbf{x})$ be the probability of being a bad if those are the characteristics.

The optimal or Bayes error rate is the minimum possible error rate if one knew these distributions completely over the whole population of possible applicants,

$$e(\text{Opt}) = \int \min\{f(B|\mathbf{x}), f(G|\mathbf{x})\} f(\mathbf{x}) d\mathbf{x}. \tag{7.1}$$

Any given credit-scoring system built on a sample S of n consumers estimates the functions $f(G|\mathbf{x})$ and $f(B|\mathbf{x})$ and in light of this defines two regions of answers A_G and A_B, in which the applicants are classified as good or bad. The actual or true error for such a system is then defined as

$$e_S(\text{Actual}) = \int_{A_G} f(B|\mathbf{x}) f(\mathbf{x}) d\mathbf{x} + \int_{A_B} f(G|\mathbf{x}) f(\mathbf{x}) d\mathbf{x}. \tag{7.2}$$

This is the error that the classifier built on the sample S would incur when applied to an infinite test set. The difference occurs because one has used only a finite sample S to build the estimator. One is usually interested in estimating $e_S(\text{Actual})$, but if one had to decide what system to use before seeing the data and adapting the system to the data, then one would take the expected error rate $e_n(\text{Expected})$, which is the expectation of $e_S(\text{Actual})$ over all samples of size n.

The difficulty in calculating $e_S(\text{Actual})$ is that one does not have an infinite test set on which to calculate it and so one has to use a sample S^* to estimate it on. Hence one calculates

$$e_S(S^*) = \int_{A_G} f(B|\mathbf{x}) f_{S^*}(\mathbf{x}) d\mathbf{x} + \int_{A_B} f(G|\mathbf{x}) f_{S^*}(\mathbf{x}) d\mathbf{x}, \tag{7.3}$$

where $f_{S^*}(\mathbf{x})$ is the distribution of characteristics $\mathbf{x}$ in the sample S^*.

The obvious thing is to check the error on the sample on which the classifier was built and so calculate $e_S(S)$. Not surprisingly, this underestimates the error considerably, so $e_S(S) < e_S(\text{Actual})$. This is because the classifier has incorporated in it all the quirks of the sample S even if these are not representative of the rest of the population. Hence it is far better to test the data on a completely independent sample S^*. This is called the holdout sample, and it is the case that the expected value of $e_S(S^*)$ over all samples excluding S is $e_S(\text{Actual})$. This means that this procedure gives an unbiased estimate of the actual error.

In credit scoring, the cost of the two errors that make up the error rate are very different. Classifying a good as a bad means a loss of profit L, while classifying a bad as a good means an expected default of D, which is often considerably higher than L. Thus instead of error rates one might look at expected loss rates, where the optimal expected loss $l(\text{Opt})$ satisfies

$$l(\text{Opt}) = \int \min\{Df(B|\mathbf{x}), Lf(G|\mathbf{x})\} f(\mathbf{x})d\mathbf{x}. \tag{7.4}$$

In analogy with (7.2) and (7.3), the actual or true loss rate for a classifier based on a sample S is

$$l_S(\text{Actual}) = \int_{A_G} Df(B|\mathbf{x}) f(\mathbf{x})d\mathbf{x} \int_{A_B} Lf(G|\mathbf{x}) f(\mathbf{x})d\mathbf{x}, \tag{7.5}$$

while the estimated rate using a test sample S^* is

$$l_S(S^*) = \int_{A_G} Df(B|\mathbf{x}) f_{S^*}(\mathbf{x})d\mathbf{x} \int_{A_B} Lf(G|\mathbf{x}) f_{S^*}(\mathbf{x})d\mathbf{x}. \tag{7.6}$$

Bayes's theorem will confirm that the expression for the actual loss rate in (7.5) is the same as that given in (4.12).

So how does one calculate $e_S(S^*)$ and $l_S(S^*)$, having built a classifier on a sample S and having a completely independent holdout sample S^* available? What we do is compare the actual class G or B of each customer in the sample S^* with the class that the scorecard predicts. The results are presented in a 2×2 table called the confusion matrix (see Table 7.1), which gives the numbers in each group.

Table 7.1. *General confusion matrix.*

		True class		
		G	B	
Predicted class	G	g_G	g_B	g
	B	b_G	b_B	b
		n_G	n_B	n

For example, we might have the confusion matrix given in Table 7.2. In this sample, n, n_G, and n_B (1000, 750, 250) are fixed as they describe the sample chosen. Thus really only two of the four entries g_G, b_G, g_B, and b_B are independent. The actual error rate is calculated as $\frac{b_G+g_B}{n} = \frac{250}{1000} = 0.25$. If the losses are $L = 100$, $D = 500$, then their actual loss per customer is $\frac{Lb_G+Dg_B}{n} = \frac{(100\cdot150)+(500\cdot100)}{1000} = 65$.

Table 7.2. *Example of a confusion matrix.*

		True class		
		G	B	
Predicted class	G	600	100	700
	B	150	150	300
		750	250	1000

One can use confusion matrices to compare systems or even to try to decide on the best cutoff score when a scorecard had been developed. In the latter case, changing the scorecard in the example in Table 7.2 may lead to the confusion matrix in Table 7.3.

Table 7.3. *Confusion matrix with a different cutoff.*

		True class		
		G	*B*	
Predicted class	*G*	670	130	800
	B	80	120	200
		750	250	1000

The actual error rate is $\frac{130+80}{1000} = 0.21$, which suggests this is a better cutoff. However, the actual expected loss rate is $\frac{(100\cdot 80)+(500\cdot 130)}{1000} = 73$, which is higher than the expected loss with the other cutoff. So perhaps the other system was superior. This difference between the two ways of trying to decide which system to go for is very common. Sometimes in credit scoring, one will find that the error rate is minimized by classifying everyone as good and so accepting them all. It is a brave—and foolish—credit analyst who suggests this and thus makes himself redundant!

One approach that is useful in comparing the difference between the way two credit-scoring systems perform on a holdout sample is to look at the swap sets. This is the group of people in the holdout sample, who are classified differently by the two systems. The swap sets for Tables 7.2 and 7.3 might look like Table 7.4.

Table 7.4. *Example of a swap set analysis.*

	True	
	G	*B*
Predicted *G* by Table 7.2, *B* by Table 7.3	50	10
Predicted *B* by Table 7.2, *G* by Table 7.3	120	40

This table cannot be calculated from the two confusion matrices, although from these we can see that 100 more in the sample were predicted *B* by Table 7.2 and *G* by Table 7.3 than were predicted *G* by Table 7.2 and *B* by Table 7.3. There could be in this case a number of people who move a different way to the overall trend. The fact that $\frac{50+10+120+40}{1000} = 0.22$ of the population change between the two scorecards suggests that the scorecards are quite different. If one looks only at the swap sets caused by changing the cutoff, then obviously no customers will have their own classifications changed against the movement. Hence one of the rows in the swap sets will be 0.

7.3 Cross-validation for small samples

The standard way to assess credit-scoring systems is to use a holdout sample since large samples of past customers usually are available. One wants the holdout sample to be independent of the development sample but to be similar to the population on which it is to be used. This causes a problem because of population drift, wherein the characteristics of consumers change over time. The new customers on whom the scorecard is to be used may be different from those of two or three years ago, on whom it was built and tested.

However, there are also situations in which not much data are available to build a scorecard. This could be because what is being assessed is a new type of loan product, or the lender is targeting a new consumer group. It could also be that it is an area like scoring of commercial customers, where the population itself is very limited. In these cases, we want to use all the data to build the system and so cannot afford to cut the size of the development sample, yet we want as true an estimate of the error rate of the classification as is possible.

The way around this is cross-validation. Subsets of the sample are used to test the performance of the scorecard built in the remainder of the sample. This process is repeated for different subsets and the results are averaged. Two ways have been suggested for choosing the subsets: leave-one-out—a single consumer is used as a test set for the scorecard built on the other $n-1$, and this is repeated for each consumer in the sample; and rotation—the sample is split into m different subsets, say, $m = 5$ or 10, and each one in turn is used as a test set for the scorecard built on the rest.

The leave-one-out method is so much the archetypal cross-validation method that it is often called cross-validation. If $\{S-i\}$ is the sample with customer i missing, one calculates

$$e_S(\text{Leave}) = \sum_{i=1}^{m} e_{S-i}\{i\}. \tag{7.7}$$

It is where the sample is used to its fullest in building the scorecard, but it has the disadvantage that n scorecards have to be built, which could be very time-consuming. Modern computing power makes this less of a burden than might have been expected, and in the linear regression and linear programming approaches to scorecard building, one can use the fact that only one sample data point is different between any two samples to cut down considerably the calculation of the new scorecard. The scorecard built on the full sample is used in practice, and the average error rate of the n scorecards built on the $n-1$ points in turn is an unbiased estimator of the actual error rate, although in practice it has quite a high variance.

The rotational approach considerably cuts down the number of scorecards that need to be built, from n to m, but the scorecards built are less similar to those built on the full sample than in the leave-one-out case. Hence if one uses the full sample scorecard in practice, the average of the error rates of the m scorecards built is not as robust an estimator of the actual error rate as the leave-one-out estimator.

7.4 Bootstrapping and jackknifing

Cross-validation approaches still split the sample into two subsets—one for building the scorecard and the other for testing it. Bootstrapping and jackknifing methods seek to estimate the bias in $e_S(\text{Actual}) - e_S(S)$ and try to remove it.

In bootstrapping, a sample, say, R, equal in size to the original sample is chosen from it, allowing repetition. Thus the same consumer can appear several times in the sample. This sample with possible repetition is taken as the development sample, and the original sample is used to test the scorecard. The procedure is repeated m times using samples chosen again with repetition allowed; call the samples $R_1, R_2, \ldots, R_m$. The idea is that one is trying to estimate the bias between the true and the apparent error rate $e_S(\text{Actual}) - e_S(S)$. For the scorecards built on the samples R, the apparent rate is $e_R(R)$ and the claim is that the true rate is $e_R(S)$. This assumes that S is, in fact, the whole population for these bootstrap estimators. Hence $e_R(S) - e_R(R)$ gives a good estimate of the bias that we are interested in, and so one takes the estimator for $e_S(\text{Actual})$ to be

$$e_S(S) + (e_S(\text{Actual}) - e_S(S)) \approx e_S(S) + \sum_{i=1}^{m} (e_{R_i}(S) - e_{R_i}(R_i)). \tag{7.8}$$

One can improve on this initial bootstrap estimation by recognizing that there are bound to be customers who are both in S and in R, and so these sets are not independent. If there are n customers in sample S, the chance a particular one does not appear in a specific R is $(1-\frac{1}{n})^n$ (i.e., $(1-\frac{1}{n})$ is the chance the customer is not sampled each time one of the n customers is sampled). Letting $n \to \infty$, this converges to $e^{-1} = 0.368$. The easiest way to see this is to take logs and to recognize that $n\log(1-\frac{1}{n})$ tends to -1. Thus there is a 0.368 chance a consumer does not appear in sample R and a 0.632 chance that the consumer does appear in sample R. It then turns out that an improved estimate of the actual error rate is obtained by taking

$$0.368 e_S(S) + 0.632 \sum_m e_{R_m}(S - R_m), \tag{7.9}$$

where $(S - R_m)$ is the set of customers who are in S but not in R_m.

The jackknife method operates very similarly to the leave-one-out estimator in that one leaves out each of the sample customers in turn and builds a scorecard on the remaining $n-1$. However, it then calculates the apparent error rate of this scorecard $e_{\{S-i\}}(S-i)$, and hence the average apparent rate is

$$\bar{e}_{S-1}(S-1) = \sum_{i=1}^{n} e_{S-i}(S-i). \tag{7.10}$$

Similarly, one calculates $e_{\{S-i\}}(S)$ and the average $\bar{e}_{S-1}(S) = \sum_{i=1}^{n} e_{S-i}(S)$, and one can show that $(n-1)(\bar{e}_{S-1}(S) - \bar{e}_{S-1}(S-1))$ is a good estimate of the bias between estimating on the true set (in this case, S) and the set on which the scorecard was built (the $S-i$ in this case). Thus the jackknife estimate $e_S(\text{Jack})$ of $e_S(\text{Actual})$ is taken as

$$e_S(\text{Jack}) = e_S(S) + (n-1)(\bar{e}_{S-1}(S) - \bar{e}_{S-1}(S-1)). \tag{7.11}$$

It can be shown (Hand 1997) that there is a relationship between the jackknife and the leave-one-out estimate of $e_S(\text{Actual})$, namely,

$$e_S(\text{Jack}) = e_S(\text{Leave}) + (e_S(S) - \bar{e}_{S-1}(S)). \tag{7.12}$$

7.5 Separation measures: Mahalanobis distance and Kolmogorov–Smirnov statistics

A number of measures used throughout statistics describe how far apart the characteristics of two populations are. If one has a scoring system that gives a score to each member of the population, then one can use these measures to describe how different the scores of the goods and the bads are. Thus these approaches can be used only for credit-scoring systems that actually give a score, like the regression approaches or linear programming. They cannot be used for the credit-scoring systems that group, like classification trees, or where a score is not explicit, like neural networks. Moreover, they describe the general properties of the scorecard and do not depend on which cutoff score is used. This is useful in that these measures give a feel for the robustness of the scorecard if the cutoff score is changed and may be useful in determining what the cutoff score should be. However, when it comes to it, people will want to know how well the scorecard will predict, and to know that one needs to have chosen a specific cutoff score so that one can estimate the error rates and confusion matrices of sections 7.2 and 7.3.

For accurate estimates, we should calculate these measures on a holdout sample, which is independent of the development sample on which the scorecard was built. However, often speed and the fact that in many statistical packages it is much easier to calculate the measures on the development sample than on the holdout sample mean that one calculates them first on the development sample. Since they are only indicators of the relative effectiveness of different scorecards, the assumption is made that if one is much better than the other on the development sample, it will remain so on the holdout sample.

The first measure, the Mahalanobis distance, appeared earlier in the discussion on the Fisher approach to using linear regression as a classification method in section 4.3. There we found the linear combination Y of the application variables so that M, the difference between the sample mean of Y for the goods and the sample mean Y of the bads, divided by the standard deviation in Y for each group was as large as possible. This M is the Mahalanobis distance and is a measure of by how much the scores of the two groups of the goods and the bads differ. Formally, if $n_G(s)$ and $n_B(s)$ are the numbers of goods and bads with score s in a sample of n, where there are n_G goods and n_B bads, the $p_G(s) = \frac{n_G(s)}{n_G}$ ($p_B(s) = \frac{n_B(s)}{n_B}$) are the probabilities of a good (and bad) having a score s. Then $m_G = \sum s p_G(s)$ and $m_B = \sum s p_B(s)$ are the mean scores of the goods and bads. Let σ_G and σ_B be the standard deviation of the scores of the goods and the bads, calculated as

$$\sigma_G^2 = \left(\sum_s s^2 p_G(s) - m_G^2 \right)^{\frac{1}{2}}, \qquad \sigma_B^2 = \left(\sum_s s^2 p_B(s) - m_B^2 \right)^{\frac{1}{2}}. \tag{7.13}$$

Let σ be the pooled standard deviation of the goods and the bads from their respective means; it is calculated as follows:

$$\sigma = \left(\frac{n_G \sigma_G^2 + n_B \sigma_B^2}{n} \right)^{\frac{1}{2}}. \tag{7.14}$$

The Mahalanobis distance M is then the difference between the mean score of the two groups, suitably standardized

$$M = \frac{m_G - m_B}{\sigma}. \tag{7.15}$$

This is indifferent to any linear scaling of the score, and as Figure 7.1 suggests, one would assume that if a scorecard has a large Mahalanobis distance, it will be a better classifier. In Figure 7.1, the dotted lines represent possible cutoff scores, and the errors in the figure on the left with the smaller M are much greater than those on the right.

The Mahalanobis distance measures how far apart the means of the goods score and the bads score are. The Kolmogorov–Smirnov statistic measures how far apart the distribution functions of the scores of the goods and the bads are. Formally, if $P_G(s) = \sum_{x \le s} p_G(x)$ and $P_B(s) = \sum_{x \le s} p_B(x)$ (or the sums replaced by integrals if the scores are continuous), then the Kolmogorov–Smirnov (KS) statistic is

$$\text{KS} = \max_s |P_G(s) - P_B(s)|. \tag{7.16}$$

In Figure 7.2, the Kolmogorov–Smirnov statistic is the length of the dotted line at the score that maximizes the separation in the distribution function. If the distribution functions are sufficiently regular, then the Kolmogorov–Smirnov distance occurs at the score where the good and bad histograms in Figure 7.1 cross. As mentioned in section 4.7, the Kolmogorov–Smirnov statistic for an attribute, rather than the score, is used to find the best splits in a classification tree.

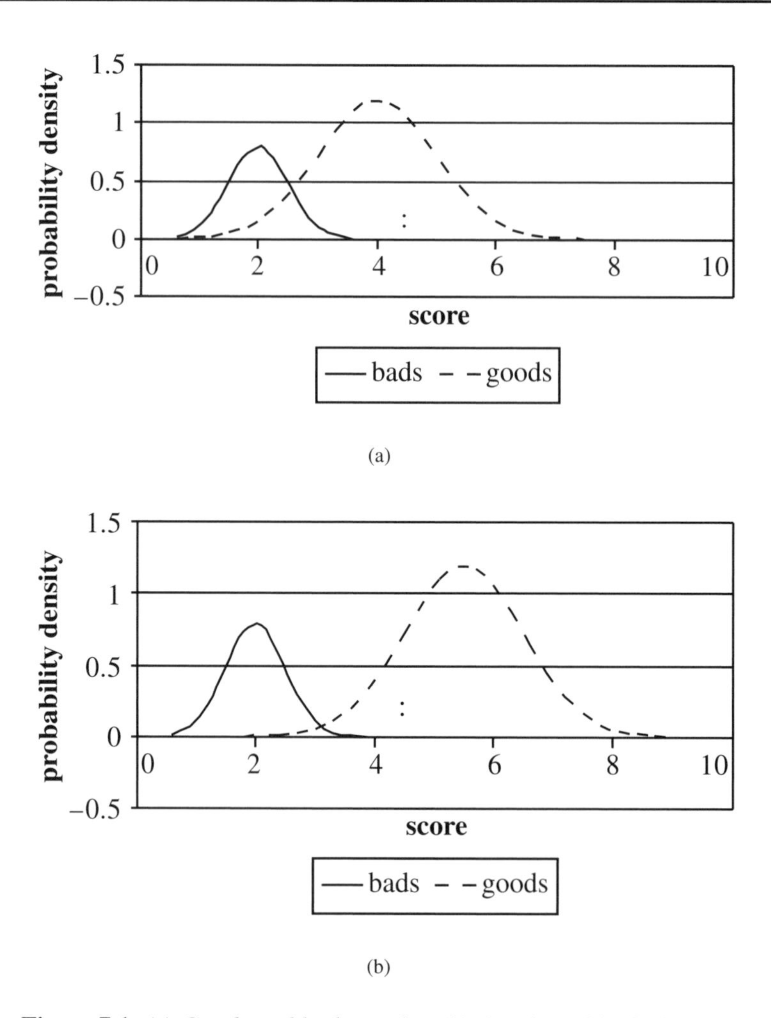

Figure 7.1. (a) *Goods and bads similar.* (b) *Goods and bads different.*

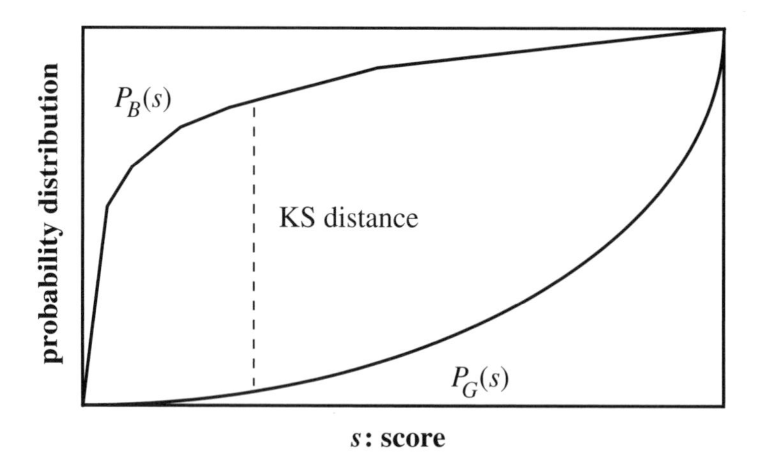

Figure 7.2. *Kolmogorov–Smirnov distance.*

7.6 ROC curves and Gini coefficients

Kolmogorov–Smirnov statistics can be displayed, as in Figure 7.2, by plotting two curves, but it is possible to display the same information on one curve by plotting $P_G(s)$ against $P_B(s)$. The result is a curve as in Figure 7.3, where each point on the curve represents some score s, and its horizontal distance is $P_B(s)$ and its vertical value is $P_G(s)$.

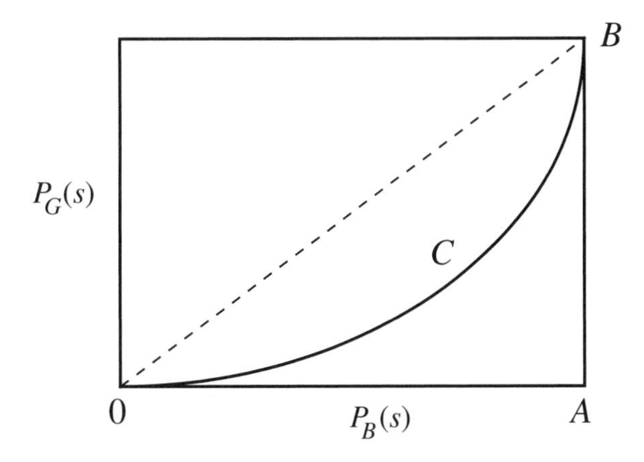

Figure 7.3. *ROC curve.*

This is the ROC curve, sometimes called the Lorentz diagram. It describes the classification property of the scorecard as the cutoff score varies. The best possible scorecard would have an ROC curve that goes all the way along the horizontal axis before going up the vertical axis. Thus point A would correspond to a score s^*, where $P_B(s^*) = 1$ and $P_G(s^*) = 0$; i.e., all of the bads have scores less than s^* and none of the goods do. An ROC curve along the diagonal OB would correspond to one where at every score $P_G(s) = P_B(s)$, so the ratio of goods to bads is the same for all score ranges. This is no better than classifying randomly given that one knows the ratio of good to bads in the whole population. Thus the further from the diagonal the ROC curve is, the better the scorecard. If one scorecard has a ROC curve that is always further from the diagonal than the ROC curve of another scorecard, then the first scorecard dominates the second and is a better classifier at all cutoff scores. More common is to find ROC curves that cross so one scorecard is a better classifier in one score region and the other is better in the other region.

The ROC curve's name suggests its origin in estimating classification errors in transmitting and receiving messages. Using classification as a tool in epidemiology has introduced two other terms that are found in scorecard measurement. If the bads are thought of as the cases one needs to detect, then the sensitivity (Se) is the proportion of bads who are predicted as bad, while the specificity (Sp) is the proportion of goods who are predicted as good. If prediction is by a scorecard, with those below the cutoff score predicted to be bad and those above the cutoff score predicted to be good, then the sensitivity Se is $P_B(s)$, while the specificity Sp is $1 - P_G(s)$. Thus the ROC curve can be thought of as plotting 1− specificity against sensitivity.

One can also use the confusion matrix to calculate sensitivity and specificity. Using the notation of Table 7.1, $\text{Se} = \frac{b_B}{n_B}$, while $\text{Sp} = \frac{g_G}{n_G}$. Recall that since n_G and n_B are fixed, there are only two variables in the confusion matrix, so specifying Se and Sp defines it exactly.

Recall that the further the ROC curve is from the diagonal the better. This suggests that the larger the shaded area in Figure 7.3 between the ROC curve and the diagonal, the better classifier the scorecard is. Define the Gini coefficient G to be twice this area. This has the useful property that the perfect classifier that goes through A in Figure 7.3 will have $G = 1$, while the random classifier with ROC curve OB in Figure 7.3 has $G = 0$. In fact, one can show that it is the probability that a randomly selected bad will have a lower score than a randomly selected good. This is related to the Wilcoxon test for two independent samples (Hand 1997). Thus the Gini coefficient gives one number that summarizes the performance of the scorecard over all cutoff scores. This is both useful in its brevity and misleading because in reality we are usually interested in the scorecard's performance over a small range of possible cutoff scores. The same complaint can be made of the Kolmogorov–Smirnov statistic and the Mahalanobis distance. They all describe general properties of the scorecard, whereas what is important in practice is how the scorecard performs at the chosen cutoff.

One can use the ROC curve to identify suitable cutoff scores. The score that maximizes the Kolmogorov–Smirnov statistic, for example, corresponds to the point on the curve whose horizontal distance from the axis is greatest. This is C in Figure 7.4. This follows since the point is $(P_B(s), P_G(s))$, so this horizontal distance is $(P_B(s) - P_G(s))$. If one assumes that the true percentages of goods and bads in the population are p_G and p_B, respectively, and that L and D are the usual loss quantities for misclassifying, the expected loss rate if one has a cutoff at s is

$$l(\text{Actual}) = LP_G(s)p_G + D(1 - P_B(s))p_B. \tag{7.17}$$

As far as an ROC curve f is concerned, this is like having to minimize $Lp_G f(x) + Dp_B(1-x)$, which occurs when

$$Lp_G f'(x) - Dp_B = 0 \tag{7.18}$$

(i.e., the derivative is 0). Hence at the minimum the slope of the tangent to the ROC curve, $f'(x)$, satisfies $f'(x) \ \frac{Dp_B}{Lp_G}$. One way to find this point is to draw the line with slope $-\frac{Lp_G}{Dp_B}$ through the point $(1, 0)$ and project the curve onto this line. The point that yields the projection nearest the point $(1, 0)$ is the point we require. This is point D in Figure 7.4.

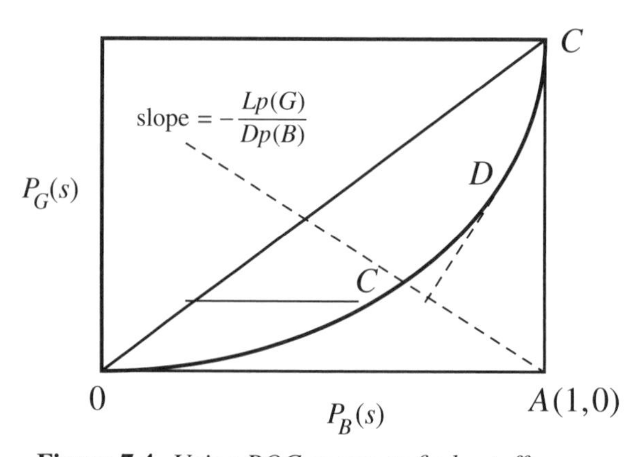

Figure 7.4. *Using ROC curves to find cutoffs.*

7.7 Comparing actual and predicted performance of scorecards: The delta approach

The previous sections of this chapter looked at ways to measure the classification performance of a scoring system and the overall characteristics of a scorecard, but they did not seek to identify where and why it was misclassifying. Those measures are used at the development of the scoring system to describe its likely performance characteristics and are used to change cutoff levels. In this section, we look at one technique that identifies the differences between the way the scorecard is performing in practice and how it was intended to perform. The different forms of monitoring reports, which also perform this function, are discussed in Chapter 9. The delta approach has the advantage of calculating a measure for the misalignment between the predicted and actual scorecard performance and splitting that to identify the misalignment in different scoring reasons. It can then be used to decide whether the scorecard is satisfactory or needs to be completely rebuilt or whether adjusting one or two scores will be sufficient. It can be used only on scoring systems based on regression methods, and we assume hereafter that the original scorecard was built using logistic regression.

The original scorecard was built assuming a model that

$$P(G|\text{Score} = s) = F(s), \quad \text{where Score} = w_0 + w_1 x_1 + \cdots + w_p x_p \quad \text{and} \quad F \text{ is a monotone function,} \tag{7.19}$$

and traditionally one wished to test whether $X_i = 1$, say, had the correct score by looking at

$$P(G|s_1 < \text{Score} \le s_2) = F(s_2) - F(s_1) = P(G|s_1 < \text{Score} \le s_2 \text{ and } X_i = 1). \tag{7.20}$$

Doing this directly can be very time-consuming since there are so many choices of s_1, s_2, i, and values for X_i. An alternative approach is to exploit knowledge of $F(s)$. In logistic regression,

$$F(s) = \frac{e^{\alpha+\beta s}}{1 + e^{\alpha+\beta s}},$$

so

$$\log\left(\frac{P(G|\text{Score} = s)}{P(B|\text{Score} = s)}\right) = \log(\text{Odds}(G : B)|s) = \alpha + \beta s. \tag{7.21}$$

What we do is plot the log of the actual good:bad odds in different score ranges for different attributes of the characteristic under consideration. Suppose that one is considering telephone ownership, with Y meaning the consumer has a telephone number and N meaning the consumer does not have a number. Figure 7.5 shows the graph of these odds when the score range has been split into four ranges, and so each graph consists of joining the four points of mean score and log of actual odds in each score range. If the scorecard were perfect, then each of the graphs would lie on the line $\alpha + \beta s$. That is not the case—the telephone owners have better odds than predicted and the nonowners have worse odds in three of the four score ranges. The delta score at each point is the amount needed to be added to the score so that the point coincides with the original graph:

$$\delta = \frac{\log(\text{Actual odds}) - \log(\text{Predicted odds})}{\beta} = \frac{\log\left(\frac{\text{Actual odds}}{\text{Predicted odds}}\right)}{\beta}. \tag{7.22}$$

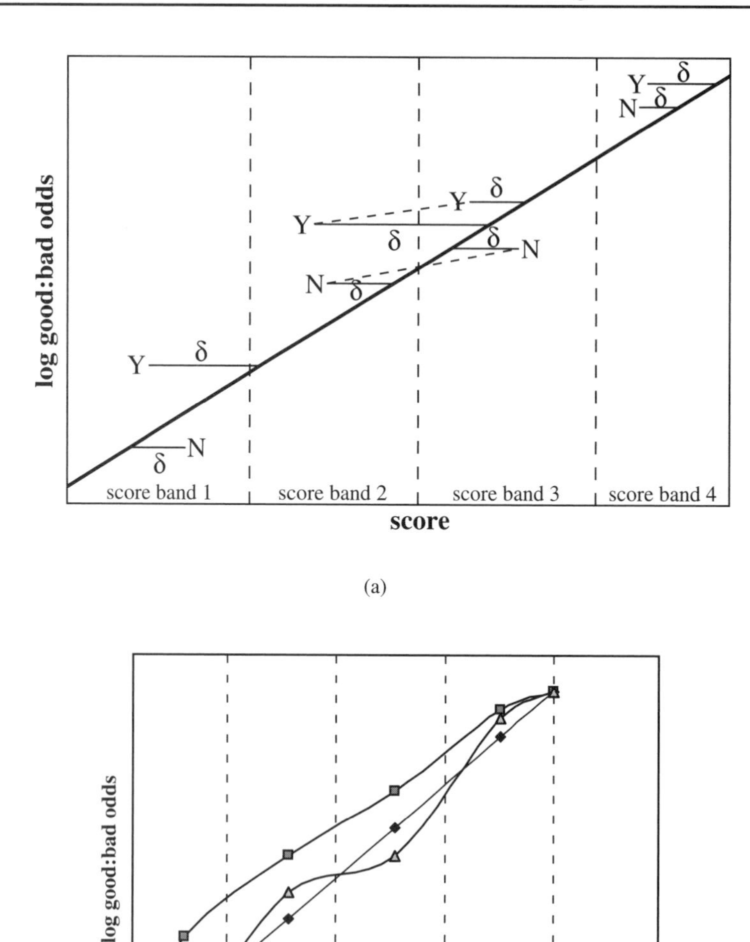

(a)

(b)

Figure 7.5. (a) *Graph of odds against score and corresponding delta distances.* (b) *Log odds against score in four score bands.*

Assume that one has a set of data of the actual performance of the scorecard, i.e., which customers are good and bad and what their original score (and the β_0 in the original logistic regression) was. Let g_{ij} (b_{ij}) be the actual number of goods (bads) in the ith score band with attribute j for a given characteristic and g_i (b_i) be the total number of goods (bads) in the ith score bands. Apply another logistic regression to this data to estimate for each original score the predicted probability of a customer with that score being bad, using the actual data; i.e., one is fitting

$$P(B|\text{Score} = s) = \left(\frac{1}{1 + \exp\left(\hat{\alpha} + \hat{\beta} s\right)} \right). \tag{7.23}$$

Let $\hat{b}_{ij}$, $\hat{b}_i$ be the sum of the predicted badness for those in the ith score band with attribute j and those in the ith score band in total.

Estimate the actual good:bad ratios by

$$r_{ij} = \frac{g_{ij} + \frac{1}{2}}{b_{ij} + \frac{1}{2}}, \qquad r_i = \frac{g_i + \frac{1}{2}}{b_i + \frac{1}{2}}. \tag{7.24}$$

(These are better estimates of $\log r_{ij}$ than taking $r_{ij} = \frac{g_{ij}}{b_{ij}}$.)

Estimate the predicted good:bad ratios by

$$\hat{r}_{ij} = \frac{g_{ij} + b_{ij} - \hat{b}_{ij}}{\hat{b}_{ij}} \quad \text{and} \quad \hat{r}_i = \frac{g_i + b_i - \hat{b}_i}{\hat{b}_i}. \tag{7.25}$$

It is possible that $\hat{r}_i$ and r_i differ considerably although all $\hat{r}_{ij}$ are consistent with $\hat{r}_i$. This means the whole curve has changed somewhat although the scores for the attributes under consideration are consistent. We allow for this by defining a scaled predictor good:bad ratio to be

$$\bar{r}_{ij} = \frac{\hat{r}_{ij} r_i}{\hat{r}_i}. \tag{7.26}$$

This adjustment ensures that the centroid of the predictor line in the ith band agrees with the centroid of the line obtained in the original logistic regression.

Then define δ:

$$\delta_{ij} = \frac{\log\left(\frac{r_{ij}}{\bar{r}_{ij}}\right)}{\beta}.$$

One can show that δ_{ij} has an approximate standard error of

$$\sigma_{ij} = \frac{\sqrt{\frac{g_i - g_{ij}}{g_i\left(g_{ij} + \frac{1}{2}\right)} + \frac{b_i - b_{ij}}{b_i\left(b_{ij} + \frac{1}{2}\right)}}}{\beta}, \tag{7.27}$$

and if one takes the weighted mean of the delta scores in region i,

$$\bar{\delta}_i = \frac{\sum_j \frac{\delta_{ij}}{\sigma_{ij}^2}}{\sum_j \frac{1}{\sigma_{ij}^2}},$$

one can apply a χ^2 test,

$$e^i = \sum_j \frac{(\delta_{ij} - \bar{\delta}_i)^2}{\sigma_{ij}^2}, \tag{7.28}$$

to check if the scorecard is badly aligned in the ith score band. Summing the e^i's gives a measure of the overall alignment of the scorecard.

If one wanted to adjust the scores to improve the alignment, one would add the aggregated delta scores for attribute j, δ_j to the current score for j, where

$$\delta_j = \frac{\sum_j \frac{\delta_{ij}}{\sigma_{ij}}}{\sum_j \frac{1}{\sigma_{ij}}}. \tag{7.29}$$

This would be a way to get an overall adjustment to that attribute, but the δ_{ij} give a more detailed insight into what are the errors caused by that attribute at different ranges of the scorecard.

Chapter 8

Practical Issues of Scorecard Development

8.1 Introduction

The previous chapters looked at the different techniques that can be used to develop application and behavioral scoring systems and ways to measure the performance of such systems. This chapter is somewhat more down to earth and concentrates on the practicalities of building a scoring system using these techniques, and it identifies some of the pitfalls that can arise.

Section 8.2 considers how to select the sample of previous customers on which to build the system. Section 8.3 discusses the usual definitions of good and bad performance for such borrowers and what one does with those in the sample who do not fall into either category. Sections 8.4 and 8.5 consider the characteristics that tend to be used in building scoring systems, the first section concentrating on what application and transactional data are used and the second on the credit bureau information available. The next three sections concentrate on how these characteristics are modified and then used in scoring systems. Section 8.6 discusses how and why one might develop a suite of scorecards—one for each of the subpopulations into which the sample is divided. Section 8.7 looks at coarse classifying the characteristics—putting several attributes into the same group so they get the same score—and splitting the values of a continuous variable into a number of different categories. Section 8.8 discusses how one chooses which variables to keep in a scorecard.

Section 8.9 addresses the thorny issue of reject inference; i.e., How can one enhance the scoring system by using the partial information available on those previous applicants who were rejected for credit? Section 8.10 discusses how and why the decisions of the scoring system may be overridden and how that affects the performance of the scoring system, while Section 8.11 concentrates on how the cutoff score in the scorecard is chosen. Section 8.12 investigates how one checks if the scorecards are correctly aligned and how to recalibrate them if necessary.

8.2 Selecting the sample

All the methodologies for credit and behavioral scoring require a sample of previous customers and their histories to develop the scoring system. There are two somewhat conflicting objectives in selecting such a sample. First, it should be representative of those people who are likely to apply for credit in the future—the through-the-door population. Second, it should incorporate sufficient of different types of repayment behavior (i.e., the goods and the bads) to make it possible to identify which characteristics reflect this behavior in the general through-the-door population. The group that must be closest to this population are the previous applicants for that lending product. The conflict arises because to get as close as possible to the future through-the-door population, we want the sample group to be as recent as possible. However, to distinguish between the good and bad repayment behavior, we need a reasonable history of repayment and thus a reasonable time since the sample group applied. This is particularly the case with behavioral scoring, where one also needs a reasonable performance period to identify the transactional characteristics as well as an outcome period. The compromise is usually an outcome period of 12 months for application-scoring systems. In behavioral scoring, one usually takes 18 to 24 months of history and splits that into 9 to 12 months of performance history and 9 to 12 months of an outcome period. These periods vary depending on the product since for mortgages the outcome period may need to be several years.

The next questions are how large the sample should be and what the split should be between the number of goods and the number of bads in the sample. Should there be equal numbers of goods and bads in the sample, or should the sample reflect the good:bad odds in the population as a whole? Normally, the latter is so strongly oriented to the goods (say, 20:1) that keeping the same odds in the sample would mean there may not be enough of a bad subpopulation to identify their characteristics. For this reason, the sample tends to be either 50:50 or somewhere between 50:50 and the true population proportion of goods to bads. If the distribution of goods and bads in the sample does not have the same distribution as that in the population as a whole, then one needs to adjust the results obtained from the sample to allow for this. In the regression approach, this is done automatically as the probability of goods and bads in the true population, p_G, p_B, is used in the calculations. In other approaches, it has to be done a posteriori, so if a classification tree built on a sample where the goods were 50% of the population (but from a true population they were 90%) has a node where the good bad ratio is 3:1 or 75% to 25%, the true odds are

$$\frac{(\text{odds in node}) \cdot (\text{odds in true population})}{(\text{odds in sample population})} = \frac{\frac{3}{1} \cdot \frac{9}{1}}{\frac{1}{1}} = 27{:}1.$$

As for the number in the sample, Lewis (1992) suggested that 1,500 goods and 1,500 bads may be enough. In practice, much larger samples are used, although Makuch (1999) makes the point that once one has 100,000 goods, there is no need for much more information on the goods. Thus a typical situation would be to take all the bads one can get into the sample and take 100,000+ goods. This sample is then randomly split into two. One part is used for developing the scoring system and the other is used as a holdout sample to test it.

The real difficulty arises when samples of such sizes cannot be obtained. Is it sensible to put applicants for different products or applicants to different lenders together to make an enlarged sample? This is discussed in more detail in section 12.2, where we look at the development of generic scorecards, but it is sufficient to say that one has to be very careful. For example, the attributes that distinguish risky borrowers for secured loans like mortgages or even hire purchase of white goods are different from those that identify the risky borrowers for unsecured loans.

If the sample is chosen randomly from an existing population of applicants, one has to be sure the choice is really random. This is not too difficult if there is a central application list kept with applicants ordered by time of application. Choosing every tenth good in the list should give a reasonably random sample of 10% of the goods. If, however, one has to go to branch levels to collect the list, one has to first randomly select the branches to ensure a good mix of urban and rural branches and a suitable spread of socioeconomic conditions and geography. Then one has to randomly select at the branch level. Again one needs to be careful. Deciding to take all the customers who apply in a given month might on the surface seem sensible, but if that is the month when universities start back, then a lot more students are likely to be in the applicant population than in a true random sample. Sometimes it may be necessary to put such a bias into the sample since, for example, the new product is more for young people than the existing one and so one wants a higher proportion of young people in the sample than in the original population. The aim always is to get a sample that will best reflect the likely through-the-door population for the new product. In fact, this is not quite correct; what one wants is a sample of the through-the-door population that the lender will consider for the scoring process. Thus those who would not be given loans for policy reasons should be removed from the sample, as should those who would be given them automatically. The former might include underage applicants, bankrupts, and those with no credit bureau file. The latter might include customers with specific savings products or employees of the lender.

All this work on the choice of sample presupposes that we can define goods and bads. In the next section, we consider what these definitions might be and what to do with those in the sample who do not obviously fall into one or the other category.

8.3 Definitions of good and bad

As part of the development of a scorecard, one needs to decide how to define good and bad. Defining a bad does not necessarily mean that all other cases are good. Often in scorecard development, at least two other types of case can be identified. The first might be labeled "indeterminates," those cases that are in between—neither good nor bad. The second might be labeled "insufficient experience."

In a scorecard development for a credit card portfolio, a common definition of bad is a case that at some point becomes three payments in arrears. This is often referred to as "ever 3+ down" or "worst 3+ down." The indeterminate cases might be those that have a worst-ever status of two payments down. Thus they have caused some problems and some additional collections activity—perhaps repeatedly if they have been two payments down on several occasions—but have never become three payments down. Then we might identify those where we have insufficient experience. Suppose that we have a development sample window of 12 months of applications and an observation point one year later, so cases have 12 to 24 months' exposure. Then we might label as having insufficient experience those with three or fewer months of sales or cash advance activity. In other words, the account is not bad, but it has been used fairly infrequently and so it would be premature to adjudge it to be good. The residue would be categorized as good accounts.

This classification is only an example. Many variations are possible. One could define bad as being ever 3+ down or twice 2 down. We could define "insufficient experience" as those that have had less than six months of debit activity. We could include in the indeterminate grouping those cases that have missed one payment.

On an installment loan portfolio, the situation might be a little clearer. Here, we might define bad as ever 3+ down or ever 2+ down or something in between, as suggested above. Indeterminates would be those that may have missed one payment, or may have missed one payment several times, or may have missed two payments. If we have chosen a sample window carefully, then there may be no cases with insufficient experience. However, if someone redeems a loan, i.e., repays a loan in a lump sum, after only a few months, we could classify the loan into this category, especially if the truncated loan means that we did not make a profit or made a very small profit. The survival analysis idea, introduced in section 12.7, is one approach to dealing with the difficulty that different outcomes are unsatisfactory as far as the profitability of a loan is concerned.

When we move into secured lending, such as a mortgage, our definitions could change markedly. Here we have some security and so our definition of good and bad may be affected dramatically by whether the case generated a loss. Thus if a mortgage defaults and the property is repossessed and we recover all our loan and our collection and litigation costs, one could argue that this is not a bad account. Some lenders would classify this as bad in the scorecard development, while others may consider indeterminate to be a more appropriate label. Similar to the installment loan, if a mortgage is redeemed fairly early in its life, we may wish to classify this as indeterminate or insufficient experience.

In the case of current (checking) accounts with an overdraft, by necessity the definitions again have to be changed. There is no fixed expected monthly repayment, so an account holder cannot be in arrears. Therefore, a completely different set of definitions needs to be introduced. We might categorize a bad account as one where there is unauthorized borrowing, i.e., borrowing above the agreed overdraft limit, if there is one.

Whatever definitions we choose, there is no effect on the scorecard methodology. (This assumes that the definitions create a partition; i.e., all possible cases fall into exactly one classification.) We would normally discard the indeterminates and those with insufficient experience and build the scorecard with the goods and bads. Of course, how we define goods and bads will clearly have an effect on the result of the scorecard development. Different definitions may create different scorecards. These differences could be in the score weights or in the actual characteristics that are in the scorecard. However, that does not mean that the results will be very different. Indeed, different definitions of good and bad may generate quite different scorecards but still result in a great similarity in the cases that are accepted and declined.

Another issue that might arise is if we use a very extreme definition of bad. Here we may end up with few actual bad cases and so our modeling robustness may suffer. At a practical level, the definition of good and bad should be considered as more than a theoretical exercise. Issues of profit or loss may have a bearing. We also want consistency in the definition of good and do not want to use a factor to help classify accounts that is highly variable. In general, however, while it is important to develop sound and reasonable definitions of good and bad, it may make only a marginal difference to the effectiveness of the scorecard actually developed based on the definitions.

8.4 Characteristics available

The characteristics available to discriminate between the good and the bad are of three types—those derived from the application form, those available from a credit bureau search, and, for behavioral scoring only, those describing the transactional history of the borrower. We deal with the credit bureau data in the next section, so we concentrate here on the application characteristics

Table 8.1 shows the characteristics that occur in three typical application forms: one from a finance house for a car loan, a U.S. credit card, and a U.K. credit card.

Several characteristics are not permitted to be used for legal reasons. The U.S. Equal Credit Opportunity Acts of 1975 and 1976 made it illegal to discriminate in the granting

Table 8.1. *Characteristics in three application forms.*

Characteristic	Finance house	U.S. credit card	U.K. credit card
ZIP/postal code	X	X	X
Time at address	X	X	X
Residential status	X	X	X
Occupation	X	X	X
Time at employment	X	X	X
Applicant's monthly salary	X	X	X
Other income	X	X	
Number of dependents	X	X	
Number children	X	X	X
Checking account/current account	X	X	X
Savings account	X	X	
Credit card	X	X	X
Store card	X	X	X
Date of birth			X
Telephone		X	X
Monthly payments	X		
Total assets	X		
Age of car	X		

of credit on the grounds of race, color, religion, national origin, sex, marital status, or age. It is interesting to note, however, that one can use age in U.S. score cards if the scores given to ages of 62+ are greater than those given to any other age. The U.K. Race and Sex Discrimination Act outlaws discrimination in credit granting on the grounds of race or sex. Other variables like health status and previous driving convictions, although not illegal, are not used because lenders consider them culturally indefensible.

Having decided on the application questions or characteristics, one also has to decide on the allowable answers or the attributes. For example, in terms of residential status, should one leave the answer open where one could have answers like "owner with no mortgage," "with friends," "barracks," "caravan," etc., or should one restrict the answers to choosing one from "owner/renting/with parents/others"? Another question is how one decides what to do when no answer is given. The usual approach is to have a category of "answer missing" for each characteristic, but there are occasions where it may be clear that no answer really means "no" or "none" (e.g., in response to the spouse's income question).

The occupation characteristics can cause real problems since it is very hard to code because there are hundreds of thousands of occupations. One approach is to specify the attributes such as executive, manual worker, manager, etc. Even with this, "manager" can cause no end of problems as one may be managing a factory employing thousands or managing a career as a part-time Santa Claus. If one goes to the other extreme and simply asks if the applicant is employed, self-employed, or unemployed, then many people can answer yes to at least two of these answers.

Income can also cause difficulty unless one is very careful in the wording of the question to clarify that it refers to income per month (or per year) and whether it is the applicant's basic income, the applicant's total income, or the total household income.

Finally, it is necessary to be able to validate the data. One needs to check that there are no ages under 18 and not too many over 100. For the different characteristics, one can analyze the distribution of the answers to check that they seem to make sense. The verification of the data is also a way of trying to identify fraud—it is amazing how often inconsistent or impossible answers are given. One of the other problems that can arise is when an applicant fills in the application form with help from the lender's staff. If the staff have incentives for signing up customers that are accepted for loans, they might be tempted to advise the applicant on suitable answers. This is why some lenders use the credit bureau data as their main source of application information, although such lenders are in a minority.

For behavioral scoring, one can add a large number of transactional characteristics. The most common would be average balance and maximum and minimum balance over the past month, the past 6 months, or the past 12 months. Other characteristics could include the total value of credit transactions and the total value of debit transactions over such periods. One would also need to include characteristics that suggest unsatisfactory behavior, like the number of times over the credit card or overdraft limit or the number of reminder letters that have had to be sent. One can produce a large number of variables from transactional data, and these are likely to be strongly correlated with each other. Deciding which to keep and which to ignore is part of the art of scoring.

8.5 Credit bureau characteristics

Credit reference agencies or credit bureaus exist in many countries. Their roles are not identical from country to country, and neither are the legislative frameworks in which they operate. Therefore, it should come as no surprise that the stored and available data vary from country to country and even within countries.

In section 2.10, we introduced, at a fairly high level, the role of the credit reference agency or credit bureau. In this section, we shall describe in some detail what is available from credit bureaus in the U.K. This is to give the reader some appreciation of the extent of the information that is collected, analyzed, and made available. Many other Western countries operate in a similar way.

In the U.K., there are two main bureaus for consumer information—Experian and Equifax. Information is accessed through name and address, although there are different levels of these being matched (see later). The information they have available on consumers falls into several types, and we deal with each one in turn:

- publicly available information,
- previous searches,
- shared contributed information—through closed user groups, many lenders share information on the performance of their customers,
- aggregated information—based on their collected information, such as data at postcode level,
- fraud warnings,
- bureau-added value.

8.5.1 Publicly available information

In the U.K., publicly available information is of two types. The first is the electoral roll or voters' roll. This is a list of all residents who have registered to vote. In the U.K., this is not a complete list of all adults since there is no requirement to vote, as there is in some other countries. This information also includes the year in which someone was registered to vote at an address. This is useful information because it can be used to validate the time that someone has stated that they've lived at an address. For example, if they state that they have lived at an address for three years but have been registered to vote there for twelve years, there is clearly a mismatch of information.

As this is being written, there is a debate in the U.K. between the Office of the Data Protection Registrar (ODPR) and the regulatory and industry bodies. This debate has arisen because the ODPR wishes voters to be able to decide whether their voter registration should be available to be used for credit or marketing purposes. It looks likely that some constraints will be implemented. However, while this is being done to provide greater protection for the consumer, it is possible that the effect will be to weaken the lenders' ability to distinguish between acceptable and unacceptable credit risks. If so, the consumer may actually suffer either by being declined or by being forced to pay more in interest rates to cover the additional risks.

There is no legal obligation on local councils to provide the electoral roll to the bureau. The electoral roll is available on paper to political agents with regard to an election and is often available for inspection in local libraries or council offices. However, in almost all cases, the electoral roll is supplied to the bureaus electronically in return for a considerable fee.

The second type of public information is public court information. These are the details of county court judgements (CCJs) or, in Scotland, court decrees. One option as part of the process to pursue a debt is to go to a county court and raise an action for a CCJ. This establishes the debt and may force the debtor into action. If the debtor then clears the debt and this information is passed to the bureau, the CCJ does not disappear but will show as having been satisfied. Similarly, if there is a dispute concerning the CCJ, which is proved in favor of the plaintiff, the CCJ may show as being corrected. This may occur, for example, where the court action was raised for a joint debt when the liabilities of one of the debtors had been extinguished. In such a case, often with a husband and wife, one of the debtors may raise a correction to show that the outstanding CCJ is not theirs but their partner's.

CCJs, etc. are information in the public domain. Some county courts can provide the information electronically; in many cases, however, the information will be entered from paper notices and records.

Perhaps the important thing to realize about this category of information—both electoral rolls and CCJs—is that it is all (or almost all) available to the public. The value that the bureaus add is to load it electronically and greatly speed up the process. Thus instead of having to contact several local authorities, which may take several days, if not weeks, one can now access the information in a few seconds.

8.5.2 Previous searches

When a lender makes an inquiry of a credit reference agency, that inquiry is recorded on the consumer's file. (There are special circumstances when it is not recorded, but we do not need to discuss them here.) When another lender makes a subsequent inquiry, a record of any previous searches will be visible. The previous searches carry a date and details of the type of organization that carried it out—bank, insurance company, credit card company, utilities, mail order, etc.

What it does not reveal is the outcome of the inquiry. Therefore, a consumer may have eight searches recorded over a two-week period by a number of companies. We cannot tell which of these are associated with offers from the lenders and in which cases the lender declined the application. For those cases where an offer was made, we cannot tell if the applicant accepted the offer. For example, the consumer could be simply shopping around several lenders for the best deal but will take out only one loan. Or the consumer could be moving house and the inquiries are to support financing the purchase of furniture, a television, kitchen units, etc. A further possibility is that the consumer is desperately short of money and is applying for a variety of loans or credit cards and intends to take everything that he can get.

Therefore, while the number and pattern of previous searches may be of interest, either in subjective assessment or in a scorecard, it requires careful interpretation. In scorecard development, the characteristics that might be included would be the number of searches in the last 3 months, 6 months, 12 months, and 24 months, as well as the time since the last inquiry. Obviously, with the likely correlation structure, it would be very rare that more than one of these would appear in a final scorecard.

8.5.3 Shared contributed information

Many years ago, both lenders and the bureaus realized that there was value in sharing information on how consumers perform on their accounts. Therefore, at its simplest, several lenders can contribute details of the current performance of their personal loans. If a consumer applies for a personal loan and they currently have one with one of the contributors, an inquiry at the bureau will provide details of whether their existing loan is up-to-date or in arrears and some brief details of the historical payment performance.

This developed in many ways but the fundamental guidelines are encapsulated in a document to which lenders and the bureaus subscribe—the principles of reciprocity. Basically, one gets to see only the same type of information that one contributes. Some lenders contribute data on only some of their products, and they should see details of other lenders' contributions only for the same products. Some lenders do not provide details on all their accounts. Rather, they provide details only of their accounts that have progressed beyond a certain stage in the collections process, usually the issue of a default notice. (Under the Consumer Credit Act of 1974, this is a statutory demand for repayment and precedes legal action.) These lenders—or, more accurately, these lenders when carrying out inquiries relating to this product—will get to see only default information supplied by other lenders. This is the case even if the other lenders also contributed details of all their accounts that are not in serious arrears or in arrears at all.

Although the adult population of the U.K. is only about 40 million, between them the bureaus have approximately 350 million records contributed by lenders. (Note that there will be some duplication and that this number will also include closed or completed accounts for a period after they have ceased to be live.)

When a lender carries out an inquiry, it will not be able to discover with which company the existing facilities and products are. However, it does get to see details of the type of product—revolving credit, mail order, credit card, etc.

The principles of reciprocity not only dictate that you get to see only the same type of information as you contribute. They also dictate the restrictions placed on access to information depending on the purpose to which it will be put. As a rough guideline, the restrictions are least when the information is being used to manage an existing account with an existing customer. Further restrictions are introduced when the data are to be used to target an existing customer for a new product, i.e., one that they do not already hold. Even further restrictions may be placed on the use of shared data for marketing products to noncustomers.

8.5.4 Aggregated information

Through having data from the electoral roll and having the contributed records from many lenders, the bureaus are in an advantageous position to create new measures that might be of use in credit assessment. With the depth of the information supplied by lenders, together with the electoral roll and the post office records, they do create variables at a postcode level. (Each individual postcode has between 15 and 25 houses allocated to it. In large blocks of flats, there may be one postcode allocated for each block.) Therefore, by aggregating this information, the bureaus are able to create and calculate measures such as

- percentage of houses at the postcode with a CCJ;
- percentage of accounts at the postcode that are up-to-date;
- percentage of accounts at the postcode that are three or more payments in arrears;
- percentage of accounts at the postcode that have been written off in the last 12 months.

Clearly, the individual lender cannot see this information in its entirety as the search is indexed by the address that is entered. However, the bureaus are able to create such measures that, in some scoring developments, prove to be of value. Also, as the ODPR tries to restrict recording and use of data pertaining to the individual, this type of information may be able to restore the lenders' ability to discriminate between good and bad credit risks.

8.5.5 Fraud warnings

The bureaus are able to record and store incidences of fraud against addresses. These could either be first-party fraud, where the perpetrator lives at the address quoted, or impersonation fraud, where the perpetrator claims that they live at the address quoted.

A credit reference inquiry carried out at the address may generate some sort of fraud warning. This does not mean that the application is fraudulent. Most lenders will use this as a trigger to be more careful and perhaps increase the level of checking that is carried out on the details supplied. In fact, many cases that generate a fraud warning are genuine applications; the warning having resulted from an attempted impersonation fraud. It would be counter to common sense, to business sense, and to the guidelines to which the U.K. industry adheres to assume that a fraud warning means the application in fraudulent. While the warning may make the lender more cautious, the lender should decline the case as a fraudulent application only if it has proof that there is some aspect of fraud involved. One cannot decline a case as fraudulent simply because another lender has evidence of an attempted fraud.

8.5.6 Bureau-added value

As can be seen from the above discussion, the credit bureaus store huge amounts of information. Using this, they are able to create and calculate measures. However, they can also develop generic scorecards. Their construction is along the lines of a standard scorecard. There is a huge volume of data and a definition of bad is taken. These scorecards do not relate to a specific lender's experience. Neither do they relate to the lender's market position. They also do not relate to specific products. Further, the bad definition may not be appropriate for a specific situation. However, they can be extremely useful in at least three environments.

The first environment is where the lender is too small to be able to build its own scorecard. To allow it some of the benefits of scoring, the lender calibrates its own experience against the generic scorecard. This will allow it to build some confidence of how the scorecard will operate and also to set a cutoff appropriate to its needs.

The second environment is when there is a new product. In such cases, a generic scorecard may be of use even for larger lenders since there would be insufficient information on which to build a customized scorecard for that product.

The third environment is in general operation. While the lender may have a scorecard that discriminates powerfully between good and bad cases, a generic scorecard allows the lender to do at least two things. The first thing is that the generic scorecard may add to the power of the lender's own scorecard as the generic scorecard will be able to reveal if an applicant is having difficulties with other lenders. The second thing, therefore, is that the generic scorecard can be used to calibrate the quality of applications the lender receives. For example, if the average generic score one month is 5% below the previous month's, the lender has a quasi-independent measure of the credit quality of the recent batch of applications. Because of the different restrictions placed on the use of shared data, each bureau constructs different generic scorecards using different levels of data.

Another example of the development and use of a generic scorecard appears in Leonard (2000). In this example, the credit bureau has built a generic scorecard but has included only cases that are three payments in arrears. Accounts are classified as good if they make payments in the subsequent 90 days. This generic scorecard is intended, therefore, only to discriminate among those accounts already three payments in arrears. It should not be used for accounts that are not in arrears nor for new applications.

In the context of understanding the data that might be available for building a credit scorecard, we have discussed the different types of information. In using credit reference information in the U.K., many other issues need to be borne in mind. However, we do not need to address them here. As we use credit reference information in other countries, other issues will arise.

One issue that we need to discuss is that of matching levels. When carrying out a credit reference inquiry, the lender submits a name and an address. However, in many cases, this does not match exactly with a record in the lender's database. For example, the applicant may have written "S. Jones" on the application form, but the record is for Steven Jones. The address may be written as "The Old Mill, 63 High Street, London," but the record on the database is for 63 High Street, London. Therefore, the bureaus need to have some way to match cases and to return cases that have an exact match, or a very close match, or a possible match. Thus either the lender specifies the level of matching that it requires or the bureaus can return elements of the inquiry with a flag detailing the level of match that was achieved.

Having discussed the information that the bureaus hold and looked at the way they might match details, it should be clear that the credit reference agencies, as they are also called, are just that. Apart from the generic scorecards that they can offer and the way they have aggregated data at postcodes, for all of the other data, they are acting as an agent of the credit industry, accelerating access to available information to support high volumes of applications.

In the U.K., the credit bureaus operate within a legal framework. For example, all consumers have the right, on payment of a small fee, to receive a print of what is held on their files at the credit bureau. The authority to take a credit reference inquiry must be explicitly given by the applicant before the inquiry is made. Also, if the lender wishes to have the authority to take repeated inquiries during the life of the account, this authority must be given in advance, although it can be given once for the life of the account. Repeated inquiries are quite often made in managing credit card limits or in assessing the appropriate action to take when a customer falls into arrears.

The discussion so far has referred to credit reference information on consumers, i.e., on individuals. However, one can also make a credit reference inquiry about a company or business. Once again, the agencies in this market provide accelerated access to public information as well as adding value.

The public information might be details of who the company directors are and with what other companies they are associated. The financial accounts of the company can also be accessed, as can any court actions, such as CCJs.

The bureaus will also build models to assess the creditworthiness of the business. This is often expressed as a credit limit. In other words, if a business has a credit limit of £10,000, that is the maximum amount that the bureau assesses that creditors should extend to the business. It is not usually possible, however, to find out easily how much credit other creditors have extended to the business toward the credit limit. Bureaus may also assess the strength of the business not only in terms of its financial performance but also in terms of the trends of the relevant industry and economy.

For small businesses, the information available is similar to that for consumers. As we move into larger businesses, the public data become more standardized in their format and in their content. For national and multinational enterprises, the value that a bureau can add diminishes and the bureaus tend toward being simply an accelerated route to public information.

8.6 Determining subpopulations

The next two sections are about deciding which of the variables should be used in the scoring systems and how they should be used. An important use of the variables is to split the population into subpopulations, and different scorecards are built on each subpopulation. There may be policy reasons as well as statistical reasons for doing this. For example, in behavioral scorecards it is usual to build different scorecards for recent customers and for long-existing customers. This is simply because some characteristics, like average balance in the last six months, are available for the latter but not for the former. Another policy might be that the lender wants younger people to be processed differently than older customers. One way to do this is to build separate scorecards for under 25s and over 25s.

The statistical reason for splitting the population into subpopulations is that there are so many interactions between one characteristic and the others that it is sensible to build a separate scorecard for the different attributes of that characteristic. One builds a classification tree using the characteristics that have interactions with lots of other variables. The top splits in such a tree might suggest suitable subpopulations for building separate scorecards. However, it is often found that although there may be strong interactions between one variable and some others, these do not remain as strong when other variables are introduced or removed.

Hence, typically, policy reasons more than statistical ones determine what subpopulations should be formed. As Banasik et al. (1996) showed, segmenting the population does not always guarantee an improved prediction. If the subpopulations are not that different, then the fact that they are built on smaller samples degrades the scorecard more than any advantage the extra flexibility gives. Segmenting the population into two subpopulations depending on whether $X_0 = 0$ or $X_1 = 1$ is equivalent to building a scorecard so that every other variable Y has two variables $Y_0 = (Y|X_0 = 0)$ and $Y_1 = (Y|X_0 = 1)$. On top of this, having two or more scorecards allows more flexibility in the cutoff scores, although one usually tries to ensure that the marginal odds of good to bad is the same at each subpopulation's cutoff score.

8.7 Coarse classifying the characteristics

Once we have divided the population into its subpopulations, one might expect that one could go ahead and apply the methodologies of Chapters 4, 5, and 6 to the variables describing the characteristics of the applicants. This is not the case. First, one needs to take each characteristic and split the possible answers to it into a relatively small number of classes, i.e., to coarse classify the characteristic. This needs to be done for two different reasons, depending on whether the variable is categorical (has a discrete set of answers) or continuous (has an infinite set of possible answers). For categorical characteristics, the reason is that there may be too many different answers or attributes, and so there may not be enough of the sample with a particular answer to make the analysis robust. For continuous characteristics, the reason is that credit scoring seeks to predict risk rather than to explain it, and so one would prefer to end up with a system in which the risk is nonlinear in the continuous variable if that is a better prediction. The following examples show these two effects.

Example 8.1 (residential status). Suppose that the distribution of answers among a sample of 10,000 to the question "What is your residential status?" was as given in Table 8.2. The overall good:bad ratio in the population is 9:1, but the ratios in the attributes range from 20:1 to 1:1. The question is whether we can allow all six types of answer (attribute) to remain. Only 20 of the populations gave "No answer"; 140 were "Other," while "Rent furnished" had only 5% of the population in that category. There are no hard and fast rules, but it would seem that these categories have too few answers for there to be much confidence that their good:bad ratios will be reproduced in the whole population. We could put all three in one category—"Other answers," with 450 goods and 200 bads—since they are the three worst good:bad ratios. There is an argument though for putting rent furnished with rent unfurnished since again their good:bad odds are not too far apart and they are both renting categories. Similarly, if we were to decide on only three categories, should it be (owner/renter) owners (6000 good, 300 bad), renters (1950 goods, 540 bads), and others (1050 good, 160 bad) or should it be (owner/parent) consisting of owners (6000 good, 300 bad), with parents (950 good, 100 bad), and others (2050 good, 600 bad).

Table 8.2. *Table of numbers in the different groups.*

Attribute	Owner	Rent unfurnished	Rent furnished	With parents	Other	No answer
Goods	6000	1600	350	950	90	10
Bads	300	400	140	100	50	10
Good:bad odds	20:1	4:1	2.5:1	9.5:1	1.8:1	1:1

Although we said that the choice of combination is as much art as it is science, one can use some statistics as guidance. Three statistics are commonly used to describe how good the characteristic with a particular coarse classification is at differentiating goods from bads. The most common is the χ^2-statistic.

8.7.1 χ^2-statistic

Let g_i and b_i be the numbers of goods and bads with attribute i, and let g and b be the total number of goods and bads. Let

$$\hat{g}_i = \frac{(g_i + b_i)g}{g + b} \quad \text{and} \quad \hat{b}_i = \frac{(g_i + b_i)b}{g + b}$$

be the expected number of goods and bads with attribute i if the ratio for this attribute is the same as for the whole population. Then

$$s^2 = \sum_i \left(\frac{(g_i - \hat{g}_i)^2}{\hat{g}_i} + \frac{(b_i - \hat{b}_i)^2}{\hat{b}_i} \right) \tag{8.1}$$

is the χ^2-statistic. Formally, this measures how likely it is that there is no difference in the good:bad ratio in the different classes, and one can compare it with the χ^2-statistics with $k-1$ degrees of freedom, where k is the number of classes of the characteristic. However, one can use it as a measure of how different the odds are in the different classes with a higher value, reflecting greater differences in the odds. Thus in the three class cases above, we get

$$\text{(owner/renter):} \quad \hat{g}_{\text{owner}} = 5670, \quad \hat{g}_{\text{renter}} = 2241, \quad \hat{g}_{\text{others}} = 1089, \quad \text{so}$$

$$\begin{aligned} \chi^2 &= \frac{(6000-5670)^2}{5670} + \frac{(300-630)^2}{630} + \frac{(1950-2241)^2}{2241} \\ &\quad + \frac{(540-249)^2}{249} + \frac{(1050-1089)^2}{1089} + \frac{(160-121)^2}{121} \\ &= 583.9, \end{aligned} \tag{8.2}$$

$$\text{(owner/parent):} \quad \hat{g}_{\text{owner}} = 5670, \quad \hat{g}_{\text{parent}} = 945, \quad \hat{g}_{\text{others}} = 2385, \quad \text{so}$$

$$\begin{aligned} \chi^2 &= \frac{(6000-5670)^2}{5670} + \frac{(300-630)^2}{630} + \frac{(950-945)^2}{945} \\ &\quad + \frac{(100-105)^2}{105} + \frac{(2050-2385)^2}{2385} + \frac{(600-265)^2}{265} \\ &= 662.9. \end{aligned} \tag{8.3}$$

One would say the owner/parent split with its larger χ^2 value is the better split.

8.7.2 Information statistic

The information statistic is related to measures of entropy that appear in information theory and is defined by

$$F = \sum_i \left(\frac{g_i}{g} - \frac{b_i}{b} \right) \log \left[\frac{g_i b}{b_i g} \right]. \tag{8.4}$$

In statistics, this is sometimes called the divergence or information value. It seeks to identify how different $p(x|G)$ and $p(x|B)$ are (which translate into $\frac{g_i}{g}$ and $\frac{b_i}{b}$) when x takes attribute value i. The motivation of the information statistic being of the form

$$\sum_i \left(\frac{g_i}{g} \right) \log \left[\frac{g_i}{g} \right]$$

when there are g observations and g_i of them were of type i is as follows. The number of ways this distribution occurs is $N_g = \frac{g!}{g_1! g_2! \ldots g_p!}$ if there are p types in total. Information is taken as the log of the number of different ways one can get the message that was seen, so $I_g = \log N_g = \log g! - \sum_i \log(g_i!) \approx g \log(g) - \sum_i g_i \log(g_i)$. The average information is $\frac{I_g}{g} \approx -\sum_i (\frac{g_i}{g})(\log(g_i) - \log(g)) = -\sum_i (\frac{g_i}{g}) \log(\frac{g_i}{g})$. This information statistic can be thought of as the difference between the information in the goods and the information in the bads, i.e., $-\sum_i (\frac{g_i}{g} - \frac{b_i}{b})(\log(\frac{g_i}{g}) - \log(\frac{b_i}{b}))$.

In Example 8.1, we get

$$\text{(owner/renter):}\quad \frac{g_{\text{owner}}}{g} = 0.667, \quad \frac{b_{\text{owner}}}{b} = 0.3, \quad \frac{g_{\text{renter}}}{g} = 0.217, \quad \frac{b_{\text{renter}}}{b} = 0.54,$$

$$\frac{g_{\text{other}}}{g} = 0.117, \quad \text{and} \quad \frac{b_{\text{owner}}}{b} = 0.16,$$

$$\begin{aligned} F &= (0.667 - 0.3)\log\left(\frac{0.667}{0.3}\right) + (0.217 - 0.54)\log\left(\frac{0.217}{0.54}\right) \\ &\quad + (0.117 - 0.16)\log\left(\frac{0.117}{0.16}\right) \\ &= 0.6017, \end{aligned} \tag{8.5}$$

$$\text{(owner/parent):}\quad \frac{g_{\text{owner}}}{g} = 0.667, \quad \frac{b_{\text{owner}}}{b} = 0.3, \quad \frac{g_{\text{parent}}}{g} = 0.106, \quad \frac{b_{\text{parent}}}{b} = 0.1,$$

$$\frac{g_{\text{other}}}{g} = 0.228, \quad \text{and} \quad \frac{b_{\text{owner}}}{b} = 0.6,$$

$$\begin{aligned} F &= (0.667 - 0.3)\log\left(\frac{0.667}{0.3}\right) + (0.106 - 0.1)\log\left(\frac{0.106}{0.1}\right) \\ &\quad + (0.228 - 0.6)\log\left(\frac{0.228}{0.6}\right) \\ &= 0.6536. \end{aligned} \tag{8.6}$$

Large values of F arise from large differences between $p(x|G)$ and $p(x|B)$ and so correspond to characteristics that are more useful in differentiating goods from bads. Thus in this case, we would take the (owner/parent) split again.

8.7.3 Somer's D concordance statistic

The Somer's D concordance statistic test assumes that the classes of the characteristic already have an ordering from low, which has the lowest good rate, to high, which has the highest good rate. The concordance statistic describes the chance that if one picks a good at random from the goods and a bad at random from the bads, the bad's attribute, x_B, will be in a lower class than the good's attribute, x_G. The higher this probability, the better the ordering of the characteristic's classes reflects the good-bad split in the population. The precise definition of D is the expected payoff of a variable, which is 1 if the ordering puts the bad below the good, -1 if it puts the good below the bad, and 0 if they are the same:

$$\begin{aligned} D &= 1.P\{x_B < x_G\} - 1.P\{x_B > x_G\} + 0.P\{x_B = x_G\} \\ &= \sum_i \frac{\left(\sum_{j<i} b_j\right) g_i - \left(\sum_{j<i} g_j\right) b_i}{bg}. \end{aligned} \tag{8.7}$$

For the three-class example of Example 8.1, we get the following calculations: For (owner/renter), the ordering is that renters have the lowest good rate, then others, then owners, so

$$\begin{aligned} D &= \left(\frac{540}{1000}\right)\cdot\left(\frac{1050}{9000}\right) + \left(\frac{700}{1000}\right)\cdot\left(\frac{6000}{9000}\right) - \left(\frac{1950}{9000}\right)\cdot\left(\frac{160}{1000}\right) - \left(\frac{3000}{9000}\right)\cdot\left(\frac{300}{1000}\right) \\ &= 0.395. \end{aligned} \tag{8.8}$$

For the (owner/parent) split, the lowest good rate is others, then parent, and then owners. Hence

$$D = \left(\frac{600}{1000}\right)\cdot\left(\frac{950}{9000}\right)+\left(\frac{700}{1000}\right)\cdot\left(\frac{6000}{9000}\right)-\left(\frac{2050}{9000}\right)\cdot\left(\frac{100}{1000}\right)-\left(\frac{3000}{9000}\right)\cdot\left(\frac{300}{1000}\right)$$
$$= 0.4072. \tag{8.9}$$

Since (owner/parent) has the higher value, this test suggests that it is a more definitive split.

Thus far, we have dealt with categorical variables, but we also need to coarse classify continuous variables. The first question is why? In normal regressions, if one has a continuous variable, one usually leaves it as such and calculates only the coefficient that it needs to be weighted by. That is because one is trying to explain some connection. If instead one is trying to predict risk, then by leaving the variable continuous, one guarantees the risk will be monotone in that variable. For example, Figure 8.1 shows the good rate at each age group from a sample of credit card holders. What is noticeable is that it is not monotone—the good rate goes up, then down, and then up again as age progresses. One can find explanations for this in terms of more responsibility and more outgoings in the 30–40 age group or the loss of a second income during the childbearing years, but the point is that the best regression line (shown in Figure 8.1) does not reflect this. It is therefore better to split age into a number of distinct attributes, 18–21, 21–28, 29–36, 37–59, 60+, and as Figure 8.2 shows, these can have scores that reflect the nonlinearity with age.

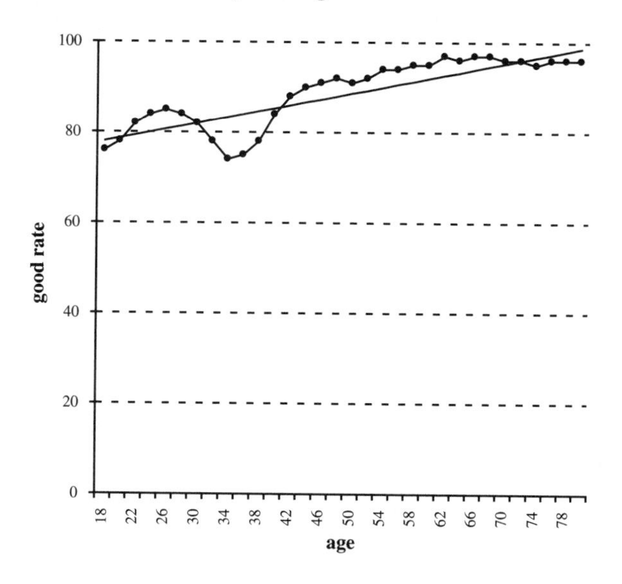

Figure 8.1. *Good rate as a function of age.*

So how should one split such continuous variables? The most obvious way is to split the characteristics first into attributes corresponding to percentiles, i.e., into 10 classes, the first having the youngest 10%, the second class the second youngest 10%, etc. There is no need to fix on 10. Some analysts use 20 groups, each of 5% of the population, or eight groups of 12.5% of the population, or even 100 groups of 1% each. Once this is done, the question is whether one should group nearest neighbors together. This depends on how close their good rates are, and one can use the statistics outlined above to help decide. Let us take time at present address as an example.

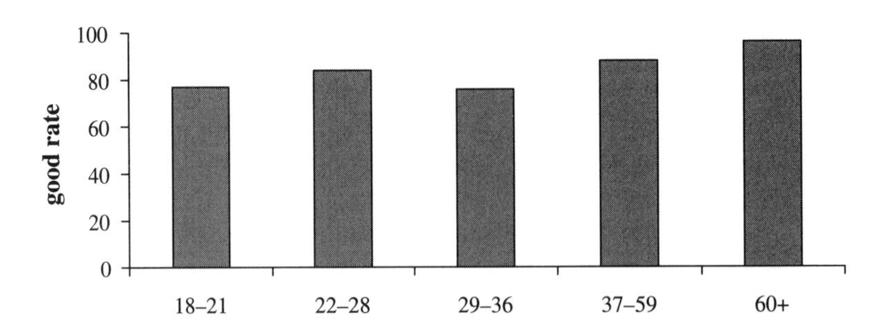

Figure 8.2. *Coarse classifying of age.*

Example 8.2 (time at present address). The data (from a hypothetical example) is given in Table 8.3, where TPA is time at present address. Choosing splits by inspection, first one might decide that the five groups of more than four years have such high good rates that they can be put together. The nonlinearities at the short times at address mean that one might keep the other five groups separate, but suppose that we want to put together at least one more pair. Some might argue for putting the under-6 months and the 6–12 months together in a <12 months group (option 1), while others might want to put 18–30 months and 30–48 months together since their ratios are close (option 2). One would not put the 12–18 months and the 30–48 months groups together because although their ratios are very similar, they are not adjacent time periods. To aid this choice between the two options, one could calculate the statistics as before.

Table 8.3. *Data for Example* 8.2.

TPA	<6 months	6–12 months	12–18 months	18–30 months	30–48 months	4–5 years	6–7 years	8–11 years	12–15 years	16+ years
Goods	800	780	840	880	860	920	970	980	980	990
Bads	200	220	160	120	140	80	30	20	20	10
Total	1000	1000	1000	1000	1000	1000	1000	1000	1000	1000
Ratio	4:1	3.5:1	5.2:1	7.3:1	6.1:1	11.5:1	31:1	49:1	49:1	99:1

Take option 1 with groups of <12 months (1580 goods, 420 bads), 12–18 months (840 goods, 160 bads), 18–30 months (880 goods, 120 bads), 30–48 months (860 goods, 140 bads), and over four years (4840 goods, 160 bads). Let option 2 be <6 months (800 goods, 200 bads), 6–12 months (780 goods, 220 bads), 12–18 months (840 goods, 160 bads), 18–48 months (1740 goods, 260 bads), and over four years (4840 goods, 160 bads). The calculations are as follows.

χ^2 statistic:

$$\begin{aligned}\text{Option 1:}\quad \chi^2 &= \frac{(1580-1800)^2}{1800}+\frac{(840-900)^2}{900}+\frac{(880-900)^2}{900}+\frac{(860-900)^2}{900}\\ &\quad+\frac{(4840-4500)^2}{4500}+\frac{(420-200)^2}{200}+\frac{(160-100)^2}{100}\\ &\quad+\frac{(120-100)^2}{100}+\frac{(140-100)^2}{100}+\frac{(160-500)^2}{500}\\ &= 588. \end{aligned}\tag{8.10}$$

$$\begin{aligned}\text{Option 2:}\quad \chi^2 &= \frac{(800-900)^2}{900} + \frac{(780-900)^2}{900} + \frac{(840-900)^2}{900} \\ &+ \frac{(1740-800)^2}{800} + \frac{(4840-4500)^2}{4500} + \frac{(200-100)^2}{100} \\ &+ \frac{(220-100)^2}{100} + \frac{(160-100)^2}{100} + \frac{(260-200)^2}{200} + \frac{(160-500)^2}{500} \\ &= 588. \end{aligned} \tag{8.11}$$

So options 1 and 2 are considered equally good under this test.

Information statistic F:

$$\begin{aligned}\text{Option 1:}\quad F &= \left(\frac{1580}{9000} - \frac{420}{1000}\right)\log\left(\frac{1580\cdot 1000}{9000\cdot 420}\right) + \left(\frac{840}{9000} - \frac{160}{1000}\right)\log\left(\frac{840\cdot 1000}{9000\cdot 160}\right) \\ &+ \left(\frac{880}{9000} - \frac{120}{1000}\right)\log\left(\frac{880\cdot 1000}{9000\cdot 120}\right) + \left(\frac{860}{9000} - \frac{140}{1000}\right)\log\left(\frac{860\cdot 1000}{9000\cdot 140}\right) \\ &+ \left(\frac{4840}{9000} - \frac{160}{1000}\right)\log\left(\frac{4840\cdot 1000}{9000\cdot 160}\right) \\ &= 0.7287. \end{aligned} \tag{8.12}$$

$$\begin{aligned}\text{Option 2:}\quad F &= \left(\frac{800}{9000} - \frac{200}{1000}\right)\log\left(\frac{800\cdot 1000}{9000\cdot 200}\right) + \left(\frac{780}{9000} - \frac{220}{1000}\right)\log\left(\frac{780\cdot 1000}{9000\cdot 220}\right) \\ &+ \left(\frac{840}{9000} - \frac{160}{1000}\right)\log\left(\frac{840\cdot 1000}{9000\cdot 160}\right) + \left(\frac{1740}{9000} - \frac{260}{1000}\right)\log\left(\frac{1740\cdot 1000}{9000\cdot 260}\right) \\ &+ \left(\frac{4840}{9000} - \frac{160}{1000}\right)\log\left(\frac{4840\cdot 1000}{9000\cdot 160}\right) \\ &= 0.7280. \end{aligned} \tag{8.13}$$

Thus in this case, the suggestion is to take the larger value—namely, option 1—but it is very close.

***D* concordance statistic:**

$$\begin{aligned}\text{Option 1:}\quad D &= \left(\frac{420}{1000}\right)\left(\frac{840}{9000}\right) + \left(\frac{580}{1000}\right)\left(\frac{860}{9000}\right) + \left(\frac{720}{1000}\right)\left(\frac{880}{9000}\right) \\ &+ \left(\frac{840}{1000}\right)\left(\frac{4840}{9000}\right) - \left(\frac{1580}{9000}\right)\left(\frac{160}{1000}\right) - \left(\frac{2420}{9000}\right)\left(\frac{140}{1000}\right) \\ &- \left(\frac{3280}{9000}\right)\left(\frac{120}{1000}\right) - \left(\frac{4160}{9000}\right)\left(\frac{160}{1000}\right) \\ &= 0.433. \end{aligned} \tag{8.14}$$

$$\begin{aligned}\text{Option 2:}\quad D &= \left(\frac{220}{1000}\right)\left(\frac{800}{9000}\right) + \left(\frac{420}{1000}\right)\left(\frac{840}{9000}\right) + \left(\frac{580}{1000}\right)\left(\frac{1740}{9000}\right) \\ &+ \left(\frac{840}{1000}\right)\left(\frac{4840}{9000}\right) - \left(\frac{780}{9000}\right)\left(\frac{200}{1000}\right) - \left(\frac{1580}{9000}\right)\left(\frac{160}{1000}\right) \\ &- \left(\frac{2420}{9000}\right)\left(\frac{260}{1000}\right) - \left(\frac{4160}{9000}\right)\left(\frac{160}{1000}\right) \\ &= 0.433. \end{aligned} \tag{8.15}$$

Thus both options have the same value under this criterion.

Note that different options are chosen by the different statistics.

Although we described that one of the reasons for coarse classifying continuous variables was to model the situation when the good rate is nonmonotonic, there may be cases in which the lenders want the good rate in a particular characteristic to be monotone if not necessarily linear. This may be because the lenders have prior beliefs about the effect of income, or time at bank on the good rate, or because they want to bias the scorecards to one or other end of the characteristic—for example, toward younger or older customers. In that case, we use the following coarse classification approach.

8.7.4 Maximum likelihood monotone coarse classifier

Assume that the bad rate is going down as the characteristic value increases. (If not, apply the following procedure starting from the other extreme of the characteristic.) Start at the lowest characteristic value and keep adding values until the cumulative bad rate hits its maximum. This is the first coarse classification split point. Start calculating the cumulative bad rate from this point until it again hits a maximum. This is the second split point. Repeat the process until all the split points are obtained.

It is an interesting exercise to show that, in fact, this algorithm gives the maximum-likelihood estimate of the fit conditional on the bad rates decreasing from class to class. That is a subtle way of saying that we will not prove it in this text but leave it for the inquiring reader to prove. We apply this procedure to the following two examples.

Example 8.2 (revisited). Using the data of Example 8.2, the calculations in Table 8.4 find the first four split points. Thus the classes are <12 months, 12–18 months, 18–48 months, and 4–5 years; doing the rest of the calculations shows that all the remaining classes remain separate, except that 8–11 years and 12–15 years clearly can be put together.

Table 8.4. *Calculations for Example* 8.2 (*revisited*).

TPA	<6 months	6–12 months	12–18 months	18–30 months	30–48 months	4–5 years	6–7 years	8–11 years	12–15 years	16+ years
Goods	800	780	840	880	860	920	970	980	980	990
Bads	200	220	160	120	140	80	30	20	20	10
Cum. bad rate	0.2	0.21*	0.193	0.175						
Cum. bad rate			0.16*	0.14	0.13	0.125				
Cum. bad rate				0.12	0.13*	0.113				
Cum. bad rate						0.0*	0.055	etc.		

The procedure can also be applied at an individual data level as follows.

Example 8.3. In a sample of consumers, the consumers have the values of a particular continuous characteristic, as shown in Table 8.5, and their good bad status is denoted by G or B. We apply the procedure described at the start of the section, data point by data point. Thus the classes are <12 (bad rate 0.71), 12–17 (bad rate 0.67), 18–25 (bad rate 0.6), 26–36 (bad rate 0.57), and 36+ (bad rate 0).

Coarse classifying characteristics is essentially the same thing as splitting the population into subclasses, which is the essence of the classification tree approach to classification. Therefore, one could also use the Kolmogorov–Smirnov statistics or indices like the Gini

Table 8.5. *Calculations for Example* 8.3.

Character value	1	3	6	7	9	10	12	14	16	17	18	20
Good-bad	G	B	B	G	B	B	B	G	B	B	G	B
Cumulative bad	0	.5	.67	.5	.6	.67	.71*	.62	.67	.7	.63	.67
								0	.5	.67*	.5	.6
											0	.5

Character value	21	24	25	27	29	30	31	32	34	36	37	38
Good-bad	G	B	B	G	B	G	B	G	B	B	G	G
Cumulative bad												
	.5	.55	.62	.55	.6	.54						
	.33	.5	.60*	.5	.57	.5						
				0	.5	.33	.5	.4	.5	.57*	.5	.44

index, which were used in section 4.7 to decide on the best splits in the trees, as a way to decide how to coarse classify a variable. The difference between coarse classifying and tree splitting is one of timing rather than approach. If one takes a variable and splits it into classes expecting thereafter to use some other scoring approach like regression or linear programming, then it is coarse classifying. If one splits the variable as part of the classifying process, then it is considered to be the splitting part of the tree-building process.

8.8 Choosing characteristics

When the variables have been coarse classified, one can end up with a large number of attributes. For application scorecards with, say, 30 to 40 variables, initially one could have more than 200 attributes, while for behavioral scorecards, which sometimes have hundreds of initial characteristics, one could be dealing with more than 1000 different attributes. If one were to take each attribute as a binary variable, this would be far too many variables for a logistic regression or a classification tree to cope with sensibly. Moreover, if one wants to end up with a scorecard that is understandable and hence accepted by managers, it should not have much more than 20 characteristics in it. So how should these characteristics be chosen?

One can rule out some variables because they are poor predictors, too variable over time, or too dependent on other variables. The first can be checked using the statistics used in the last section to decide on the coarse classification. If one calculates the χ^2-statistic, the information statistic F, or the concordance statistic D for each characteristic, these statistics give crude orderings of the importance of the variables as predictors.

Consider the following example.

Example 8.4. Consider the characteristics X_1, X_2, and X_3, all of which are binary variables. Table 8.6 gives the data and the values of the various statistics. This suggests that one would choose X_1 rather than X_2 or X_3. It is preferred to X_2 because it has more discrimination and to X_3 because there is a more even split between the numbers in the different characteristics. However, χ^2 does not show this, while F and D do not pick up the difference between X_2 and X_3 in terms of the size of the populations splits.

Blackwell (1993) pointed out the limitations in these characteristics in that the χ^2 test is looking at the statistical significant difference from all attributes having the same bad rate, which is not the same as discriminatory power. The information statistic F, on the other hand, does measure discriminatory power well, but it is very insensitive to sample size.

Table 8.6. *Data and calculations of Example* 8.4.

	X_1		X_2		X_3	
	Good	**Bad**	**Good**	**Bad**	**Good**	**Bad**
$X_i = 1$	2000	3000	4000	3000	5800	3600
$X_i = 0$	4000	1000	2000	1000	200	400
χ^2	1.611		5.480		245.9*	
F	0.746*		0.0337		0.078	
D	0.416*		0.083		0.067	

Blackwell suggests an efficiency measure $H(p)$, where p is the decision rule that accepts those whose probability of being good exceeds p. $H(p)$ takes into account $\frac{D}{L}$, the relative loss of accepting a bad to rejecting a good. He defines $g(p)$ to be the number of goods in attributes of the characteristic where the good rate is larger than p and defines $b(p)$ to be the number of bads in attributes of the characteristic where the good rate is below p. Let g and b be the total numbers of good and bad in the population; then

$$H(p) = \begin{cases} \dfrac{Lg(p) + Db(p) - Lg}{Db} & \text{if } \dfrac{g}{g+b} > p, \\[2ex] \dfrac{Lg(p) + Db(p) - Db}{Lg} & \text{if } \dfrac{g}{g+b} \leq p. \end{cases} \tag{8.16}$$

If a characteristic can be split into attributes that are either all good or all bad, $g(p) = g$ and $b(p) = b$, so $H(p) = 1$, while if all the attributes have the same proportion of goods, then if $\frac{g}{g+b} > p$, $g(p) = g$ and $b(p) = 0$; so $H(p) = 0$, with a similar result if $\frac{g}{g+b} \leq p$. If one wanted to use this to rank characteristics, one would have to decide on a specific value of p or else integrate $H(p)$ over all p, i.e., $\int H(p)dp$. However, it is more usual to use χ^2, F, and D to measure the discrimination of the characteristics.

One can check for the robustness of a characteristic over time by splitting the sample into subsamples depending on the date the account was opened and checking whether the distribution of the characteristic was consistent over these time-dependent cohorts of accounts. Dependence can be checked using the correlation between variables, but common sense also plays its part. If, for example, there are several income variables in an application score data set or several variants of the average balance variables in a behavioral score data set, it is likely that the scorecard will want to have only one of the group in it. One way to choose is to run a regression of the good-bad variable on this group of variables and chose which one first enters the regression under forward selection or which one is the last one left under backward selection.

If one is using dummy variables to define the characteristics after coarse classification, then this stepwise linear-regression approach is also a good way to cut down the number of variables at that stage. The advantage of using linear regression over the other techniques, including logistic regression, at this choice stage is that it is much faster to run and there could be lots of variables still around. It also produces more poorly aligned scorecards, and this makes it more obvious which attributes are unsatisfactory predictors.

An alternative approach that can be used after performing the coarse classifying of the previous section does not increase the number of variables from the original number of characteristics. In this approach, we replace the value of attribute i of a characteristic by the good:bad odds or the weight of evidence for that attribute. This has the advantage of not increasing the number of variables at all and so is particularly useful for the regression approaches. It ensures that, provided the coefficient of the modified variable is positive, the scores for the attributes will reflect the ranking of their good:bad odds. The problem with it is that there is a bias because these modified independent variables are now not independent of the dependent good:bad variable in the regression. This could be overcome by calculating the good:bad odds or the weight of evidence on part of the sample and calculating the regression coefficients on the remainder. However, this is rarely done in practice. It is also the case that since setting the weight of evidence attribute score is a univariate exercise, the results will be poor if there are still strong correlations between the variables. It certainly is a way to cut down on the number of variables produced, but even in this case one would still use stepwise, forward, or backward regression to remove some of the variables.

8.9 Reject inference

One of the major problems in developing scoring models in credit granting is that only the accounts that were approved in the past have performance data, which allows them to be classified as good or bad. For the customers that were declined in the past, one only has their characteristic values but not their good-bad status. If these customers are ignored and dropped out of the sample, then it no longer reflects the through-the-door population that one is after. This causes a bias (the "reject bias") in any classification procedure built on this sample. A number of techniques have been proposed to overcome this bias, and they come under the name of reject inference. How can one use the partial information available on the rejects to improve the scoring system? Since this has been an area which scorecard builders have sought commercial advantage, it is not surprising that it is an area of some controversy.

The cleanest way to deal with reject bias is to construct a sample in which no one was rejected. Retailers and mail-order firms traditionally do this. They take everyone who applies in a certain period with the intention of using that sample for building the next generation of scorecards. The culture in financial organizations does not accept this solution. "There are bads out there and we just cannot take them" is the argument—although, of course, the greater loss involved in someone going bad on a personal loan, compared with not paying for the two books ordered from a book mail-order club, may have something to do with it. However, this approach has considerable merit and can be modified to address the concerns that it will cost too much in losses. Traditionally banks take probability of default as their criterion because they assume that the losses from defaulters do not have huge variations. In that case, one can diminish these losses by not taking everyone but taking a proportion $p(x)$ of those whose probability of defaulting is x. One lets this proportion vary with x; it is very small when x is almost 1, and it tends to 1 when x is small. By allowing the possibility of picking everyone, reweighting would allow one to reconstruct a sample with no rejection without having to incur the full losses such a sample brings. Although this approach is not finding much support among credit granters, some are trying to get information on reject performance from other credit granters who did give these consumers credit.

If one has rejects in the sample on which the system is to be built, then there are five ways that have been suggested for dealing with rejects: define as bads, extrapolation, the industry norm—augmentation, mixed populations, and the three-way group approach.

8.9.1 Define as bad

The crudest approach is to assign bad status to all the rejects on the grounds that there must have been bad information about them for them to have been rejected previously. The scoring system is then built using this full classification. The problems with this approach are obvious. It reinforces the prejudices of bad decisions of the past. Once some group of potential customers has been given a bad classification, no matter how erroneously, they will never again get the opportunity to disprove this assumption. It is a wrong approach on statistical grounds and wrong on ethical grounds.

8.9.2 Extrapolation

Hand and Henley (1993), in their careful analysis of reject inference, pointed out that two different situations can occur depending on the relationship between the characteristics X_{old} of the system, which was used for the accept-reject decision, and X_{new}, the characteristics available to build the new scorecard. If X_{old} is a subset of X_{new}, i.e., the new characteristics include all those that were used in the original classification, then for some combination of attributes (those where X_{old} rejected the applicant), we will know nothing about the good-bad split because they were all rejected. For the other combinations, where X_{old} accepted the applicants, we should have information about the proportion of goods to bads but only among the accepted ones. However, X_{old} will have accepted all the applicants with these combination of characteristics. Then one has to extrapolate, i.e., fit a model for all the probability of being good for the attribute combinations that were accepted and extend this model to the combinations that were previously rejected. As was pointed out forcibly by Hand and Henley (1993), this method works far better on methods that estimate $q(G|x)$ directly, like logistic regression, rather than ones that estimate $p(x|G)$ and $p(x|B)$, like linear regression–discriminant analysis. This is because when estimating $p(x|G)$, the sampling fraction is varying with x and so will lead to biases when one tries to estimate the parameters of $p(x|G)$ (e.g., the mean and variance of the normal distribution in discriminant analysis). However, for $q(G|x)$, the fraction of the underlying population with that value of x which is sampled is either 0 or 1, so there is no bias in the model's parameter estimation. What happens is that the model gives a probability of being good to each of the population that was rejected and the scoring systems are then built on the whole population with the rejects having this value. This would not work for methods like nearest neighbors, but it might work for logistic regression if one can believe the form of the model.

8.9.3 Augmentation

If X_{old} is not a subset of X_{new}, so there were unknown variables or reasons for the original rejection decisions, then the situation is even more complicated. The usual approach is the augmentation method, outlined by Hsia (1978). First, one builds a good-bad model using only the accepted population to estimate $p(G|x, A)$, the probability of being good if accepted and with characteristic values x. One then builds an accept-reject model using similar techniques to obtain $p(A|x) = p(A|s(x)) = p(A|s)$, where s is the accept-reject score. The original approach of Hsia (1978) then makes the assumption that $p(G|s, R) = p(G|s, A)$—the probability of being good is the same among the accepteds and the rejecteds at the same accept-reject score, where

$$p(G|s, A) = \sum_{x;s(x)=0} p(G|x, A)p(x|s(x) = s). \tag{8.17}$$

This is like reweighting the distribution of the sample populations so that the percentage with a score s moves from $p(A, s)$ to $p(s)$. A new good-bad scorecard is now built on the full sample including these rejects. The rejects with accept-reject score s are given the probability $p(G|s, A)$ of being good.

Other methodologies for assessing the $p(G|s, R)$ have also been suggested. One approach assumes that $p(G|s, R) \leq p(G|s, A)$ and chooses this probability subjectively. The discount could depend on the type of account, whether there have been CCJs against the person, and when the account was opened. Others have assumed that $p(G|s, R) = kp(G|s, A)$, where k might be obtained by bootstrapping using a subset of the variables to build a good-bad score, which is then used to obtain the accept-reject decision. However, all these variants of augmentation have some strong assumptions in them concerning the form of the distributions or the relationship between $p(G|s, R)$ and $p(G|s, A)$. In practical applications, these assumptions are not validated and are rarely true.

8.9.4 Mixture of distributions

If one is making assumptions, an alternative approach is to say that the population is a mixture of two distributions—one for the goods and one for the bads—and that the form of these distributions is known. For example, if $p(\mathbf{x})$ is the proportion of applicants with characteristic $\mathbf{x}$, one says that

$$p(\mathbf{x}) = p(\mathbf{x}|G)p_G + p(\mathbf{x}|B)p_B, \tag{8.18}$$

and one can estimate the parameters of $p(\mathbf{x}|G)$ and $p(\mathbf{x}|B)$ using the accepts and, by using the EM algorithm, even the rejects. A usual assumption is that $p(\mathbf{x}|G)$ and $p(\mathbf{x}|B)$ are multivariate normal despite the fact that so many characteristics are categorical or binary. A halfway approach between this approach and augmentation is to assume that the good-bad scores and the accept-reject score for the goods are a bivariate normal distribution with a similar assumption for the bads. In this approach, one initially estimates the parameters of these distributions from the accepts and uses these parameters to estimate the probability of each reject being a good. Using these new estimates for the probability of a reject being a good, reestimate the parameters of the two distributions and iterate the process until there is convergence.

8.9.5 Three-group approach

A final approach is to classify the sample into three groups: goods, bads, and rejects. This was proposed by Reichert, Cho, and Wagner (1983), but the problem is that we want to use our classification system to split future applicants into only two classes—the goods, whom we will accept, and the bads, whom we will reject. What one does with those who are classified as rejects is not clear. If one rejects them, this approach reduces to classifying all rejects as bad. Its only saving grace is that in classical linear discriminant analysis, when one classifies into three groups, it is assumed that all three groups have a common covariance matrix. Hence this is a way to use the information in the rejects to improve the estimation of the covariance matrix.

To sum up, it does seem that reject inference is valid only if one can confidently make some assumptions about the accepted and rejected populations. It may work in practice because these assumptions may sometimes be reasonable or at least moving in the right direction. For example, it must be sensible to assume that $p(G|s, R) < p(G|s, A)$ even if one cannot correctly estimate the drop in probability.

In all the above approaches, we have concentrated on the rejects—the customers to whom the lender decided not to give credit—but exactly the same problem (and hence the same procedures) would need to be applied to deal with the bias of these who withdraw after being made the offer of a loan. Withdrawal might mean exactly that in the case of personal loans, or it might mean never using the loan facility in the case of credit cards.

8.10 Overrides and their effect in the scorecards

Overrides are when the lender decides to take an action contrary to the recommendation of the scoring system. High-side overrides (HSOs) are when the applicant is not given a loan although the score is above the cutoff; low-side overrides (LSOs) are when the applicant is accepted for a loan although the score is below the cutoff. Such overrides can be because the lender has more information than that in the scorecard or because of company policy, or it may be a subjective decision by the credit manager or underwriter.

Informational overrides are usually rare, but it could happen that an accompanying letter or knowledge by the branch suggests one of the characteristics in the scorecard does not tell the whole story. A raise in salary has been authorized but not paid, for example. The facility should be available for the decision to be reversed in such a case.

Policy overrides are when the lender has decided that certain types of applicant will be treated in certain ways irrespective of their other characteristics. For example, it may be decided that students are good long-term prospects and so should be given an overdraft no matter what. It may also be politic for all employees of the company to be accepted for the credit facility. An apocryphal story says that when a major international retailer first introduced store cards, the scorecard rejected the applications from the wives of its directors. An override was immediately instituted.

Many LSOs are justified because the possible loss of lucrative business if the applicant is upset by being rejected. The same argument is used when third parties (mortgage brokers, for example) demand that a loan for one of their clients be agreed. It is important to have the information to make a sensible decision in such cases. How much business and, more important, how much profit could be lost, and how likely is it that the loan will default? One needs to weigh these factors against one another. This means that it should be standard policy for the lender to keep track of the subsequent performance of overrides so that these estimates of losses can be calculated. Sangha (1999) looked in some detail at what information needs to be collected and the pitfalls, including possible noncompliance with the fair-lending laws, that need to be watched. One obvious point is to ensure that if a policy override is known, then those affected by the override should be removed from the sample on which the scoring system is developed.

When underwriters or credit evaluators are running the scoring system, subjective overrides are common but unjustifiable. Either the underwriter or the scoring system is wrong. If it is the underwriter, then he should not have interfered; if it is the scoring system, then the underwriter obviously has generic information that can improve the scorecard and so the scorecard should be redeveloped to include this. The way to find out is again to keep a record of how such overrides perform and get whoever is wrong to learn from this experience. The problem usually is human nature. An underwriter or credit evaluator wants to be seen as useful to protect his position, and accepting all the decisions of an automatic system does not do this. The trick is to redefine the evaluator's role into one of verifying and validating the system, i.e., checking that it is doing what it is supposed to be doing and that the population being assessed is still sufficiently close to the population the system was built on for the decisions still to be sensible.

8.11 Setting the cutoff

When introducing a new scorecard, there are several ways to select a cutoff.

A simplistic approach, especially for the first period following live implementation, is to adopt a cutoff that produces the same acceptance rate as the existing scorecard (on the current quality of applicants). This can be adopted until we have more confidence in the scorecard and its programming and in the other implementation issues that may arise. Even if the objective of the new scorecard is to increase acceptance and maintain the bad rate, it might still be worthwhile to retain the current acceptance rate for a few weeks at least. This can instill confidence in the programming and implementation, and it allows one to flush out any anomalies without many things being changed. Anomalies that might occur would be in the swap sets—those cases that previously were declined and are now approved, and vice versa. For example, we may have had an old scorecard where the acceptance rates were 74% of existing branch customers and 53% for noncustomers. With the new scorecard, these percentages may be 66% and 58%. If we have tried to have the same acceptance rate, then the cause of this shift is either a change in the relative quality of applicants from these two sources or the change in scorecard. If at the same time we have tried to increase acceptance rates and also reduce our expected bad rate by a little, it becomes much more difficult to identify the causes of this shift.

(Of course, at the end of the scorecard development phase or during the validation phase, analysis should reveal, for a given cutoff, the swap sets in terms of individual characteristics. These are important not only for scorecard variables but also for other variables since they may have a major effect on marketing strategy, staff bonus payments, or even internal politics.)

Implementing with the same acceptance rate is recommended only for a few weeks. In any case, having spent time and money building a new scorecard, one will be keen to begin to accrue the benefits of the development. (Certainly, maintaining the same acceptance rate will usually accrue benefits in terms of a reduced bad rate. However, this benefit will really be seen and be proved only many months later. In many organizations, pressure will be brought to bear to increase the acceptance rate.)

If we can ignore that a new scorecard will usually be replacing an old scorecard, then one approach to setting a cutoff is by assessment of the breakeven point. Let us assume for the moment that we have perfect information on the future performance of groups of accounts in terms of repayment, early settlement, arrears, ultimate loss, and the timing of these events. In such a case, we should be able to assess how much money this group of applications will make for the organization if we choose to accept them. We can produce these data on income and loss for each possible score—or for each score that is a candidate for the cutoff. A simplistic—but not necessarily incorrect—view is to accept all applications that will generate a profit, however small. One could also argue that we could accept those applications whose profit will be zero. This implicitly at least assumes that cases above this point will also generate a profit, although with some products—for example, a credit card—cases that score very highly and are unlikely to generate bad debts also may be unlikely to generate much income.

As was stated, this is simplistic but not necessarily incorrect, at least as a starting point. This falls down mainly because of accounting reasons.

Rather than consider profit, we certainly need to express income and losses in terms of their NPV. This equates future income and expenditure to what they are worth today. For example, is £1000 today worth more or less than £1200 in three years, or a monthly income of £32 for the next three years? The NPV tries to take into account such factors as the interest one could earn with the money if received earlier as well as the possible opportunity cost of having money tied up for a period.

Rather than consider profit or net income or NPV, we should consider the return on our investment. In lending someone, say, £5000, we are making an investment in that person and in their future ability and likelihood to repay the loan with interest. However, there is a cost in raising the £5000 to be able to lend it. We need to consider how well we have managed the money. For example, if we can lend £10,000 and make a £1000 profit and also lend £5000 and make a £600 profit, in a very simple form, the former represents a 10% return and the latter a 12% return. Clearly, if we could have a portfolio of £1 million of either the former or the latter type of loan, the latter is preferable. (This evaluation of return is not intended to be comprehensive. Indeed, it is extremely simplistic. Many finance and accounting books deal with the matter in much more detail and depth.)

What is important from the point of view of scoring is that we may wish to set a cutoff using return rather than profit. Therefore, we would set our cutoff at a point where all applications will meet some required minimum return threshold. Again, we may be making the assumption that cases with a score higher than this threshold will also meet the minimum return threshold.

In assessing both profit and return, we may run into another accounting issue, the allocation of fixed costs. For example, suppose that our fixed costs are £11 million per annum. These fixed costs may be for the building, staff, marketing, etc. With a scorecard cutoff of, say, 245, we expect to grant 50,000 loans. The fixed costs equate to £220 per loan. However, at a cutoff of, say, 235, we expect to grant 55,000 loans, equating to a fixed cost allocation of £200 per loan. If at a scorecard cutoff of 245 we meet our required profit or return, when we consider whether to lower the cutoff to 235, do we reallocate the fixed costs? Alternatively, do we assume that they have all been taken care of and that we can consider the marginal business from 235 to 244 using only the variable costs in the profit-return calculation? There are complex variations of this, and there is no single answer. Rather, the answer usually rests within the organization's accounting procedures and business objectives.

Another challenge that arises in assessing the profit or return for an application if granted is that, once again, we need to assume that the future will be like the past. We cannot know if the loan will be settled early or prepaid. We can only make some assumptions based on past performance of similar loans. We also have to make some assumptions about the recovery performance of future problem cases—indeed, of problem cases arising from applications that we have not yet even approved.

Therefore, in assessing which cutoff to use, several interrelated credit and accounting issues may need to be considered.

From the scorecard development, we should be able to construct a graph similar to Figure 8.3. For example, at an acceptance rate of only 50%, we have a bad rate of 2%. If we accept 70% of applications, the bad rate climbs to almost 4%. At an acceptance rate of 90%, the bad rate is approximately 7.5%. The triangle represents the current position before implementation of the new scorecard. Clearly, we can take up a position to the right of this, thereby maintaining the bad rate but increasing the acceptance rate. We can also take up a position below it, maintaining our acceptance rate but reducing the bad rate. In fact, any combination above or on the curve and to the right or below (or both) of the triangle represents an improvement on current practice. However, common sense dictates that the only positions worth considering are those actually on the curve.

Often this data is given in the form of a table reporting the results of applying the scorecard to the holdout sample. This table, the run-book, gives the likely effect of each

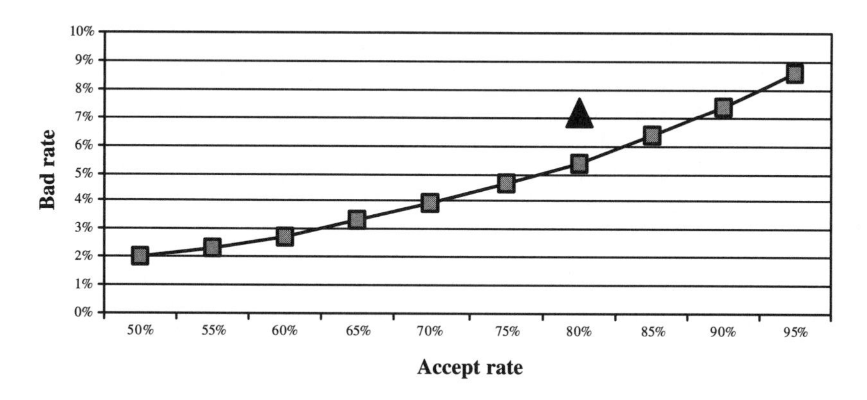

Figure 8.3. *Strategy curve.*

cutoff score on the whole population. Table 8.7 shows a typical run-book with a holdout of 10,000 (9000 goods and 1000 bads). We could also construct alternative graphs that look at score versus profit or score versus return.

Table 8.7. *Example of a run-book.*

Score	Cum. goods above score	Cum. bad above score	Cum. total	Percent of total	Bad rate percent of acceptance	Marginal good:bad odds
400	2700	100	2800	28	3.6	—
380	3300	130	3420	34.3	3.8	20:1
360	3800	160	3960	39.6	4.0	16.7:1
340	4350	195	4545	45.4	4.3	15.7:1
320	4850	230	5080	50.8	4.5	14.3:1
300	5400	270	5670	56.7	4.5	13.7:1
280	5900	310	6210	62.1	5.0	12.5:1
260	6500	360	6860	68.6	5.2	12:1
240	7100	420	7520	75.2	5.6	10:1
220	7800	500	8300	83.0	6.0	8.7:1
200	8600	600	9200	92	6.5	8:1
180	8900	800	9700	97	8.2	15:1

While profit or return is often the driving force behind a scorecard cutoff decision, there are operational factors to be considered. For example, if we greatly increase our acceptance rate, do we have sufficient funds to lend the increased sums of money? Also, have we the operational capacity to deal with more cases being fulfilled and funds being drawn down, as well as the fact that there will be more live accounts and so more customer queries and more redemptions, either early on or at the end of the scheduled term?

If we implement a behavioral scorecard, similar issues arise. Are we about to greatly increase the number of telephone calls or letters that will be required, and do we have the capacity? On the financial side, are we about to greatly increase our credit limits, and do we need to get approval for this, or are we about to increase our provisions for arrears cases?

8.12 Aligning and recalibrating scorecards

If one has a suite of scorecards each built on a separate subpopulation, then it would seem sensible to align them so that a score has the same meaning on all the scorecards. What this means is that at each score the marginal good:bad odds should be the same on all the scorecards built on the different subpopulations. This is the only part of a run-book that is specific to a scorecard since the acceptance rate and the bad rate are needed for the whole population and their values on individual subpopulations are of little interest. One way to do the recalibration is to run a logistic regression of good-bad outcome against score for each scorecard separately. For scorecard i, this leads to the equation

$$\log\left(\frac{p_G^i(s)}{p_B^i(s)}\right) = a_i + b_i s, \tag{8.19}$$

where s is the score, $p_G(s)$ and $p_B(s)$ are the probabilities of an applicant with score s being a good and bad, respectively, and a_i and b_i are the coefficients of the regression. If this logistic regression is a good fit, one can recalibrate the scorecard for population i by multiplying the score by $\frac{b_i}{b_1}$ and adding $\frac{a_i - a_1}{b_i}$ to all the attributes of the scorecard. This guarantees that all the scorecards have good:bad odds satisfying $\frac{p_G}{p_B} = e^{a_i + b_i s}$. Even if the logistic regression does not fit very well, one should multiply all the scores in the scorecard by $\frac{b_i}{b_1}$ and then use a new variable (to which subpopulation an applicant belongs) in a regression to find the best constants to add to the scorecard to recalibrate it.

What this has done is to align scorecards belonging to different subpopulations in terms of their marginal good:bad odds so that at each score the good:bad odds is a known constant. This idea of recalibrating the scorecard so that it has certain properties can be applied to individual scorecards as well as suites of scorecards. The properties one might want in a scorecard are as follows:

(a) The total score is positive.

(b) The score for each attribute is positive.

(c) There are reference scores that have specific marginal good:bad odds.

(d) Differences in scores have the same meaning at all points on the scale.

However, recalibrating a scorecard to satisfy some of these properties makes it lose others. For example, we have a scorecard where all the attribute scores (and hence the total score) are positive and we want it to have marginal good:bad odds of o_1 at s_1 and o_2 at s_2. If s_1^* and s_2^* are the scores on the existing scorecard that have these marginal odds, then the linear transformation

$$s^* = \frac{s_2^* - s_1^*}{s_2 - s_1} s + \frac{s_1^* s_2 - s_2^* s_1}{s_2 - s_1} \tag{8.20}$$

ensures that the reference scores s_1 and s_2 have the required marginal odds. Since this is a linear transformation, it can be applied to each of the attribute scores in turn, and the result is that the total score is transformed in the correct way. However, it may turn positive attribute scores into negative ones and even introduce negative total scores. Trying to correct these problems will then destroy the two reference score properties.

Thus given an existing scorecard, it is not possible to ensure that it has all the properties (a)–(d). Condition (d) is very like condition (c) in the following way: Suppose that one wanted an increase in the score of s to double the good:bad odds; then this is equivalent to requiring the odds at s_0 to be o_0, at $s_0 + s$ to be $2o_0$, and at $s_0 + ks$ to be $2^k o_0$ for $k = 2, 3, \ldots$. One way to seek a new scorecard that approximately satisfies all these properties and also keeps the relative rankings in the sample used to build the original scorecard as unchanged as possible is to use linear programming.

Assume that there are N accounts in the sample and they are ordered in increasing original score. Let there be p characteristics and x_{ij}, $i = 1, 2, \ldots, N$, $j = 1, 2, \ldots, p$, are the values of the characteristics given by the ith consumer in the sample. Assume that $x_{ij} \geq 0$ for all i, j. Then the requirement that the ordering is maintained under a new scorecard with weights $w_1, \ldots, w_p$ is that we seek to minimize e_{ir} for $i, r, i < r$, where

$$w_0 + \sum_j w_i w_{ij} < w_0 + \sum_j w_j w_{rj} + e_{ir}. \tag{8.21}$$

This is trying to ensure that the score for the ith consumer is below that for the rth consumer. This has $\frac{N(N-1)}{2}$ conditions, which for reasonable-size data sets make it too large a problem to deal with. We can approximate this requirement by asking that it hold only for nearest neighbors, i.e., that the score for the ith consumer is below that for the $(i + 1)$st consumer. This can then be incorporated into a linear program, which also seeks to satisfy the other constraints, namely,

$$\begin{aligned}
&\text{Minimize} \quad a_1 \sum_{i=1}^{N-1} e_i + a_2 \sum_{v=3}^{m} (d_v^+ + d_v^-) + a_3 \sum_{l=LM}^{UM} (f_l^{d+} + f_l^{d-}) \\
&\text{subject to} \quad w_0 + \sum_j w_j x_{ij} \leq w_0 + \sum_j w_j x_{i+1j} + e_i \quad \text{for } 1 \leq i < N, \\
&w_0 + \sum_{j=1}^{p} w_j x_{i_k j} = s_k \quad \text{for } k = 1, 2 \\
&\qquad (i_1, i_2 \text{ accounts have properties required at scores } s_1, s_2), \\
&w_0 + \sum_{j=1}^{p} w_j x_{i_k j} = s_k + d_k^+ - d_k^- \quad \text{for } k = 3, \ldots, R, \\
&\qquad (i_k, k = 3, \ldots, R, \text{ accounts have properties required at scores } s_k), \\
&w_0 + \sum_{j=1}^{p} (w_j) x_{i_t j} = s_0 + ts + f_t^{d+} - f_t^{d-} \quad \text{for } 0 \leq t \leq T \\
&\qquad (\text{where } i_t \text{ is the account at whose score the marginal odds are } 2^t o_0), \\
&e_i \geq 0 \quad \text{for } 1 \leq i < N, \qquad w_j \geq 0 \quad \text{for } 1 \leq j \leq p \\
&\qquad (\text{attribute score is positive}), \\
&d_k^+, d_k^- \geq 0 \quad \text{for } 3 \leq k \leq R, \qquad f_t^{d+}, f_t^{d-} \geq 0 \quad \text{for } 0 \leq t \leq T.
\end{aligned} \tag{8.22}$$

For a fuller description of how this linear program can recalibrate the scorecard, see Thomas, Banasik, and Crook (2001). It is clear, however, that there are psychological and operational reasons to adjust a scorecard so that the scores tend to be in the range 100 to 1000.

Chapter 9

Implementation and Areas of Application

9.1 Introduction

This chapter deals with some of the issues involved in implementing a scorecard, either an application scorecard or a behavioral scorecard. It also discusses some related issues. In much of the chapter, there is little difference between application and behavioral scorecards; where this difference is significant and is not obvious it is indicated.

The chapter also moves on to deal with monitoring and tracking, defining these separately rather than treating them as interchangeable terms as is often the case. We also look at setting and testing strategies and then try to bring some of the matters together in answering questions such as, "When is my scorecard too old and in need of redevelopment?"

9.2 Implementing a scorecard

For whatever reason, an application scorecard has been commissioned and built, but how do you implement it successfully? To do this, several questions need to be considered.

- Do we have access to all the scorecard elements in the new scorecard?

In theory, one would assume that the answer to this is yes. After all, if we did not have access to the information, how could the data element have been included in the model building? However, several possibilities can arise. For example, it is possible that a data element was collected at the time of the development sample but has since been removed or dropped. In another example, the data elements in some questions may have been collected as part of a retro (i.e., a retrospective analysis) at a credit bureau but are not part of a standard return from the credit bureau.

- Are there any definitional questions as a result of the elements in the scorecard?

One needs to be clear about how the customer or operator will interpret each question. For example, are the questions and answers as used in the development the same as those used now? If we ask about the applicant's marital status, common options are single, married, widowed, divorced, and separated. However, people may consider themselves both married and separated. Other people may consider themselves to be both single and divorced. Whether applicants record themselves as married or separated or single or divorced may change according to a variety of influences, including the region of the country. What is recorded now may be subtly different from what was recorded in the development sample. The difference may be less subtle, caused perhaps by a change in the order or layout of the options. A member of the staff involved in the application—in a branch- or telephone-based environment—may have an influence on what is actually recorded, by answering the applicant's queries or by the way the questions are phrased. To reduce this effect, a clear policy relating to a wide range of possible questions is needed, and those colleagues who deal with customers should be trained carefully, thoroughly, and repeatedly and their performance continually monitored.

Another area that should be checked is whether the data received from the bureau has changed. For example, in the U.K., from October 2001 some changes were implemented regarding what data are available from the bureau. These changes may lead to fewer searches being recorded against individuals. Therefore, a scorecard characteristic looking at the number of searches in the last six months may change its value.

- What programming is required?

This may seem a trivial point, but even with the advanced state of computer systems used for processing applications and accounts, this often needs consideration. Clearly, if the programming of the scorecard requires the support of a systems department, then the size and complexity of the task may have a bearing on the time scale for its full implementation. One also needs to be very careful with how the implementation and scoring is actually tested. The alternative implementation route is to utilize one of the many pieces of software that are commercially available or, in some organizations, that have been built on to the standard systems. In these cases, the scorecard is parameter driven and can usually be programmed by a user or scoring analyst. However, one should never underestimate the effort required to key in the characteristics and their relative weights and to test that this has been completed successfully. One also should never take shortcuts, for example, to achieve implementation a few days earlier.

- How are we going to treat "pipeline" cases?

In many environments, there will be cases that are in the pipeline at the point we implement the new scorecard. This will generally happen when we have processed an application using the old scorecard but have not yet taken the application to approval or the funds have not yet been drawn down. One reason for this is that we are running a telephone- or Internet-based operation and we have issued a credit agreement for the applicant to sign and complete. Another possibility is where we have approved a mortgage but the property has not yet changed hands and so the funds have not been drawn. In many of these cases, we may be able to adopt the original decision. (In some of these cases, we are in a very weak position if we make someone an offer and then rescind it without their circumstances changing.) However, if, for example, we needed to reprocess the application, we may find that, because of the change in scorecard, the application now fails. Clearly, we need to adopt a policy. One option is to go with the original decision and live with a few marginal cases being accepted for the first few weeks. (This may require some systems solution to override a case that fails and to force it through the system.) Another option is to adhere to the new scorecard rigidly and reverse an original decision, communicating this to the applicant.

9.3 Monitoring a scorecard

In most scoring operations, analysts and managers refer to *monitoring* and *tracking*. In many cases, these terms are used either generically or interchangeably. In this text, we treat these as two separate functions, each with their separate but linked purpose.

Monitoring is considered to be passive and static. It is analogous to the traffic census taker who sits beside the road and logs the numbers of different types of vehicle passing a point in a period of time.

When carrying out monitoring, we should remember that a scorecard provides a prediction of risk for individual accounts but is also used to manage portfolios.

Therefore, some key questions can and should be addressed, including the following:

- Are applications being scored properly?
- Is the profile of current applicants the same as previously, e.g., last quarter, and the development sample?
- Is the current acceptance rate the same as previously and the development sample?

The first question should be required less often than it used to be. Earlier in this chapter, we dealt with ensuring that the scorecard was accurately implemented into our systems. Here we are referring to that area of the application that requires human input, whether by a member of the staff or by the applicant. One common area where some scrutiny might yield dividends is where there is a list of options. This might relate to the classification of occupation, for example. Suppose that the development process used 12 categories of occupation, the 12th one being "other." Suppose also that the percentage of cases scored as other was 5%. We now find that in a current batch of applications, there are 12% scored as other. We need to examine this difference and understand how this change might have happened. There are several possible explanations:

- The 12% is correct and was the comparative figure for the development sample. However, in the development process, some additional effort was spent in trying to allocate these cases of other to one of the other 11 categories.

If this is what has happened, then we could argue that the additional effort was wasted. Our general assumption of the future being like the past has been invalidated by our extra investigation.

- The operators who are classifying occupation are not being as effective as they should and, if they do not immediately know the category into which an employment should fall, they classify it as other and move on.

In this case, the issue becomes one of training and perhaps motivation. We have seen this type of thing happen in one organization split into three regions, where the percentage of cases classified as other in the three regions was 8%, 12%, and 26%. The major reason for these differences was the different emphases on data quality and process efficiency placed by different regional directors.

- There has been a change in our applicant profile.

If this is true, then we need to investigate the cause of it, what can be done about it, and the effect on the scorecard. One possible cause is that new occupations are developing all the time and our processes do not always keep pace with these. For example, 5 or 10 years ago, there were far fewer website developers than there are now. Some further comments on this area can be found in (Edelman 1988).

Another possible cause of a change in our applicant profile is that we have seen a change in the type of person who is applying. This may be caused by a change in the organization's marketing strategy, which may employ mailing, television advertising, or press campaigns—or a mix of these. Alternatively, the organization could have changed policy or strategy on a whole subset of the potential applicant population. For example, there could have been a marketing drive to attract students or a loan promotion aimed at gold card customers. Despite many years of recognizing the mutual advantages of communication between credit and marketing strategists, it still happens too infrequently and to too little effect. Most managers of scoring operations will be able to recount tales of discovering changes in marketing strategies after the event, by which time they have declined a high percentage of the additional business that was attracted. Worse is that if a poor-quality profile of applications is generated by a campaign, even after the high decline rate, the average quality of the accepted propositions will be poor. Therefore, while some of the marketing department may be content with a high response, most of them will be unhappy that the credit department declined many of them. Of course, the credit department is also unhappy because the quality of the book of accepted cases is being diluted by a poor-quality batch of accounts.

A further possibility is that there has been a change in the marketing strategy of a competitor. This can happen if the competitor is relatively large or where their offering or positioning is similar.

A change in profile may be caused by a change in the economic environment and outlook. If the economy suffers a recession or an expansion or if forecasts are that either of these will occur, we may see changes not only in the numbers of applicants for credit but also in the quality.

Having discussed at length some of the causes of a change in our applicant profile, let us move on to consider the measurement and the possible resolutions.

Table 9.1 is a common type of report. The characteristic analysis report takes each attribute of a characteristic and considers the differences in the proportions with each attribute between the development sample and a current sample. It then calculates the effect on the score for the characteristic.

Table 9.1. *Characteristic analysis report.*

Attribute	Development sample percentage	Current sample percentage	Score	Difference percentage	Difference percentage x score
Employed full time	52	42	37	–10	–370
Employed part time	9	16	18	7	126
Self-employed	18	23	15	5	75
Retired	11	10	28	–1	–28
Houseperson	6	4	11	–2	–22
Unemployed	1	3	3	2	6
Student	3	2	8	–1	–8
					–221

Thus for this characteristic, the average score has fallen by 2.21 points. We may find other characteristics where the change in average score is greater or less than this, and it is the total of the changes in the scorecard characteristic that determines the total change in score between the development sample and the current sample.

Therefore, we should do this type of analysis in three ways. The first is to examine each of the characteristics in the scorecard. The second is to do a similar analysis for the total score, and the third is to look for significant changes in nonscorecard characteristics. Significant changes here may reveal a significant change in our applicant profile or a characteristic that should be included in the scorecard, especially if it is highly correlated with performance. Clearly, we cannot construct a meaningful characteristic analysis report for a characteristic not in the scorecard (as the score weights will all be zero). However, we can construct a population stability report, as in Table 9.2.

Table 9.2. *Population stability report.*

Score	Development sample percentage	Current sample percentage	$B - A$	$\frac{B}{A}$	$\ln\left(\frac{B}{A}\right)$	$C \times D$
	A	B	C		D	
<200	27	29	0.02	1.074074	0.07146	0.0014
200–219	20	22	0.02	1.1	0.09531	0.0019
220–239	17	14	–0.03	0.823529	–0.19416	0.0058
240–259	12	9	–0.03	0.75	–0.28768	0.0086
260–279	10	11	0.01	1.1	0.09531	0.0010
280–299	8	6	–0.02	0.75	–0.28768	0.0058
300+	6	9	0.03	1.5	0.40547	0.0122
					Stability index =	0.0367

We need to consider how to interpret this.

Statistically speaking, the stability index is an X^2-type of measure, and therefore the interpretation of the stability index should incorporate some indexing. We may wish to consider using degrees of freedom, especially because the number of cells used may affect the value of the index.

Some have suggested a rule of thumb to the extent that a stability index of less than 0.1 indicates that the current population is fairly similar to that from the development sample. An index of between 0.1 and 0.25 suggests that we should pay attention to the characteristic analysis reports to see if we can identify shifts within the characteristics. However, an index value above 0.25 indicates some significant changes in the score distribution.

In general, however, this is a tool that is fairly easy to use but lacks some sophistication and consistency. Other measures could be used that would counteract these objections. To measure the significance of any difference between the score distributions at the time of the development and the current distributions, one could calculate a Kolmogorov–Smirnov test statistic and test its significance. Alternatively, one could calculate the Gini coefficient.

However, of utmost importance is that, on a regular basis, one goes through the process of measuring the stability of the characteristics and the score distribution. Even if one has no notion of significance, if the index or measure is growing, this alone is a sign that there is some population drift.

We need to consider one other point here. When developing a scorecard, we can and do use exactly the same measures although then we are looking for large significant differences to support the development and implementation of a new scorecard as being better than the old one. Once we have implemented the new scorecard, we are looking for insignificant values to suggest that there has been little change in applicant profile and so forth since the development.

9.4 Tracking a scorecard

While monitoring is passive and static, tracking can be considered to be active and dynamic. If we consider the two analogies mentioned above, tracking is more like following an animal to see where it has its young or like a police car following another car to assess its speed or quality of driving.

Tracking principally allows us to assess whether the scoring predictions are coming true. Following that, we need to consider whether the predictions are coming true for realistic subsets of the population. Also, if the predictions are not coming true, we need to consider what action to take. When tracking, as with monitoring, we should remember that a scorecard provides a prediction of risk. Therefore, one of the key reports is to examine how accurate the scorecard's predictions are. Thus, the key questions are the following:

- Is the scorecard ranking risk properly?
- Is the portfolio bad rate what we would expect given the development sample or the distribution of accounts booked?
- What changes can we detect between different subsets of the account population, defined by segments, or cohorts, or by different account strategies?

To assess if the scorecard is ranking risk properly, we need to examine whether at each score the percentage of cases expected is approximately what happens. In Figure 9.1, we plotted our expectations from our development sample along with data from three different samples.

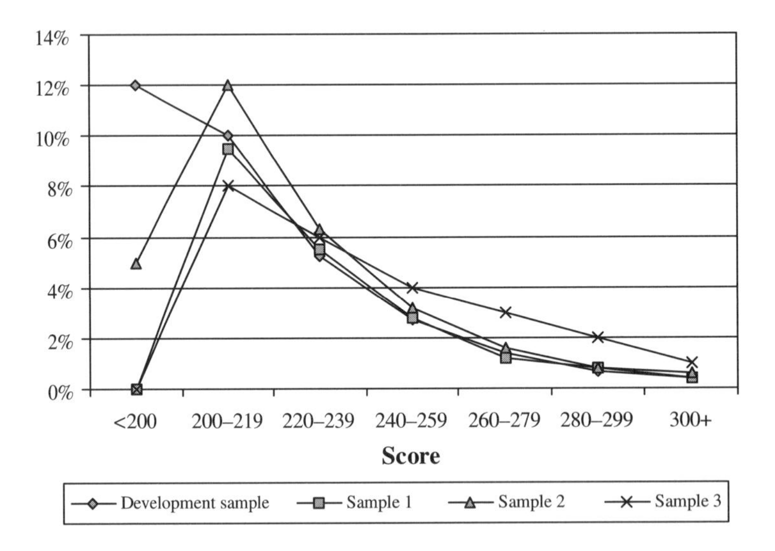

Figure 9.1. *Bad rates.*

Clearly, we are operating this portfolio with a cutoff of about 200. With the first sample, our bad rates for each scoreband are approximately what the scorecard predicted. There are some fluctuations around the development sample, but on the whole it is well aligned.

The second sample conveys some different messages. First, note that in this sample we took on some business below our cutoff. The bad rate in this region is below that of the first region above cutoff, so we might conclude that there has been some intelligent overriding in this area. Above the cutoff, the bad rates are consistently above expectations. This suggests that the economic environment in which this sample is operating is worse than that in which the development sample operated. Note, however, that the scorecard is still working well since the curve has a similar shape. On the other hand, if the second sample comes from a different type of business, e.g., newspaper advertising rather than branch-based applications, then we need to consider whether we should amend our process. Another possibility is that we raise the cutoff so that the profit or return at the cutoff is the same. Of course, if the applications from newspaper advertising are cheaper to acquire, it may be that we can accommodate a slightly higher bad rate at each score and still produce the same return.

With the third sample, we see lower bad rates than expected just above the cutoff and higher bad rates than expected at higher scores. As this curve is beginning to move away from the curve we expected and tending toward a flat horizontal line, we might suspect that the scorecard is not working as well as it should.

We should consider two other points here. First, sample 3 might actually be making more money. For example, a 50% reduction in bad accounts in the 200–219 scoreband and a 50% or even 100% increase in bad accounts in the 260–279 scoreband, say, might mean that overall we have fewer bad accounts. Obviously, one of the key factors in this calculation is the number of accounts in each scoreband, although with careful marketing and targeting, we should be able to exert some control over the number at each scoreband. Nevertheless, the flatter shape of the curve does suggest that the scorecard is not working well on this sample of our business.

The other point worthy of consideration here is that we ought to place greater focus on what is happening at and around the cutoff or at and around candidate cutoffs. If we are operating this portfolio with a cutoff of 200 and find that cases scoring in the 260–279 scoreband have a higher bad rate than was expected, while that may be a nuisance, we are never going to avoid these accounts. First, the actual bad rate for this high scoreband is still likely to be lower than for the scoreband just above cutoff, even if the actual bad rate at that point is below expectations. Second, if an application scores in the 260–279 band, a redeveloped scorecard will most likely still score it well above any feasible cutoff.

If we move away from application scoring to behavioral scoring (or even to collections scoring, attrition scoring, etc.), the principles are very similar. If we can wait for the outcome to appear, which in these cases may be as soon as 6 months later, rather than 18 to 24 months for an application scorecard, we can assess how well the scorecard is performing on different parts of the portfolio.

Now the problem is that if we wait 24 months to discover that we have a weakness, we have taken a lot of business during that 24-month period that we would now rather not have taken on. Therefore, one key objective would be to make some sort of interim assessment.

These interim assessments can be carried out using some simple tables and graphs that are known by a variety of names: dynamic delinquency reports or delinquency trend reports, or vintage analyses or cohort analyses. The fundamental issue is that for each cohort of business, usually one month's business or one quarter's business, perhaps split also by another variable—customer-noncustomer, branch-telephone, score—we compare the performance at similar stages in that cohort's life.

Therefore, suppose we take as a cohort a month's business and record its performance at the end of each subsequent month. We can build up pictures of how this cohort is performing by recording the percentage of accounts that are one down, two down, or three or more down. We can also record the percentage of accounts that by the end of the month in question have ever been one down, two down, or three down. We can do the same analysis by value rather than by number, taking either the loan balance or the arrears balance as a percentage of the total amount advanced. This could lead to 12 graphs or reports for each subset. However, these graphs or reports encapsulate the whole history of the portfolio from one particular point of view.

An example of this appears in Figure 9.2. In this graph, we plotted the performance of 12 cohorts of business from April to the next March. For the oldest cohort—business from April—we examined its performance at the end of each month from the third month of exposure until the 15th month of exposure.

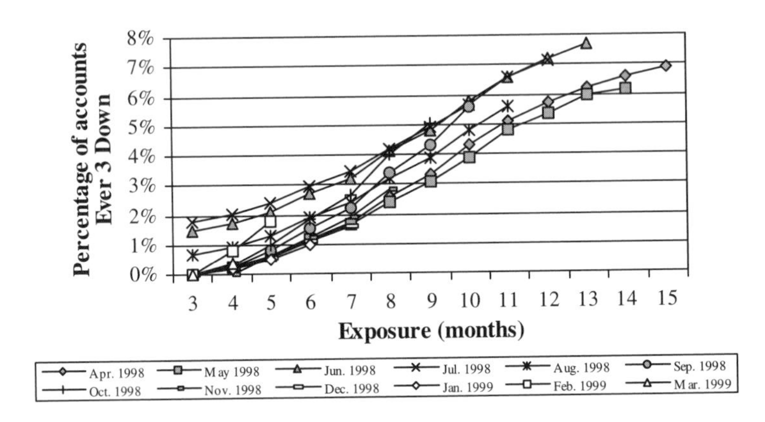

Figure 9.2. *Dynamic delinquency report.*

(One needs to be careful about how a cohort is defined. This may vary from product to product. One could define a cohort by the month or quarter in which the application was made, or the application was approved, or the money was drawn down, or when the account was activated. Further definitions include the period in which there was first financial activity or the period in which the first payment was due. It is obviously important that the definition is applied consistently. It is also important that the definition is consistent with the objectives of the analysis and with the product. For a credit card, one could use the month in which the account was activated, if it requires activation prior to use, or the month of the first debit activity, i.e., transaction, or the month the card was issued. For a loan, the better choices would be the month or quarter the funds were drawn or the month or quarter when the first payment is expected.)

Let us assume that the definition used in Figure 9.2 is from the month in which a loan payment is first due. Therefore, April 1998 business is a tranche of loans that have a first payment due date in April 1998. We have 15 months of exposure for this cohort; i.e., we looked at this group of loans at each month until June 1999.

For the next cohort of business—May 1998—we also looked at its performance every month until June 1999, but that makes only 14 months of exposure, and so on.

This graph looks at the percentage of accounts in each cohort that ever got to being three payments down. This is one of the more common pictures as this is often the scorecard's definition of bad.

We can see a number of things from this graph:

- The graph starts at only the third month of exposure. Accounts cannot miss three payments until three payments have been due.
- All curves are either flat or increasing. Because we are looking at a worst-ever status, the percentage of accounts reaching this stage cannot ever decrease as we lengthen the

exposure. (This relies on our systems retaining a record of accounts even if they are closed.)

- By examining the curves, we can establish a common pattern and then are able to identify deviations from this.

- If a curve is below the general trends, this cohort of business is performing better than the norm. Such a deviation from the norm should be investigated and, if possible and appropriate (taking into consideration profit, etc.), we may wish to attract more of this business. On the other hand, a more radical approach here might be to increase the level of risk by, for example, reducing the level of checking a part of the credit assessment. In other words, bring the business back into line by relaxing our credit assessment process, increasing the risk of each case, but accruing some financial benefit from the relaxation of the credit assessment. For example, if the organization asks the applicant to supply proof of their income, we could increase our risk and reduce our costs by dispensing with this requirement. Another option is to reduce the level of credit reference checking, perhaps not checking on the previous address, irrespective of the length of tenure at the current address.

- If a curve is above the general trend, this cohort of business is performing worse than the norm. Such a deviation should be investigated and, if possible and appropriate, we may wish to try to attract less of this business. If we can identify this very early on, then we can take early corrective action. In Figure 9.2, July 1998 is performing worse. If we identify this through the percentage of cases one payment down and can identify this after three months, i.e., at the end of September, then the change will begin to take effect on some of October's business and certainly November's business. The longer we leave it, the more cohorts of business may also perform poorly. Of course, we must also be careful that we do not act prematurely without proper analysis and consideration of the cases of the poor performance. Alternatively, we could introduce additional steps into the process to improve the credit quality.

- April 1998 and May 1998 business look similar. We then have a worse pattern for June 1998 and July 1998 business. However, these curves are not steeper than the previous two; rather, there is a parallel shift upward. What appears to have happened is that this tranche of business is about 1.5% higher than the norm. A possible reason for this, especially since the shift occurred from the beginning of the plotted history of the performance of the cohort, is that we were hit by some increased fraud. Once this increment is removed, the credit quality of the rest is similar. By August 1998, we appear to have begun to sort this out as the parallel shift is reduced.

- September and October 1998 business is clearly worse than the norm. The curves are steeper, suggesting a poorer quality of business or poorer performance. November 1998 is almost back in line with April and May 1998, suggesting again that we have sorted things out. This could be due to a change in the process, or a change in the scorecard cutoff, or a change in marketing strategy. However, February 1999 again looks to have a worse performance, perhaps caused by the post-Christmas rush for further credit among many consumers.

As was stated earlier, the same type of graph can be constructed for different definitions. We can clearly use Ever One Payment Down and Ever Two Payments Down. We can also use Currently One Payment Down, Currently Two Payments Down, Currently Two or More

Payments Down, etc. In these cases, obviously, the curves can go down as well as up as accounts in arrears can improve from one month to the next. In fact, later on in the life of the cohort, once most of the bad cases have been flushed out, we might expect to see these curves fall as our collections and recoveries activity corrects more accounts each month than the maturity process pushes into arrears.

Clearly, the reports and graphs looking at earlier stages of delinquency can be used to identify at an early stage if a cohort of business appears to be performing worse than the norm. For example, if we see a higher rate of one-down cases after three months and can take corrective action immediately, we have only three months of business taken on with the current process and procedures.

We can also use percentages of values rather than the percentages of numbers. We should recognize that moving into current arrears or value may represent a step away from tracking the performance of the scorecard because scorecards in general work on the percentage of accounts going good and bad, rather than the percentage of the monetary value of the accounts. However, it is a step toward managing the portfolio.

To return to scorecard tracking, we can also produce these graphs for a particular scoreband, thereby controling one of the key factors—the quality of business. Clearly, we can get a better or worse tranche of business by changing our cutoffs or our marketing. To some extent, if we make a conscious decision to do this, we would expect the effect to show up in a dynamic delinquency report. However, if we were to produce a report for, say, scoreband 220–239, then irrespective of the number of accounts that fall into that band in each cohort, we would expect the performance to be similar. Any deviation from the norm could be a sign that the scorecard is deteriorating. Of course, we might also find that for one cohort in this scoreband, the average score is 231, while for another it is 227, so we may need to control further by cutting the the scorecard into even narrower intervals, provided we have enough data to make analyses and conclusions reliable.

We could also produce the same reports and graphs for specific attributes. For example, we might produce one graph for homeowners, or one for customers aged 25–34, or one for customers whose loan is between £5000 and £8000. Obviously, the more we cut the data, the fewer cases there are, leading to more volatile results and less certainty in our conclusions. If we have so little data that we cannot be certain about our conclusions, then we might decide that it is not worthwhile to carry out the analysis.

These graphs allow the user to easily identify trends that happen at the same point in a cohort's exposure. They will not easily allow one to identify trends or events that happen at the same point in real time. For example, if we expect arrears to rise post-Christmas 1998 and then fall in early spring, this will affect each tranche at a different point in its exposure. For April 1998 business, January 1999 is month 10, while for August 1998 business, it is month 6. A much clearer example of this happening occurred in the late 1980s in the U.K., when there was a six-week postal strike. Although cardholders could make their monthly credit card payments through the banking system, many payments were received late, if at all. Cardholders claimed—and some believed—that if they did not receive a monthly bill, they were not due to make a payment. In terms of current arrears, the percentage of cases one payment down on some portfolios rose from, say, 10% to 30%. At the end of the postal strike, the wave had been started and the percentage of cases two payments down rose from 4% to 7%. However, three to four months later, the wave had dissipated and there was minimal effect on ultimate write-offs. (Indeed, any adverse effect on the ultimate write-offs was more than compensated by the additional interest income.)

One may also wish to identify whether there are any seasonal effects. For example, it may be that the February business, because of lower quality, always performs worse.

In some portfolios, the cutoff has been raised for a short period at the start of the year to counteract this.

Once again, the above focusses on application scoring, so how does this type of analysis work for behavioral scoring? To a great extent, the application of the analysis is exactly the same. The score provides a prediction of the percentage of cases of a particular type that will behave in a particular way within a period of time. Therefore, we should track the performance of these accounts after they have been scored and dealt with accordingly to assess the accuracy of the scorecard and the consistency of their behavior. What we might need to be more careful about is the definition of a cohort of business, although factors commonly used in such definitions include time on books and the size of the limit or balance.

We also need to bear in mind that behavioral scoring is often used in conjunction with adaptive control strategies. With these, score may be only one of a set of factors that are used to determine how to manage the account and how to react to the customer's behavior and the account performance. We deal with this more in section 9.6. The key point here is that not all accounts scoring, say, 500–539 will be dealt with identically. Also, the range of actions for them could be quite different from those accounts scoring, say, 580–619 and so we should not necessarily expect a smooth curve, even in theory.

9.5 When are scorecards too old?

Through tracking and monitoring, we can begin to assess when the applicant population has changed significantly from our development sample, in terms of both the demographics and other data available at application stage and in the performance.

There is no simple answer or simple statistical or business test that can be performed to decide when corrective action is required and what it should be. Scorecard performance degrades over time for a number of valid reasons. There is then a business decision to be taken to weigh up the advantages and disadvantages of a new scorecard, bearing in mind the feasibility of building a new one. For example, do we have enough data with enough mature performance and a sufficient number of bad accounts? For example, if our scorecard has degraded because of a change in marketing strategy, will a new scorecard suffer the same fate?

Before we set about redeveloping the scorecard in its entirety, there are a number of possible modifications that will either extend the life of the scorecard or improve its effectiveness through the rest of its life.

The first thing we can do is examine any potential for realignment. Suppose we have a scorecard where we have a characteristic of time with bank, as in Table 9.3. We have been tracking the performance of accounts based on their total score but looking at the different attributes of this (and other characteristics) separately. This produces a graph similar to Figure 9.3.

Table 9.3. *Time with bank.*

Time with bank (years)	
Attribute	**Score**
0	5
<1	7
1–3	18
4–6	31
7–10	38
11+	44

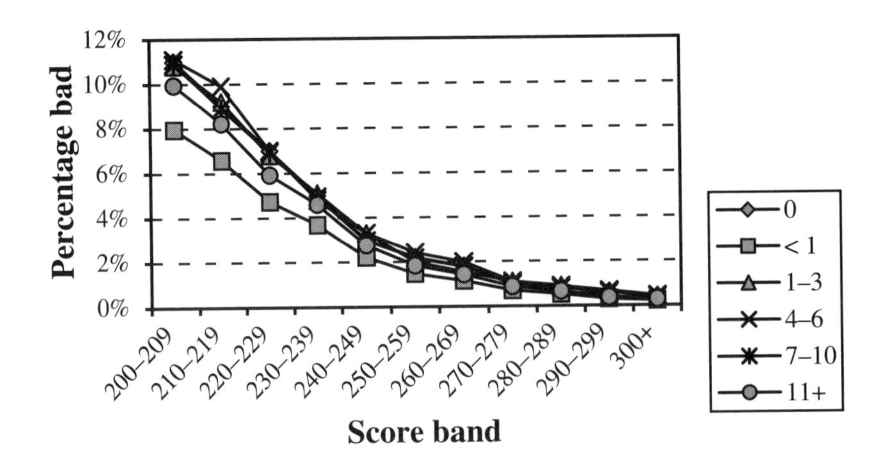

Figure 9.3. *Alignment analysis: Time with bank.*

What we would expect to see if the scorecard is working is that at each scoreband, the percentage of bad cases is the same whatever the time with bank. In other words, the scorecard has adjusted adequately for the different levels of risk represented by the different tenure of the existing banking relationship.

What we see here is that the pattern is similar except for applicants who have been at their current bank for less than one year. What could be causing this persistent anomaly?

A possible reason is that we are more stringent in taking on these applicants as customers, and we may find that we filter out some of them. For example, if we use some other step in the credit assessment—a credit reference search, for example—we may find that for those that score 200–209, say, 88% of them, are accepted. However, for those who have been at their current bank for less than one year, this percentage is only 70%.

Another possibility is that there has been a shift in their level of risk as measured by the scorecard itself. This might be the case here if we find out that banks are being more selective in taking on new customers. Therefore, anyone who has recently changed banks has also recently passed through the new bank's assessment procedures and therefore is better, on average, than the corresponding group in the development sample.

Other reasons could be hypothesized, and some may even be capable of being proved or disproved. However, from the point of view of managing the lending business, all we need to do is to understand that the effect is happening and that it is happening fairly consistently. Once we have established that, we can take steps to deal with it. In this case, applicants with a time with bank of less than one year are performing similarly to other applicants scoring approximately eight points more. Therefore, one simple adjustment is to change the scorecard weight for this attribute from 7 to 15 points. This not only reclassifies all of the existing accounts but also introduces some accounts that previously scored 192–199 and will now score eight points more and achieve the cutoff of 200. We have not seen how these will perform, but with such a smooth pattern, one may be comfortable assuming that they should perform in line with the new score calculated for them.

If there is some concern about taking on new business, this can be introduced on a trial basis, for a month, say, or one could take on a random 1 in 10 of these additional accounts until confirmatory evidence has been gathered. On the other hand, if one does not introduce additional business, then the scorecard realignment becomes a hypothetical and philosophical exercise, i.e., if we are not prepared to act, why bother doing the analysis?

Similarly, if an attribute is underperforming—accounts are performing worse than expected—we can reduce the score weight and, as a result, will now decline cases that previously would have been approved.

(When looking at percentages of bad cases, there are many situations for which we need to consider carefully of what it is that we are taking a percentage. Not only may we wish to look at numbers and financial values separately, but depending on the product, we may have to look separately at accounts opened, accounts still open, active accounts—debit active or credit active or either. We also need to ensure that our computer and accounting systems include accounts written off as bad and carry the actual loss somewhere—although the account balance has been set to zero by the write-off.)

Therefore, it is possible to realign the scorecard, reflecting gradual or developing changes in performance. In general, one should not be too alarmed. It is recognized that scorecards do not suddenly change their performance. Changes happen gradually. However, eventually a decision to redevelop has to be taken. The principle reasons for this are the following:

- Minor realignment either is impossible or has become too complex to understand what is and has been going on.
- Better quality data are now available.
- There are new information sources.
- The goals and practices of the organization or the marketing strategy have changed.

Monitoring and tracking, each in their own way, will provide clues to when the point has been reached that a redevelopment is required. Other factors also need to be considered. For example, the economics of the scorecard and its redevelopment may have a bearing on the timing decision. For example, the redevelopment could take three to nine months (or even longer) to complete and be running live. For example, if the strategy of the organization is likely to change again soon or has changed fairly recently so that there is little performance information available, this may affect the timing decision.

On the other hand, Lewis (1992) puts forward the view that "only rarely are credit scoring systems replaced due to failure to perform as expected" (pp. 115–116). Indeed, he viewed the replacement of the scorecard as being a task of medium-term to long-term planning. "Credit management are well advised to give thought in advance to the degree of degradation of the performance of their scoring system that would cause them to replace it. Since system performance degrades slowly, if at all, if internal policies have not changed, management can predict when the replacement should take place and can plan the assembly of the sample data that will be needed for the development of the new system."

9.6 Champion versus challenger

The use of champion and challenger strategies became much more widespread through the 1990s and is now more accepted through credit and other areas. The principle is quite simple. There is an accepted way of doing something. This is known as the champion. However, there are one or more alternative ways to achieve the same (or a very similar) objective. These are known as the challengers. However, there is no evidence of the effect of the challengers. Therefore, on a random sample of cases, we try the challengers. This trial not only will test effectiveness in comparison with the champion but also will allow us the opportunity to identify the existence and extent of side effects. Eventually, we decide that one of the challengers is better overall than the champion, and this challenger becomes the new champion.

This occurs in medicine. There are established ways to treat specific ailments. These established ways might be by drugs, counseling, etc. Suppose that we are developing a new toothpaste. We have passed a number of clinical trials in which the chemical composition of the toothpaste was assessed, and we also tried it out on laboratory animals. Eventually, we need to try it out on humans. The key factors in assessing the toothpaste's effectiveness include the taste, how clean the teeth feel, the color of the teeth after brushing, and the ability of the toothpaste to prevent tooth decay. Of these four factors, the first three are fairly subjective, while the fourth takes some time to establish, and we must test the toothpaste on large numbers of humans to establish average effectiveness. (We cannot easily, for example, brush half of each person's teeth with one toothpaste and half with another.) The existing toothpaste is the champion while the new toothpaste is the challenger. In carrying out a trial, we try to capture meaningful figures on the key factors as well as record other measures. For example, was there a change in the incidence of mouth ulcers? Did the subjects endure more headaches or nausea?

Television advertisements abound with similar cases, not necessarily to do with toothpaste. Laundry detergent is a common example. Within medical research, there are other experiments, for example, in cancer treatments. Within the credit industry, we also carry out such experiments. Indeed, a person who has a credit card is almost certain to have formed part of a challenger group at some point without knowing about it.

In the credit environment, we often refer to the whole process as adaptive control. By recording and analyzing customer reactions to our management and control actions, i.e., experiments, we can learn how best to control their behavior to maximize some objectives, such as profit and retention. One might expect a behavioral psychologist to use similar approaches and terminology in controlling animal behavior.

We can carry out such experiments at the application stage. For example, for loan applicants, we could assess their affordability and, on a random sample, select those with higher affordability and offer them either an increased loan or a credit card as well. The desired benefit would be increased income. However, it might be that those who are more susceptible to such an offer are also those who are more likely to end up in trouble, and so our increased bad debt might wipe out the increased income.

Another example is in a telephone-based lending operation, where, after the customer has received an agreement in principle, the customer is sent an agreement to sign and is asked to return this, completed, along with some documentary proof of income or employment or residency. Not everyone completes this process. Some people do not because they change their minds about borrowing the money or about the planned purchase. Others do not because they find another lender with less-stringent requirements for proofs. A challenger strategy is to remove the requirement for the documentary proof. This can be trialled for a short period to see the effect on the percentage who complete the process. Of course, removing the requirement for proof will increase the risk of fraud, and so this will have to be monitored and taken into consideration when assessing the effectiveness of the challenger.

Another challenger strategy would be to offer the customer an incentive to complete the loan—i.e., complete the loan documentation so that the loan can be drawn—or to complete it within a specific period. Here, running strategies in parallel will allow us to measure the relative cost-effectiveness of offering different incentives.

There are other potential impacts from such a challenger. It may be that if customers who are asked to send in proof of their income receive their agreement in principle a few days before they are due to be paid, they wait until they have an up-to-date payslip available. If this proof is not required, they may return their agreement, not only in greater numbers, but by return of post. Therefore, it is possible that the flow of returned agreements would be affected.

Probably the most common use of champion and challenger strategies within credit is in the area of managing accounts. There are several areas in which we can operate experiments, and the systems for managing accounts often differentiate between these. Trivially, in most cases, when we talk about account management, we are talking about accounts with something to manage. These are, therefore, revolving credit accounts. They may be overdrafts, or current accounts, or flexible mortgages, but the most common area is credit cards. Within credit cards, there are at least six areas with which account management deals. These are as follows:

- Overlimit—-how to deal with accounts where the balance is over their agreed credit limit.

- Delinquent—how to deal with accounts that have missed one or more due payments. These may include accounts that are also overlimit.

- Credit limit—how to manage the cardholder's agreed credit limits.

- Authorization—how to deal with requests for authorization from merchants. These will be generated either because the proposed purchase is above the merchant's limit or because the proposed purchase will take the customer above some threshold, e.g., their credit limit, or a system warning on the number of transactions in a period.

- Reissue—whether to reissue a card to a cardholder at the expiration date of the card (or when it is reported as lost) and, if so, for what period.

- Marketing—how to maximize retention or cross-selling opportunities.

As an example, we can consider how we deal with cases that miss a payment and become one month delinquent. The credit card organization will have an established way to deal with these. Years ago, this might simply have been to send the cardholders a letter reminding them of their responsibility. However, we can construct a challenger to this strategy. We have a range of options open to us:

- We can send the customer a letter. Indeed, we can have different strategies using a range of letters varying in the severity of the tone and what we are threatening to do in terms of further action.

- We can telephone the customer, either at work or at home.

- We can send someone to the customer's home to talk.

- We can generate a message on the statement.

- We can ignore the missed payment.

How do we decide which action to take for which customer? Well, this decision would depend on a range of other factors, including the following:

- The arrears history. If this is the first time this has happened, we might wish to take a firm but friendly stance. On the other hand, if the customer regularly misses a payment and then makes it up the next month, it might be worth ignoring it and seeing if the customer makes the payment. Similarly, the time since their last delinquency (and the severity of it) might have a bearing.

- The value of the balance or the arrears. In terms of cost-effectiveness, action that might be appropriate for a £4000 balance where there is a £200 payment that has been missed might not be appropriate for a £200 balance where the missed payment was for £5.

- The relative value of the balance compared with their normal usage patterns, following on from the above. If the customer normally has a £1000 month-end balance, but this month, when the payment was missed, the month-end balance has risen to £4000, this could be a sign that there is an increased risk, and so appropriate, firmer action is required. On the other hand, it could also be a sign that the customer is away from home—on holiday, perhaps—and is using their card, having forgotten to arrange for a payment to be made.

- The behavioral score, which represents the probability that an account in a certain current state will reach some other state at some future point. For example, for accounts that are up to date or two or fewer payments down, the behavioral score might predict the probability that the account will reach three payments down within the next 6 (or 12) months. Therefore, even for accounts that are one payment down or are even up to date, if the customer has a low behavioral score, we ought to treat them differently from those with a high behavioral score.

- How long one has been a customer. This is similar to consideration of arrears history. We may wish to deal with relatively new customers in a different way from more established customers.

- Other relationships. We may wish to take into consideration other accounts held by the customer.

- The overlimit status. If the customer is also overlimit, this will probably call for a more stringent approach.

- The card is due for reissue. If the card is due for reissue, we may be able to use the promise of reissue to encourage the cardholder to get their account back in order. Alternatively, we can take the opposite view and decide not to reissue the card.

Account management packages allow the user to select from a large number of factors in setting a strategy. Each strategy, therefore, can have several nodes. What might happen in a collections area is that there is a champion strategy that operates on 80% of the accounts and that we run with two challengers on 10% each, one that is more stringent and one that is less stringent. As with the applications strategies (and the medical experiments), there are several factors to be assessed with the three strategies. However, usually the overriding consideration is of profit or return. Therefore, we need to understand the cost-effectiveness for each strategy. To do this, we need to consider several areas:

- The cost—of letters, telephone calls, visits.
- The effectiveness: What ultimately happens to the accounts? Do the accounts get back in line? Do we end up writing off bad debts? Do customers get annoyed at our approaches and take their business elsewhere?
- The positioning: Have we generated customer complaints? Has there been an effect on our public position?

We can then run these strategies for a few months and monitor the effect on many measures. We need to monitor income, and also how many cases pass on to the next stages of delinquency. We need to record customer complaints (and even customer praise). Costs will need to be allocated quite carefully. However, after a few months, we should be able to assess if any of the strategies is coming out on top or at the bottom. Clearly, if one is coming out poorly, we remove it and either move back to using the champion on that group of accounts or develop a new strategy.

If one of the challengers is coming out on top, we should expand the size of the group of accounts being managed using this strategy. However, this requires that we consider carefully the resource implications. For example, the best challenger may involve making telephone calls instead of sending out letters. We may have reached our capacity of outbound calls until we recruit more staff. Alternatively, the number of telephone lines the system will support might restrict us. Also, when making outbound calls, if we are unable to contact the cardholder and leave a message, we may greatly increase the number of inbound calls, with the obvious resource implications.

Running these competing strategies and allocating cases randomly and being able to monitor the outcomes and allocate costs to assess the effectiveness places great strains on our computer systems. Account management packages provide support in these areas. However, they are quite expensive and so are probably of use only when the portfolio is of significant size. How large it has to be to justify the cost will obviously depend on how much additional profit can be squeezed out of the accounts by better management. In addition, there is the view that if we can manage existing accounts much better and maximize our profit from each one, then we can take on much higher risks at the front end. Therefore, a good strategy management system, together with good strategy managers, will not only increase profit from the accounts but will allow more applicants to be approved.

Systems will also allow the user to randomly select accounts. To do so, most systems have a random number allocated to each account. This might be a two- or three-digit number. However, the best systems will have more than one random number allocated. This is useful in reallocating test groups.

In the example above, if one challenger proves to be ineffective, we really want to randomly reassign these cases to the other strategies. If we replace the ineffective strategy with a new strategy, then we are not, strictly speaking, testing the new strategy. Rather, we are testing a combined process of using the now-discarded strategy followed by the new strategy—in statistical terms, there may be a carry-over effect—and accounts managed in this way may not perform in the same way as accounts managed by the new strategy alone. It should also be recognized that, if the strategies are significantly different, this may generate customer confusion and, perhaps, complaints.

Chapter 10

Applications of Scoring in Other Areas of Lending

10.1 Introduction

In the previous chapter, we dealt with the most common applications of scoring or of a scoring approach in lending. In this chapter, we look at some other areas of lending where a scoring-type approach can be used. In Chapter 11, we look at some other areas where a scoring-type approach can or has been used but which fall beyond the normal confines of a lending environment.

The topics we touch on in this chapter are addressed in order of the process, but not necessarily in the order of their priority or importance. These topics are prescreening, preapproval, fraud prevention, mortgage scoring, small-business scoring, risk pricing, credit extension, transaction authorization, aspects of debt recovery, and provisioning for bad debt.

In almost all of these cases, there is a lending decision to be made and there are either two possible outcomes or we can reconstruct the situation so that it looks as if there are two possible outcomes. In a few cases, there are clearly more than two possible outcomes, and we try to identify these and comment as appropriate.

10.2 Prescreening

In principle, prescreening is where we make some or all of the lending and associated decisions before the applicant is aware of what is happening. In some cases the applicant will never be aware of what has happened. For example, before sending out a mailing, we may be able to match the name and address against a file of previous bad debtors and remove those who have a poor credit history. Clearly, in such a case, those who receive the mailing are not explicitly aware that they have passed through some form of initial test. Of course, those who are excluded from the mailing are certainly not aware that they were initially considered and then filtered out. In this example, the file of previous bad debtors may be an internally held file of the institution's previous borrowing customers. Alternatively, it could be a list from a credit bureau. (However, the reader should be aware that different restrictions exist in different countries regarding exactly how this information is used.)

Before approaching the customer with an offer, we can carry out a wide range of prescreens. Perhaps we merely wish to match or confirm the address. On the other hand, we can carry out several prescreen tests and, if we have enough information, we can even carry out a full assessment and maybe make a firm offer (preapproval). Another option open to us is that, on the basis of the results of the prescreen, we may elect to make the customer a different offer.

When considering using prescreening, one needs to carefully consider the cost and benefits. Using third-party information comes at a cost. Also, if using previous serious arrears as a prescreen rule, it may be that the population one is about to approach has few members who would be removed from the list. Therefore, the information is useful in theory but not on the particular target population. It may also be that the credit-assessment process would identify those higher-risk cases that respond. Therefore, one needs to consider the trade-off between removing a large number of high-risk cases from the mailing list and trying to identify the smaller number of high-risk cases among those who respond.

One other point may need to be borne in mind. In building a traditional scorecard, we carry out reject inference on those who applied but either were not approved or chose not to take out a loan or credit card. When we do some prescreening of a mailing list, for example, we now have two other sets of potential applicants whose performance we may need to infer—those who were excluded from the mailing and those who were mailed but did not respond. This is particularly important when we try to build response models; see (Bennett, Platts, and Crossley 1996) and (Crook, Hamilton, and Thomas 1992b).

In many countries, there are regulatory and self-regulatory rules concerning the use of information for prescreening. In the U.K., for example, the credit bureaus adhere to the principles of reciprocity regarding the sharing of credit information among lenders. Within this, there are rules about what levels of information a lender can use, and these vary, for example, depending on whether the person being considered is an existing customer. The rules also vary depending on whether the credit facility being considered is a product they already hold or is a new product. All of this varies also depending on the level of information the institution is supplying.

In the U.S., credit reference information is more readily available and reveals more about a potential borrower. On the other hand, restrictions are placed on a lender to the effect that, if they carry out a prescreen, an offer to a potential customer must then be an unconditional offer.

Another example of where we might use prescreening is if we have a response model and we can exclude from a mailing those who are unlikely to respond, thereby reducing the cost but not the effectiveness of the mailing. Indeed, it will generate a higher response rate.

Another factor to bear in mind with a response model or response scorecard is that it often works counter to a credit scorecard. At its simplest, those who are credit hungry are those who are most likely to respond, while those who are more creditworthy are less likely to respond. For example, we may find that younger people are more likely to respond to a loan offer or mailing, while older people are more likely to be deemed acceptable risks. The same might be true of other attributes. Therefore, we often have to reach a compromise so that we can attract sufficient volumes of applicants with a reasonable chance of being accepted and turning out to be good performers.

In a traditional scoring approach, our binary outcomes are that the customer will perform and not perform. With the first example of prescreening, this is also the case. We also discussed the binary outcomes of respond and not respond. Ultimately, when trying to maximize profit, the binary outcomes are composite ones of "will respond and will perform" and "will not respond or will respond but will not perform." (There are other occasions when we might use a prescreening approach. In one of these, we put applicants through an initial scorecard, and on the basis of the results, we could decline the application or approve it or carry out further assessment. Another is where the applications have to pass through a scorecard using only application data and a scorecard using credit reference data. In section 12.3, we discuss how we might combine two scorecards operated in such a fashion.)

10.3 Preapproval

Preapproval is prescreening taken to the next stage. Here we have carried out a sufficiently extensive prescreen that we are prepared to make the applicant a firm offer. This offer could be guaranteed or conditionally guaranteed, subject to some mild terms. One way to do this is with one's existing customers. If a loan customer has made their last 12 payments on a £5000 loan on time, one may be able to assess that they are a good risk and in control of their finances. One could even build a model that predicts the probability of such a customer later defaulting. In general, it is not too extreme to offer the customer another similar loan—say, for £5000 or £7500—but with a guarantee. Alternatively, one could make them a conditional offer of a much larger loan—say, £15,000—provided they could produce some evidence that they could afford it. Many lenders nowadays would not even require such evidence.

Another area in which this happens is with the management of credit limits on a credit card portfolio. In general, the higher the credit limit, the higher the balance. Therefore, credit card companies regularly review customer limits and amend them, usually upward, although if someone's performance has deteriorated, they may move the limit downward.

Many years ago in the U.K., the card companies reviewed the limits and then offered them to the cardholders and had to wait for them to reply stating that they wished the higher limit. Nowadays, the higher limits are effected without any customer interface. While discussing this, one might ask why a higher limit produces a higher balance. There are several reasons but one key factor to consider is that good cardholders do not want to be embarrassed when using their card. Therefore, they will use a card up to a margin below the limit. Once the balance reaches that point, they may start to use another card. If your card is the one they prefer, i.e., the one that is used first, the higher the limit, the longer it is before a second card is called into play. Thus in this way, a higher limit will not necessarily generate a higher expenditure from the cardholder, just concentrate that expenditure on one card.

10.4 Fraud prevention

In many ways, using scoring for fraud prevention is the same as any other use. We have experience of past cases and a binary outcome—genuine or fraud. The key difference is that good fraudsters will make their applications look very genuine. Therefore, some scoring developments for fraud prevention have not proved worthwhile because they are unable to differentiate between genuine applications and fraudulent applications. On the other hand, if we use scoring as a fraud check in addition to using a different scoring model as a credit risk check, any improvement, i.e., any success in identifying fraudulent applications, will add value. However, the value of this additional check relies on it not presenting too many false-positive cases.

Because of the specialized nature of fraud prevention, two other approaches may offer some help. These are neural networks and cross-matching. Neural networks were discussed in section 5.4, and an interesting application appears in Leonard (1993a, 1993b).

Cross-matching of applications takes a completely different approach and does not use statistical models. It works on the premise that once someone has been successful in perpetrating a fraud, they will attempt to repeat their success with another lender. Therefore, some lenders have begun to send details of applications into a central data bank, where some matching algorithms operate to identify common features. Many matching rules will be applied and it is acknowledged that many false-positive cases will be identified. However, this approach has also been useful in identifying common addresses and telephone numbers.

For example, an applicant uses 14 Main Street as their home address on a personal loan application but uses 14 Main Street as their lawyer's address on a mortgage application and 14 Main Street as their employer's address on a credit card application. Even if these applications were made to three different lenders, only the first one would stand any chance of not being identified. Now it is possible, especially if the applicant is self-employed, that their home and work address and telephone numbers are the same. Also, if they are a lawyer, it is possible at they have named one of their associates as the lawyer for the transaction, and so the work and lawyer's address and telephone number are the same.

An applicant needs to use a genuine telephone number on an application in case the lender telephones as part of the application process. Also, in many cases, and even more so in the U.K. with checks to prevent money laundering, the telephone number will be checked against some directory. Therefore, if a fraudster has set up one or even a few dummy telephone numbers, these will be used repeatedly, either as home or business telephone numbers, and again cross-matching will identify these cases.

Scoring and other techniques for fraud prevention are continuing to be developed. The need for ongoing development arises because they have not yet reached a stage of offering significant chances of success. Also—and perhaps this is the reason for the previous comment—applicants attempting to commit fraud are changing their techniques and ways and, to some extent, are getting cleverer. However, given the annual amount lost on fraud, this is an area that will continue to attract attention.

10.5 Mortgage scoring

Scoring for mortgages has the same binary outcomes of performing and nonperforming. However, because it is for a mortgage, it may be quite different from scoring for a personal loan or a credit card in a number of ways, which we touch on in this section. First, there is some security or collateral—the property being purchased. Second, we may need to consider more carefully the interest rate schedule. Third, there is the eventual repayment to consider. Also, we need to consider different costs and processes—lawyers to take a legal charge over the property, valuations of the property, etc.

On a more positive note, scoring for mortgages may be simpler in some ways. Because there is a property secured that can be repossessed should the worst happen, we may be able to relax some of our concerns. We may be able to be less concerned about how the borrower is to repay the mortgage or whether the borrower will try to repay the mortgage, either on an ongoing basis or at the end of the term.

From as early as Chapter 1, we looked at what we are often trying to assess in application scoring. One aspect is the stability of the applicant, and one of the key factors or variables that might be used for this is the applicant's time at their present address. In mortgage borrowing, if a house purchase is involved—rather than a remortgage, where the borrowing moves from one lender to another while the borrower remains owning the same property—then we know that there is some instability involved as the borrower is about to move house. Therefore, if we wish to assess stability, we may need to do so using different means.

Another key factor in mortgages is the interest rate schedule and the income. Some mortgages are variable-rate mortgages. Here the rate varies and is set by the lender. The rate will be changed by the lender typically in response to a change in the interest rate set by the central bank. As a central bank lowers its rates, these reductions are, for the most part, passed on to the customer. Similarly, as rates rise, the mortgage rate will rise.

Some mortgages are fixed-rate mortgages. With these mortgages, the lender buys in funds from the capital market and lends it out to borrowers at a margin for a fixed term at a fixed rate. If the borrower continues with the mortgage until the end of its duration, the lender can assess the interest margin and income and the costs incurred at the application stage and at the end of the mortgage. However, if the customer chooses to repay the mortgage at the end of the period of the fixed rate, or perhaps even earlier, the lender may not have enough income to cover the costs. To reduce the effect of this, many mortgage lenders impose restrictions on mortgage redemptions. For example, they may impose penalties if the mortgage is redeemed not only within the period of the fixed rate but also within a period after, known as the tie-in period. During the tie-in period, the mortgage interest rate applying to the mortgage will revert to some standard variable rate that typically will produce a wider margin for the lender. If the mortgage is redeemed during the period of the fixed rate, a penalty may apply, which is partly to recompense the lender for the funds that they have bought for the fixed-rate period and which they may now not be able to lend. Some approaches have been developed to address this, and in section 12.7 we look at survival analysis to model early repayment patterns.

There are other variations of mortgage interest rate schedules such as mortgages with capped rates, where the rate is variable but is guaranteed not to rise above a specified maximum. From a scoring point of view, the interest rate schedule affects our ability to decide if a mortgage is good or bad. This is partly because the profitability equation is altered dramatically. However, it is also affected by the fluidity of the market for mortgages and remortgages. In a fluid market, many mortgages do not reach a mature-enough point for us to decide whether they are good or bad.

While considering the profitability of a mortgage, it should also be realized that the asset and the repossession of the property in the event of default puts a notional, and perhaps actual, limit on the downside risk. Of course, the lender may also consider what to do at the end of the term of the mortgage. Specifically, should the lender insist on repayment or allow the customer to continue to make their regular monthly payments? The costs and the legal and valuation processes involved also affect the profitability of a mortgage. The proposition will need to consider who meets these costs and how effectively these processes have been carried out. Another factor that may have a bearing on the profitability of a mortgage portfolio is the use of mortgage-backed securities; these are discussed in section 14.6.

10.6 Small business scoring

If we can use scoring to assess lending propositions from consumers, then many people believe and almost as many have used scoring to assess lending propositions from small businesses. The key features are the same. We have a large volume of seemingly similar transactions. We can examine the information we had available to us at the point we had to reach a decision. We can review, with hindsight, which pieces of information would have allowed us to differentiate between loans performing well and those performing badly. Eisenbeis (1996) not only supported these views but also reviewed some of the modeling methodologies in use. However, there are many differences that should be recognized and considered.

Some of this information is subject to interpretation. For example, if a business produces audited financial accounts, these are produced for the specific purposes of taxation and statutory requirement at a given point and are out of date by the time an auditor approves them. In a very small business, it is up to the owners how much cash they withdraw from the business and how much the business retains. Also, whether the business makes an accounting profit or loss in a given year is not particularly indicative of the medium-term strength of the business and of its ability to generate cash and to repay any loan. The business may, in a normal course of events, change its strategy and its mix of operations. This can happen for even the smallest business. This may have an affect on our view of the likelihood of the business to succeed or to be able to support the lending being considered.

One part of the assessment of a lending proposition from a small business is an assessment of the owners. However, this may be a matter of interpretation. The person with the largest share in the business may not actually be the person running the business. (As we move up the corporate scale, the people who run large businesses are not the people who own them. These businesses are mostly owned by investment and pension funds.) With a small business, we need to define who actually affects the success of the business or is affected by the success of the business, because it is their stability and propensity to repay that we should try to assess.

In looking at a small business proposition, there is a greater need to consider the business environment. For example, a highly profitable and well-run business making circuit boards, say, may run into trouble if a major customer who is a PC manufacturer is hit by falling sales or adverse currency movements. However, once we consider these and many other factors, the conclusion is still that, with a lot of care and proper definition of subpopulations that are genuinely similar, it is possible to use scoring for small business lending.

10.7 Risk-based pricing

In risk-based pricing, we move the focus away from risk and closer to profitability. Risk-based pricing—or differential pricing, as it is sometimes called—is where we adjust the price or interest rate we offer to the customer to reflect our view of their risk or potential profit to us. While the binary outcomes continue to be whether the loan will be repaid, we now segment our population into different levels of expectation of performance or profitability and apply a different price to each. (Of course, it is possible that we adjust items other than the price. For example, we may reduce our requirement for security or collateral. In effect, this makes the proposition riskier but may be an acceptable business alternative to lowering the price for a low-risk proposition. We can also adjust requirements for documentary proofs, e.g., to substantiate income. With mortgage lending, we can also adjust our requirement for a property valuation to require either a more stringent or a less stringent valuation or none at all.)

In principle, we calculate the risk attributed to a proposition or customer and use this to assess the income necessary to make a profit or to achieve a target return. We then offer the customer the product or service at the interest rate to produce the required income. In effect, it means that very good customers will get to borrow money at lower rates and riskier customers will be charged a higher rate.

This is contrary to the standard way of doing retail business in western society. When we shop, the price of an item is displayed and either we choose to purchase at that price or we choose not to. From the lender's perspective, risk-based pricing usually increases the ability to sell loans and also means that the income more closely matches the expenses, at least in terms of the risk and the bad debt costs. However, there are also some potential challenges with this:

- Adverse selection. In scoring, analysis is carried out on past experience. Therefore, a consumer can be made different offers by different lenders. With risk pricing, this is even more widespread. The analysis to support risk pricing will include some assumptions about the take-up rate for marginal customers. However, we need to consider the fact that some or most of those marginal customers take out a loan at a higher rate because rates offered by others lenders are even higher. For example, if our standard loan is at an annual interest rate of, say, 12% and we offer marginal customers a loan at 14%, those who take up the offer are not a random sample of those offered. Rather, they will be on average higher risk since the 14% offer will be lower than other offers they have received. To accommodate this, we should increase the rate offered, but this only aggravates the situation and we could end up chasing our own tail. Understanding the marketplace and the competitive environment is just as important when pricing to risk than with a standard offer.

- Good customers. At the low-risk end, we may be making offers at interest rates lower than the customer would deem acceptable. Therefore, for low-risk propositions, we may be undercutting ourselves and throwing away income. Again, an understanding of the marketplace and of the competitive environment is important.

- One needs to be able to explain risk-based pricing to customer-facing staff and to applicants. The main question that usually requires an answer is, "How did you calculate the rate you are now offering?" This is especially the case when the rate being offered is substantially different from an advertised rate.

Models for risk-based pricing are discussed later in section 14.5.

10.8 Credit extension and transaction authorization

We have referred to behavioral and performance scoring at several points. At this point we can explain some of the scenarios where we might use it. When an applicant applies for a credit card, credit or application scoring will be used to make the decision to issue a card, and the credit score may also have some impact on the actual credit limit to be granted. This credit limit is, however, an initial credit limit, and this limit needs to be managed throughout the lifetime of the account.

Once the account as been running for a few months and once some activity has been seen on the account, the customer's performance on the account gives us a much stronger indication of the ongoing likelihood of the customer's failure to operate the account within the terms set out. Even if we were to update the application data, the performance data will still be more powerful. So in what types of situation might we wish to use the score?

As stated earlier, the credit limit needs to be managed. If the limit is too low, the customer will concentrate their expenditure and usage on one or more other credit cards. Of course, if the limit is too high, then the customer may be tempted or encouraged to achieve a level of activity that they cannot afford to maintain. The skill of the area of credit limit management is to walk this tightrope between no income and high bad debts. Therefore, credit card companies will regularly review the credit limits of their cardholders and increase those that appear to be able to maintain a slightly higher level of expenditure.

There also will usually be an ongoing assessment that generates a shadow credit limit. This is not known to the customer but is in the background. It can be used if the customer either goes overlimit or is referred for a transaction that would take them overlimit.

Now, a credit card product is merely one example of a revolving credit type of product. In the U.K., there are other types of revolving credit. One is the overdraft, i.e., the account with a checkbook and a credit (overdraft) limit. In a similar way, such accounts will have a shadow limit so that when a customer requests an overdraft limit or a higher limit or when a check is presented that takes them over their current limit, the system can decide what action to take. Another revolving credit account is a budget account, where the customer receives a credit limit equivalent to 24 or 36 times an agreed monthly payment. Mail-order facilities will also involve some credit limit so that as a customer or an agent builds up their history, they are also able to build up the value of goods they can deal with on a credit basis.

We may also decrease credit limits. As the customer is not aware of their shadow limit, this can fall as well as rise. For example, should a credit card customer miss a payment, this will tend to reduce their score and therefore reduce their shadow limit. If adverse behavior continues, the credit card company may reduce the actual customer's credit limit and inform the customer accordingly. This can also happen with an overdraft. This may be effective in reducing the lender's exposure to the customer. It may also generate a change in customer behavior that, in many cases, is what is desired. Of course, in some cases it may be too late and is merely reducing or even only capping the eventual loss.

We may also use the behavioral score to allocate strategies for account handling. Once behavioral scores are available, we can move into the area of adaptive control. This involves managing accounts differently according to a number of key factors and learning from their behavior. Therefore, if a customer misses a payment, we may react differently to this depending on their score, the age of the account, the security held, and the past payment behavior. Many other factors can be used, but the principal aim is to manage accounts both effectively and efficiently.

For example, based on score, we may be able to predict that the customer who misses a payment after making the previous 12 payments is highly likely to catch up with their payments in the next month. Therefore, one possible course of action is to do nothing and wait for them to resolve the situation. Should the probability of their recovery be slightly less, another option is to send them a letter, and we can utilize a whole range of letters within this general strategy, from the soft to the hard reminder. An even lower probability may generate a collections telephone call. Of course, at the other extreme, should it be highly unlikely that the customer will recover, it may be that no action should be taken as it would be a waste of time and money. This last strategy, however, is rarely implemented.

With a revolving credit facility, the decision process may differ depending on how the credit application arose. For example, we may have different processes for credit limit increases requested by the customer and those generated automatically. We may have another process for credit limit increases generated by a customer writing a check or attempting a credit card transaction that takes them over their credit limit. Each different process could simply be a different allocation of score (and other variables) to limit. Alternatively, it could be a different scorecard if a large enough data set is available and supports such a difference.

10.9 Debt recovery: Collections scoring and litigation scoring

In the previous section, we touched on using the score in deciding what collections activity to carry out. This is part of behavioral scoring and, as mentioned above, the usual way to implement this is by means of adaptive control, where the lender implements various strategies and through them tries to learn what is most effective and efficient.

At a later stage in the process, a scoring approach has occasionally been used to decide what action to take. The available options include writing off the account, pursuing it, taking it through the legal process, sending it to a debt collection agency, and selling it to a third party. The outcomes in which we are interested are the likelihood of recovering the money. Of course, these are not simple binary outcomes. For example, one course of action may have an 80% chance of recovering all of the money and a 20% chance of losing a further 25%, say, to cover costs. Another course of action may have no chance of recovering all the money but is almost certain to recover 75% in each case. Both courses of action expect to lose 25% of the balance, and so there may be other issues involved in deciding which is the better course of action, such as resource availability or costs.

10.10 Provisioning for bad debt

Bad debt provisioning is not really a feature of lending. Rather, it is a feature of our accounting systems and of the control mechanisms laid down by the central banks and other regulatory bodies. It arises because it is deemed important that a lender has sufficient funds set aside to cover the bad debts that are likely to occur. However, the scale of these losses may be larger than expected, or the timing of them may cause them to occur sooner than expected. Funds set aside should be able to cover these as well.

When we ultimately finish with the debt collection and litigation process, we may have a balance remaining that we write off our books. A provision is, in essence, a prediction of a future write-off. Therefore, once the provision has been made or set aside, we do not need to concern ourselves with the timing of the eventual write-off since funds have already been set aside to meet this eventual loss. (This assumes that we have made a sufficiently large provision.)

In the U.K. environment, we refer to general provisions and specific provisions. At a simple level, specific provisions are those that are set aside for specific cases of bad debts that have already been recognized. For example, most lenders would set aside a specific provision once an account becomes three payments in arrears, and many will set aside provisions earlier than this.

A general provision is a provision set aside for cases of bad debts that we have not yet recognized but which we believe are in the current lending book. Suppose that we lend a new tranche of business with a score distribution such that we expect 2% of them to go bad. Suppose further that we expect 30% of the balances on the bad accounts to be recovered. We could set aside a general provision of 1.4% of the value of this new tranche of business as a general provision. Since we do not know yet exactly which accounts will fall into the 2%, we cannot set a specific provision. As these bad cases become evident, we can reduce the general provision and increase the specific provision accordingly.

How much we set as a specific provision and when we set this aside may be based on scores, especially behavioral scores, where we can distinguish between cases that are two down or three down but are more or less likely to recover.

10.11 Credit reference export guarantees

In the U.K., there is a branch of the government called the Exports Credit Guarantee Department (ECGD). It is not the function of this department to lend money. Its function is to support export trade, and one of the ways it does this is by guaranteeing an exporter that it will be paid. For them to understand their liability under such a guarantee, they need to assess the risk that the ECGD are running, and they do this with a form of scoring. Part of the scoring algorithm will be a factor related to the country into which the goods are being imported. In principle, however, the scoring methodology used is similar to those already discussed in this chapter. The ECGD's decision is in the form of a risk price, i.e., a cost for providing the guarantee.

Chapter 11

Applications of Scoring in Other Areas

11.1 Introduction

In the previous two chapters, we dealt with how scoring is used to make lending decisions. In this chapter, we look at some other areas where a scoring-type approach has been used but that are either at the periphery of lending or fall completely beyond the boundaries of a lending environment.

The first two topics have some relation to lending. We first deal with how we might use scoring in direct marketing. Then we look at profit scoring. After that, we move further away from the lending environment and touch on auditing in a variety of guises and then onto the parole process. The common features are that in almost all of these cases, there are two possible actions—analogous to whether to lend money—as well as data available on past cases—both at the point when a decision has to be made and ultimate performance.

11.2 Direct marketing

The previous chapter addressed direct marketing under the heading of prescreening. However, it is worth revisiting the topic to cement some of the ideas and to expand on the topic.

In many environments—and the lending environment is only one example—there are, in principle, two possible actions. Either we make someone an offer or we do not. (In practice, of course, there are many possible actions as we can choose from a range of offers or products, and a score could determine which offer to make.) A score might predict who is likely to respond and who is not. A score might also predict who is likely to be loyal and who is likely to move to a better offer from another organization as soon as one appears. A score might also predict who is likely to trade up to a better product. It should be obvious that these three examples would use different scores, i.e., different models.

We might also be able to build a model and produce a score to predict which marketing channel would be better to use. For example, we might be able to segment our target population into several segments according to whether we should market them by mail—cold mailing—or, for existing customers, statement inserts, or telephone, or even e-mail. (One useful model for such a development would be a multinomial logit model.) Clearly, the score becomes part of a larger business decision because it is much more feasible to mail 250,000 prospective customers than to make telephone calls to them, especially when one considers the necessity for repeat attempts to make contact with people unavailable on the first attempt.

If we are scoring prospects from our own list of customers or from a mailing or membership list, then there are clear restrictions. For example, we are limited to the information available. This may seem an obvious point, but if you haven't actually received a loan application yet, you cannot rely on purpose or term of loan as a scorecard characteristic (although you could make only a specific offer, e.g., for a 48-month loan for a car). Also, there are strict rules and regulations about what credit reference information is available and how it might be used. If the decision is whether to target someone for an offer, if it is decided not to select someone, they will not even be aware that they have been considered.

Thus far in this section, nothing is specific to lending. Indeed, models and a scoring methodology can work, in general, where we are trying to predict someone's propensity to do something. It may be that we are trying to offer them a visit to a timeshare complex and wish to predict, among a list of prospects, who is most likely to attend and to purchase. Similarly, we may be interested in offering people a chance to test drive a new model of car. Here we may be interested in those people who will make a purchase, whether they take up the opportunity offered. The clear message here is that there is no reason why a scoring methodology will not work well.

However, there are two warnings. First, with many uses of scoring in direct marketing, it is often the case that a scorecard trying to predict response will work in the opposite direction from one trying to predict sales or even performance. For example, the people who are most likely to respond to a loan offer are often those hungriest for credit and are least likely to repay satisfactorily. For example, the people most likely to respond to a cheap weekend at a timeshare resort as part of a sales promotion may be the people who are least likely to buy or least likely to be able to afford to make a purchase.

The second warning is that we should also consider the benefits of preselection. In a lending environment, if the target population is of fairly good quality, it may be cheaper to mail to all of them and to live with the few bad cases that apply and are approved than to prescreen all of them to remove the few bad cases. This is especially so once we realize that many of the bad cases would not apply anyway. In a sales environment, it might be better to make a blanket offer of a test drive rather than deselect some people, removing them from the prospect list. This might be the case where the cost of a test drive is minimal.

Whether we use scoring in lending or in direct marketing of other goods and services, the usual objective is to maximize some measure of profit. Often, the stated objective for the marketing department is response rate or cost per response. However, in terms of running a business overall, profit is the key measure.

11.3 Profit scoring

In most areas of scoring, we consider two outcomes—good and bad. Sometimes these are polarized by removing from consideration those cases in between—indeterminates. Often, good and bad refer to some performance definition. However, in most lending organizations, the ultimate objective is profit. Therefore, while we generally use performance definitions to classify accounts as good or bad, some scoring developments or implementations classify accounts as being good or bad depending on the profit they make.

In principle, this is simple. An account that is expected to make a profit is good and one that is expected to lose money is bad. In this section, we discuss two different aspects of this simple statement. The first is what to include in the calculation of profit or loss and the second is what profit measure to use. Both of these will affect scorecard development and implementation, although they rely on the introduction of some basic concepts in finance. (One or two of the finance concepts were discussed in section 8.11, where we discussed using profit as a means of deciding where to set the scorecard cutoff.)

Consider for a moment a simple example. An applicant applies for a 36-month loan with a fixed payment each month. If the customer makes all their repayments, we should be able to calculate our administrative and funding costs and subtract them from the interest income, and the resultant figure is profit.

As it stands, this is a simple and simplistic scenario. Let us consider some of the sources of complexity. The profit will clearly be affected by how we fund the loan. Indeed, if we fund it on a rolling basis rather than buying three-year money when we grant the loan, then at the beginning of the loan, we do not know our costs exactly. On the other hand, if we fund the loan on a rolling basis but have a variable interest rate applied to the loan, we may be able to assume a fixed margin.

In assessing profit and loss, we also need to consider our marketing acquisition costs. It is possible to debate this issue at length, but there are at least two different points of view that can be held. The first is that we should assume that acquisition costs are fixed overheads and should not be included in any consideration of profit or loss of individual cases. This might be valid in cases where we have a fixed marketing cost and we can apply a portion of this to a new loan retrospectively only when we can tell how many loans were generated by the marketing expenditure. The second is that we should factor an allocation of the acquisition cost of each loan into the calculation.

Now, if the customer repays the loan early, we may lose some of our interest income. On the other hand, we have the money returned to us and can lend it to someone else. We may need to consider timing of early payments.

Thus far, this is still relatively simple. However, especially if this is a new customer to our organization, we may also have to consider the cross-selling opportunities. While the customer may take out a small loan that barely breaks even, if at all, even if they make all their payments on time, we could include in our assessment of profit and loss the opportunities to sell them other products—savings account, insurance, etc. Clearly, once we have an existing customer and a relationship with them, there may be a greater propensity for them to buy other products. In any case, there is greater income to be set against the acquisition cost.

The debate about acquisition costs can carry over into a debate about other overhead-type costs. How much of our branch costs should we include in our calculation of profit and loss? For example, if the customer had not taken out the loan, the branch costs would still be there. On the other hand, if we used that argument across the whole spectrum of products and services, we would never have any money set aside to pay for the branch network. Moving to telephone-based processes or to Internet-based processes merely changes the size of the overhead. It does not change the argument, although in the case of the Internet, one could argue that there is minimal manual cost. In other words, once we have set up and developed a site and advertised its existence and address, handling 1000 inquiries may cost the same as handling 1 million inquiries. This leads to the next issue, as we also need to consider how we account for and recover our system development costs.

Another finance issue is that we need to convert everything into the value of money today, typically using an NPV calculation, as outlined in Chapter 3. In calculating an NPV, we need to know not only the probability of a loan not being repaid but also the timing of the likely default. We also need to know the timing of an early settlement and the timing and size of any other income streams arising from cross-selling.

A further finance issue is the fact that we may be able to measure some costs and factors very accurately, while some will be measured very roughly. In some cases, there will be no measurement, just an allocation of costs. These and other related issues are discussed by Hopper and Lewis (1991).

Thus we have seen that a consideration of the elements in our profit and loss calculation is not simple. However, we also need to decide on a profit threshold, and this introduces the issue of what profit measure to use. If we use a performance definition of good and bad—e.g., bad = ever three payments in arrears—then we might set the cutoff at the point where our marginal cases break even (in some sense). We could do the same when we consider profit once we decide what constitutes breaking even. Once we have definitions of profit and breaking even, we can then adopt a number of acceptance strategies, including the following:

- Accept all loans where the expected income is greater than the expected cost. (If this ignores overheads, the accountants might say that these cases are making a positive contribution.)

- Accept all loans where the expected income is greater than the expected cost by a fixed amount. An example of this would be to accept loans only if we expect to make at least £100 on them. This margin could be introduced to cover the uncertainty in the calculations.

- Accept all loans where the expected income is greater than the expected cost by a fixed percentage. An example of this would be to accept loans only where we expect income to exceed costs by at least 10% (of the costs). This is getting us a little bit closer to calculating a return.

- Accept all loans where the expected profit is at least a fixed percentage of the amount of the loan. This is a form of return. In other words, we are risking, say, £10,000 and will do so only if we expect to make a profit of at least, say, 20%. We could also have a loan proposal of £30,000 with an expected profit of £5,000 but would reject it as three loans of £10,000, each with an expected profit of £2000, would produce more profit for the same amount exposed.

Now, once again, we need to turn to our finance colleagues to get a calculation of return. Often it is the return on equity or return on capital employed or risk-adjusted return on capital (known in some circles as RAROC). These take into account the fact that when we lend money in aggregate, we do not have all that money in the organization. We have to borrow some of it from investors—shareholders—who require a considerable return for their money to compensate them for their risk.

Once we have a calculation of our return, we still need to resolve how we introduce our overheads, what minimum return we should set, and whether we are looking at cases that reach this required return marginally, i.e., in their own right, or in aggregate. Specifically, we could take on cases using a cutoff that gives the maximum return. On the other hand, we should probably set our cutoff to select those cases that maximize our expected profit, subject to the minimum return being achieved, either in aggregate or in each case.

In considering the overheads, we need to allocate them but may run into some difficulties until we have set the cutoffs. For example, with a cutoff of, say, 240, we may have 7,000 loans accepted per month, so we can spread our fixed costs across these 7,000 loans or the 84,000 loans per annum. In this case, our fixed costs might include the cost of our building and the salaries of senior management and our systems staff, as these are the same whether we have 7,000 loans per month or 8,000 per month. However, if we have a cutoff of, say, 225, we may end up taking on 8,000 loans per month and can spread the same fixed costs over more loans, although the quality of the loans taken on will decrease, in terms of the average profit per loan. One way to get out of this potential trap is to take a pragmatic business or accounting decision. Another way is to try to iterate to an optimal position, i.e., set a cutoff, calculate the marginal profit, identify declined cases (i.e., below cutoff) that would be profitable, and change the cutoff to include them, and so on.

Once we go down the route of scoring to assess profit, we then are faced with the issue of what to do with applicants whose performance is predicted to be good in terms of having low arrears but whose transactions may not be sufficient to generate a profit. We could consider several actions, including

- declining them, because they are, in some sense, too good;
- try to convert them to another product, better suited to their needs, that will generate a profit;
- accept them and accept the loss as either a cost of being in business or as the cost of a marketing error;
- rather than risk price them, profit price them; in other words, offer them the product at a higher price that will generate a profit.

None of these is completely satisfactory. Often the approach taken is a mixture of the second and the third options, although some companies, especially those trying to break into a market, may be better placed to adopt the first option, although even for them the third option looks likely. Whatever stance an organization takes, improved prescreening and targeting may help to reduce the scale of the problem.

Of course, some lending organizations do not adhere as strictly to the profit ethos as others. Cooperative societies, and the Coop Bank in the U.K., may be prepared to take a longer-term view of profitability. Similarly, credit unions may be more willing to lend to applicants who represent marginal profit, or even likely loss, as they may wish to achieve a high acceptance rate and a high penetration among their members. With these organizations, there may be some social or political capital to be made in lieu of financial profit.

In summary, from the viewpoint of the scoring methodology, using a profit-loss definition of good and bad rather than a repayment performance definition should present no technical difficulties. The main difficulties arise in the calculation of profit and loss, in handling applicants who are creditworthy but unlikely to generate sufficient profit, and in the (probably greater) number of assumptions that are required for the profitability assessment. In Chapter 14, we begin to look at modeling approaches that might allow us to improve our profit scoring.

11.4 Tax inspection

For many years in the U.S. and over the past few years in the U.K., there has been in place what we might refer to as an honor system of tax assessments for consumers. In essence, each consumer or each taxpayer is responsible for completing a tax return and calculating the total amount of tax they owe for the tax year in question. (In the U.K., if one submits by September 30, Inland Revenue will calculate the tax due.) If one is due to pay more tax than has been withheld through salary payments, etc., then one either encloses a check with the return or makes the payment by a specified date. If more tax has been withheld than one owed, one is due a refund, and one may be able to opt to receive this either as a lump sum from the tax authorities or as a reduction in the amount withheld in the current and subsequent tax years.

Now with such a system, the key to being able to manage the volumes is that not all tax returns are fully audited by the tax authorities. In effect, they are likely to check more fully the returns of those people who have had gross inaccuracies in past returns, especially if these inaccuracies were in the favor of the individual. For the others, they check less fully and may indeed sample among them to decide which ones to audit. Some authorities use a form of scoring to decide which ones to audit.

They can use past experience and past cases to identify those characteristics of the returns that are associated with cases where the tax reported has been a gross underestimate. In scoring terminology, they have a large set of previous cases on which to build a model. The allocation of good and bad can be fairly arbitrary and can generally be changed around, but let us consider a bad case to be one where an audit reveals that the tax due is greater than that on the return. A good case is one where the tax due is either correct or is overstated. (This book is not the appropriate place, but one could enter into a debate concerning the responsibility of the tax authorities, as a public sector body, to identify cases where the individual pays too much tax.)

This could be considered to be akin to application credit scoring, to behavioral scoring, and to fraud scoring. In some cases, the difference between the tax assessment on the return and the tax properly due is caused by errors, misunderstandings of the taxation rules, and allowances, i.e., honest mistakes. In some other cases, there is planned deception to evade large portions of tax.

As we discussed above, the definition of good and bad could depend on whether there is an underpayment of tax. On the other hand, we could adopt a different definition that defines a bad to be a case where the expected underpayment is at least the expected cost of further investigation. If we compare this to scoring in the lending environment, we find that the scoring methodology would be similar if not identical and the monitoring and tracking reports could also be utilized in the same way.

In fact, almost any decision on whether to carry out an audit and the depth of that audit could be supported by scoring methodology.

11.5 Payment of fines and maintenance payments

In the U.K., one of the tasks of the Child Services Agency is to supervise the maintenance of payments by nonresident parents (NRPs) to the parent charged with care for the child or children, often after the breakdown of a marriage. There has been some investigation into whether scoring models will help. The investigations have looked at whether it is possible to build robust models predicting NRP noncompliance with court orders, etc., and, based on these models, whether it is possible to build predictive application scoring–type models.

The first type of model would be a behavioral type of model and uses characteristics about the NRP such as gender, age, employment, and whether they were in debt when the court order was granted. It also includes whether the money comes direct from the employer, whether they have a mortgage, their income, and the number of nights each week they spend with the child. Other factors are included, some of which relate to the parent who has custody of the child, including whether they are receiving state benefits.

The application-type scorecard development is intended to be able to predict, at the outset, from the characteristics known to the agency, how much resource will be required to ensure that the NRP is compliant with their maintenance schedule. If we contrast this with a lending situation, here a bad case is one that will require considerable amounts of time to make the NRP compliant. It is akin to assuming that all moneys lent will be recovered in full but that some will take more collections effort than others to achieve full recovery.

Thus far in this section, we have considered maintenance payments. However, scoring can be used in a very similar way when we consider the payment of fines. We need to get payment of a fine from an individual, and there are several methods of doing so. At one extreme, we hold the individual in a court building until the fine is paid in cash. At the other extreme, we agree to their making payments of the fine over several years. It is possible to build a scorecard that can be used to determine the appropriate level of risk, bearing in mind the objectives of maximizing receipts of fines, minimizing the effort in chasing fines, and dealing fairly with members of the public. (With regard to the last point, it might not be considered fair to hold an elderly person in the court or an adjacent building for several hours until a friend or relative arrives to pay a small fine.)

11.6 Parole

While use of a scoring approach for parole has not become widespread, scoring has been used in reaching the decision on whether to allow a prisoner to be released on parole. Here we can consider two variants of parole. First, the prisoner is let out of prison for a short period, perhaps 48 or 72 hours, as part of the Christmas period, or as a step in rehabilitation, or on compassionate grounds such as a family funeral. Second, the choice is whether to release the prisoner from their sentence early, perhaps for good behavior, either freeing them completely or in terms of a conditional release. Common conditions are continuing good behavior, no reoffense, or regular reporting to a police station.

Once again, the methodology is more or less the same. There is a large set of historical cases, each with the data that were available at the time the decision was made. With the first type of parole, a good case is a prisoner who is let out on parole and returns as agreed. A bad case is one who does not return as agreed. With the second type of parole, a good case is one where the prisoner is released and does not reoffend, either ever or for a specific period. Bad cases are where there is some reoffense. In both cases, rejects are those who are not granted parole.

Similar to many lending environments, one of the main objectives is to minimize the percentage of bad cases, as a released convict may be bad for public safety as well as for public relations.

It is possible to build a scorecard using as predictive data such pieces of information as the prisoner's sentence and crime, how long they have to go until the end, and how long they have already served. Other pieces of information include the proximity of their home from the prison, the frequency of visits from their family, their previous parole history, their age, their recent behavior, and an assessment of their current mental health.

As was stated earlier, standard approaches to monitoring and tracking apply to determine how well the scorecard is working. Perhaps, the only major difference here might be that if there is any doubt about a prisoner, the prisoner be refused and, also, if there is any doubt about the parole process, parole can be withdrawn. However, the prison system continues to operate. This contrasts with the lending environment, where a lender cannot simply decide not to lend any more money because it is too risky. To do so for more than a short period would mean that the lender is withdrawing from this line of business.

11.7 Miscellany

If we can use scoring to assess the probability of recovery of a debt, we can also use it to assess the probability of recovery from a medical operation. This assessment might be used to decide whether to treat a patient or how to treat a patient.

Government security agencies may use a scoring methodology to assess the risk represented by a potential recruit. This could be extended to a general employment position.

To use scoring to assess the risk associated with recruiting an individual as an employee, we clearly need to define the risks with which we are dealing. A similar approach could also be used in selection for university acceptance.

Chapter 12

New Ways to Build Scorecards

12.1 Introduction

Part of the excitement of being involved in credit scoring is the constant search for new ways to improve scorecard development and for new areas where scorecard building can be applied. The former is motivated by the realization that with the volume of consumer borrowing, even a 0.25% drop in the bad debt rate can save millions. The latter is motivated by the success of credit and behavioral scoring and the recognition that using data on customers and their transactional behavior to identify target groups is generic to many areas of retail business. It is the raison d'être for much of the current investment in data warehouses and data-mining tools.

In Chapters 4, 5, and 6, we discussed not only the industry standard methods for developing scorecards, such as logistic regression, classification trees, and linear programming, but also those that have been piloted in the last decade, such as neural networks and nearest-neighbor methods. In this chapter, we discuss approaches that are still in the research stage. They involve ways to combine existing credit-scoring methodologies and to use approaches and models that have proved successful in other areas of statistical and operational research modeling.

In section 12.2, some of the work on generic scorecards and building scorecards on small samples is discussed. If the sample available to a lender is not sufficient to build a robust scorecard, what can one do? Can one put together samples of different populations to construct a scorecard that is robust on any one of the populations? If not, are there other ways to deal with small samples? Sections 12.3 and 12.4 look at the problems of combining classifiers. Section 12.3 investigates simulation work on what happens if there are two different scorecards that a customer has to pass to get a loan. This describes the situation in mortgage scoring, when consumers have to pass a loan-to-value condition and be credit scored. Section 12.4 describes methods that have been suggested to combine all types of credit scoring classifiers, not just scorecards.

An alternative approach to combining credit-scoring classifiers is to use the statistical procedures to estimate intermediate objectives like outstanding balance and amount in excess of overdraft limit and then to use these intermediates to estimate likelihood of defaulting. Section 12.5 discusses this approach. This idea of estimating a number of interdependent quantities rather than just default risk leads one to consider the use of Bayesian networks and statistical graphical techniques in section 12.6. These calculate the strength of the relationships between different variables. Some of these can be the independent variables, like the application-form questions, which are in the usual credit scorecards. Others are variables that one is seeking to estimate, both the final outcomes, like default probability or profitability, and the intermediate outcomes, like credit balance. Finally, there are exogenous variables that describe the current economic conditions, like interest rate and unemployment rate. Bayesian networks give a picture of how these variables are connected, which in turn suggests the variables (and the way they should be interrelated) that are needed to produce a successful credit prediction system.

These approaches not only increase the number of outcome variables but could also allow them to be continuous rather than the standard binary outcome of whether the customer defaults in the next x months. One obviously useful extension is to estimate the distribution of when the customer will default—so instead of estimating if the customer will default, one estimates when they will default. Section 12.7 discusses how survival analysis techniques used in reliability modeling can be translated to deal with this extension.

Not all these ideas will prove to be useful in practical credit scoring, but they give an indication of the way the subject is moving. Analysts are always seeking ways to extend the scope and improve the accuracy of credit scoring.

12.2 Generic scorecards and small sample modeling

It is always assumed that for scoring to be successful, one needs a homogeneous application population. If the population is not homogeneous, then in practice one often segments the population into more homogeneous subpopulations and builds a separate scorecard on each subpopulation. However, at times one might deliberately put together different populations to build a generic scorecard. This might be because the numbers in the individual populations are too small to build a sensible scorecard or when for legal or economic reasons the same scorecard is needed for all the populations. In this section, we describe one such experiment.

U.S. credit unions lend only to their members, who usually have common employment or live in a common locality. They tend to have memberships in the hundreds or low thousands—often not enough on which to build a scorecard. Overstreet, Bradley, and Kemp (1992) looked at what would happen if a generic scorecard was built using the data from 10 such credit unions from the southeastern U.S. They took the scorecards that had been built for the five largest unions and averaged the scores for each attribute in the five scorecards to define a generic scorecard. Their results showed that although the generic scorecard was not as good a discriminator on each population as the scorecard built on that population, in all but one case the results were not too disappointing. At the same acceptance rate, while the individual scorecard identified 70% of the bads, the generic scorecard identified 62% of the bads.

Another way to build a generic scorecard is to pool the populations and use the pooled sample to build the generic scorecard. In a subsequent analysis, Overstreet and Bradley (1996) employed this approach with similar results. At the level where 10% of the goods were misclassified, the generic scorecard correctly identified 53% of the bads, while the individual scorecards correctly classified an average of 58% of the bads.

If the sample population is small and there are no other similar populations to augment it with, other methods may have to be used. The cross-validation, jackknifing, and bootstrapping approaches of section 7.4 allow one to use the whole sample for development rather than leaving some of it out for validation. In some cases, the sample population is still too small for one to have confidence in the results. In that case, one has to resort to a series of ad hoc approaches to combine all the knowledge one has about the situation and to compensate for putting too much emphasis on statistical results derived from very little data.

One example of this was described by Lucas and Powell (1997). They had a sample of only 120 customers with which to build a scorecard for credit cards for the self-employed. These were part of a larger sample who had been given a behavior score and so they took the definition of good to be a behavior score above the median of the larger sample and bad to be a behavior score below the median. There were 84 characteristics available on each applicant in the sample, but using expert opinion, the discrimination tests of section 8.7, and correlations, these were cut to 13 characteristics. For each attribute of each of these characteristics, the sample good rate $\frac{g_i}{g_i+b_i}$, where g_i and b_i, the numbers of goods and bads with attribute i, were adjusted in two ways—one to limit the importance of the statistical results obtained from such a small sample and one to include the subjective opinion of experts on credit lending to the self-employed. First, the estimates were shrunk to allow for the fact that in small data sets, the variation is too high and one wants to shrink the estimate towards the overall mean. The James–Stein estimate (Hoffmann 2000) is one way to do this and is essentially a combination of the original estimate and the overall mean. This estimate of the good rate is

$$\tilde{r} = \lambda \left(\frac{g_i}{g_i + b_i} \right) + (1 - \lambda) \left(\frac{g}{n} \right), \tag{12.1}$$

where g_i, b_i, and n_i are the numbers of goods and bads and the total in the sample with attribute i and g and n are the total of goods in the sample and the total in the sample. The choice of λ is discussed in statistical books (Hoffmann 2000).

Second, an expert on credit lending to the self-employed was used to give a prior estimate of the good rate for each attribute i. His belief was given in the form of a beta distribution with parameters (r, m) (recall section 6.7). This was then adjusted to include $\tilde{r}$, the estimate of the good rate obtained from the sample, so that the posterior belief for the good rate is a beta distribution with parameters $(r + n_i\tilde{r}, m + n_i)$. The score for the attribute is then taken as its weight of evidence with this adjusted good rate, i.e.,

$$\log \left(\frac{r + n_i\tilde{r}}{m + n_i} \cdot \frac{n}{g} \right), \tag{12.2}$$

and the scorecard is taken to be the sum of these scores. Thus this approach used subjective estimates in two different ways and used a shrinkage of the estimate to compensate for the small sample sets. Each of these ideas makes sense in its own right, but there is no rationale for the overall methodology. It reflects the ad-hoc nature of scorecard building with very small samples.

12.3 Combining two scorecards: Sufficiency and screening

There are several examples in consumer lending where more than one scoring system is available to the lender. For example, many lenders will build a scoring system based only on the application-form variables and then another one using credit bureau data. The reason is that the lender has to pay a fee to acquire the latter information and so may hope that the results of the application-only score are sufficiently clear that this expense is not necessary. In mortgage lending, lenders often use a loan-to-income ratio as an initial rule of thumb and then use a mortgage score later in their credit decision process. A number of investigations looked at whether these scoring systems should be combined and how.

Zhu, Beling, and Overstreet (1999a) looked at the problem of an application score S_1 and a credit score S_2 for a data set of 600,000 automobile loan applications. They calculated a combined score $s_c = a_0 + a_1 s_1 + a_2 s_2$ using logistic regression. They wanted to show that whether the two individual scorecards S_1 and S_2 were better than each other and if the combined scorecard was better than both. In Chapter 7, we discussed ways to measure the performance of scorecards, but here we want to introduce a new way to look at the performance of these three scorecards, which originated in measuring probabilistic forecasts (Clemen, Murphy, and Winkler 1995, Zhu, Beling, and Overstreet 2001)

Assume that the credit-scoring system gives a score s and that if necessary a nonlinear transformation has been applied to the score so that the probability of being good is the score s, $P(G|s) = s$. Suppose we have a number of such scorecards $A, B, C, \ldots$ based on the same sample; then we can think of the scores s_A, s_B, s_C as random variables and define $f_A(s_A|G)$, $f_A(s_A|B)$, $f_A(s_A)$ to be, respectively, the conditional probability of a good (bad) having a score s_A under A and the distribution of scores under A.

Definition 12.1. *If A and B are two scorecards, then A is sufficient for B if there exists a stochastic transformation h so that*

$$0 \leq h(s_A|s_B) \quad \textit{for all } s_A, s_B, \qquad \sum_{s_B} h(s_B|s_A) = 1 \quad \textit{for all } s_A, \quad \textit{and}$$
$$\sum_{s_A} h(s_B|s_A) f_A(s_A|G) = f_B(s_B|G) \quad \textit{for all } s_B, \tag{12.3}$$

This means that B can be thought of as A with extra uncertainty added and so is dominated by A in that A will be superior under any of the normal business measures. Sufficiency is useful because, as Zhu, Beling, and Overstreet (1999b) pointed out, there is a relationship between sufficiency and the actual loss rates defined by (7.5). Scorecard S_1 is sufficient for scorecard S_2 if and only if for every expected default cost D, every expected profit L, and every score distribution $f(x)$ in (7.5), the actual loss rate l_{s_1}(Actual) is less than or equal to the actual loss rate l_{s_2}(Actual). However, in many pairs of scorecards neither is sufficient for the other.

The definition of sufficiency in (12.3), although making things clear, is very difficult to check. A much easier condition was given by De Groot and Eriksson (1985), namely, that A is sufficient for B if and only if, when $F_A(s) = \sum_{t \leq s} f_A(t)$, $F_B(s) = \sum_{t \leq s} f_B(t)$,

$$\int_0^p F_A(s)ds - \int_0^p F_B(s)ds \geq 0 \quad \text{for all } 0 \leq p \leq 1.$$

This implies that the distribution f_A has a second-order stochastic dominance over f_B.

When these measurements were applied to the scorecards developed by Zhu, Beling, and Overstreet (2001) (recall that $s_c = a_0 + a_1 s_1 + a_2 s_2$), the results were as in Figure 12.1. What these show is that scorecard 1—the application score—essentially dominates scorecard 2—the pure bureau score, but the combined scorecard is again much better than the application score. These results can be confirmed using the other performance measurements of Chapter 7, and hence we have the not-unexpected result that the combined scorecard is superior to the individual ones. One can show (Zhu, Beling, and Overstreet 1999) that scorecard S_1 is sufficient for scorecard S_2 if and only if the ROC curve of S_1 lies above that of S_2 everywhere. The results in Figure 12.1 do give a warning about sufficiency being used to decide whether to combine scorecards. S_2 is essentially dominated by S_1 and so one might feel that the second scorecard cannot enhance the first one, and yet the combined scorecard is superior to S_1. These are not startling results, but the results give a warning about the use of domination as a way to dismiss scores. Just because one scorecard dominates another does not mean that a combined scorecard may not be better still.

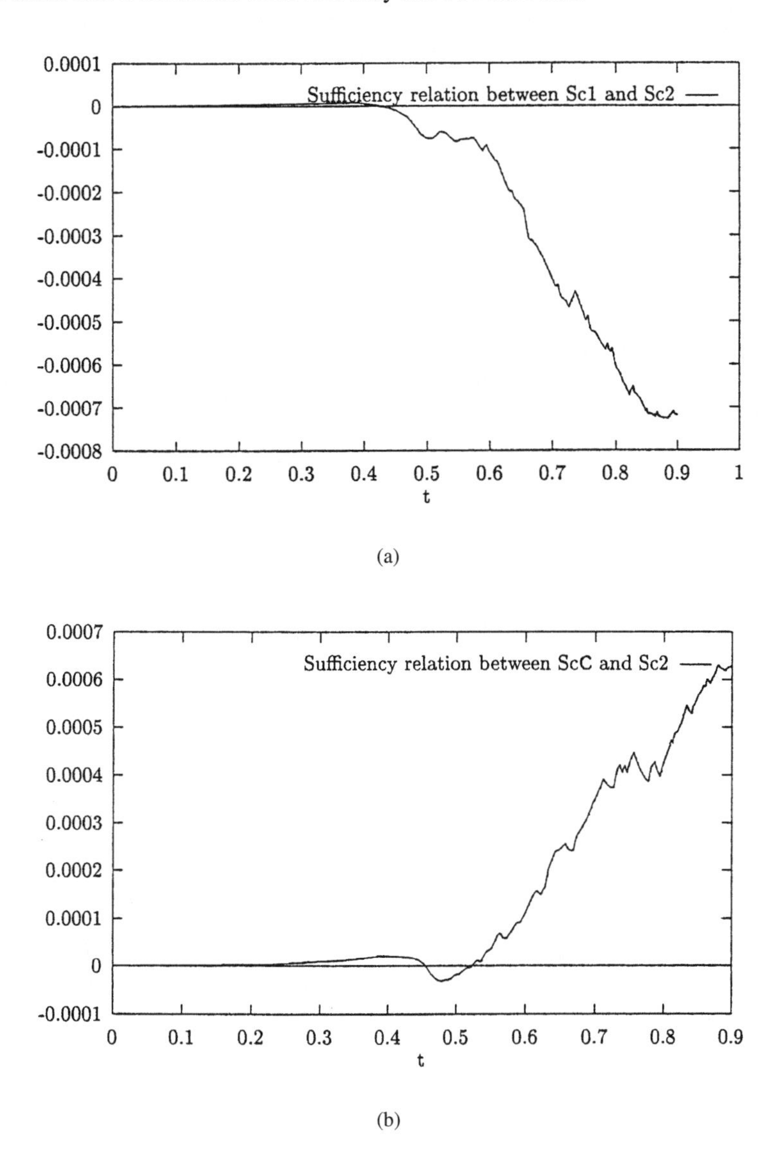

Figure 12.1. (a) *Graph of* $\int_0^p F_1(s)ds - \int_0^p F_2(s)ds$. (b) *Graph of* $\int_0^p F_c(s)ds - \int_0^p F_2(s)ds$.

Another approach to investigating whether to combine two scorecards is to use the analogies with screening methods suggested by Tsai and Yeh (1999). Screening models are used in quality control, psychology, education, and medicine to select items whose performance variables are within specifications by observing correlated screening variables rather than the performance variable. In credit scoring, the performance variable would be whether the loan defaults, while the screening variables are the application characteristics and the resultant scores. The screening models assume that all the variables form a multidimensional normal distribution, and the operational characteristics of using the screening—the scorecards in this case—can be obtained partly analytically and partly by simulation.

Tsai, Thomas, and Yeh (1999) look at the problem where there are two scorecards with S_1 and S_2 being the resultant score variables and Y the propensity to be good. It is assumed that (Y, S_1, S_2) have a standard trivariate normal distribution with covariance matrix:

$$\begin{pmatrix} 1 & \rho_{01} & \rho_{02} \\ \rho_{01} & 1 & \rho_{12} \\ \rho_{02} & \rho_{12} & 1 \end{pmatrix}. \tag{12.4}$$

Two different approaches for using the two scorecards in credit decision making are considered. The first is to have two separate scorecards and accept an applicant only if $s_1 \geq k_1$ and $s_2 \geq k_2$; i.e., they pass the cutoff on both scorecards. Thus the acceptance rate is $a_1 = \text{Prob}(S_1 \geq k_1, S_2 \geq k_2)$, and if we assume that bad applicants are those whose propensity to be good is below k_0, then the overall default rate is $d_1 = \text{Prob}\{Y \leq k_0 | S_1 \geq k_1, S_2 \geq k_2\}$. One can then choose k_1 and k_2 to minimize d_1 subject to a_1 equal to some agreed constant.

The alternative is to have a combined scorecard of $w_1 S_1 + w_2 S_2$ and assume that $Y = w_1 S_1 + w_2 S_2 + e$. Using standard regression analysis, one can show that

$$\begin{aligned} w_1 &= \frac{\rho_{01} - \rho_{02}\rho_{12}}{1 - \rho_{12}^2}, \qquad w_2 = \frac{\rho_{02} - \rho_{01}\rho_{12}}{1 - \rho_{12}^2}, \qquad \text{and} \\ \sigma^2 &= \frac{1 + 2\rho_{01}\rho_{02}\rho_{12} - \rho_{01}^2 - \rho_{02}^2 - \rho_{12}^2}{1 - \rho_{12}^2}, \end{aligned} \tag{12.5}$$

where σ^2 is the variance of e. Assuming that the cutoff score is k, so the acceptance rate is $a_2 = \text{Prob}\{w_1 S_1 + w_2 S_2 \geq k\}$, one can show that $k = \Phi^{-1}(z_{1-a_2})\sqrt{1 - \sigma^2}$, where Φ^{-1} is the standard inverse normal distribution. Similarly, one can get an expression for the default rate by

$$d_2 = \text{Prob}\{Y \leq k_0 | w_1 S_1 + w_2 S_2 \geq k\} = \frac{BN(-k(1-\sigma^2)^{-\frac{1}{2}}, k_0, -(1-\sigma^2)^{\frac{1}{2}})}{1 - \Phi(k(1-\sigma^2)^{-\frac{1}{2}})},$$

where BN is the standard bivariate normal distribution function, where values can be compiled using subroutines in the International Mathematical and Statistical Library. This allows one to use simulation to compare the effect of the different ways of using two scorecards in credit decisions.

Using an example where $p_{01} = 0.6$, $p_{02} = 0.4$, $p_{12} = 0.1$, and the overall default rate is 20%, so $\text{Prob}\{Y \leq k_0\} = 0.2$, Tsai, Thomas, and Yeh (1999) compared four different systems. The first used only the scorecard S_1, the second used only the scorecard S_2, the third was where customers were accepted only if they were above the cutoff score on both scorecards, and the last was a new scorecard whose score was a linear combination of S_1 and S_2. Using the analysis above, they found the default rates for different levels of acceptance described in Table 12.1. Again the results showed that combining the scorecards into a single scorecard seemed best.

In this section, we looked at two theoretical approaches that underpin the idea that combining scorecards is in general a sensible thing to do. They are useful tools in trying to understand when such combining of scorecards makes sense and when it does not, but more tools may be necessary. In the next section, we look at other ways to combine the outcome of two scoring systems apart from taking a linear combination of the underlying scores.

Table 12.1. *Default rates for four different scorecards.*

Acceptance rate	S_1	S_2	S_1 and S_2	Linear combination of S_1 and S_2
0.1	.011	.048	.007	.003
0.2	.022	.067	.017	.010
0.3	.035	.082	.029	.018
0.4	.049	.096	.040	.030
0.5	.064	.110	.055	.044
0.6	.082	.124	.076	.061
0.7	.102	.139	.097	.083
0.8	.126	.155	.123	.111
0.9	.156	.173	.155	.147

12.4 Combining classifiers

Suppose that one has n different credit scoring classifiers built on a population where the prior probability of being good (bad) is p_G (p_B). Can we combine these classifiers to produce something more powerful than the best of the individual classifiers? Zhu, Beling, and Overstreet (1999b) showed that this is possible and that one can construct a combined classification that is sufficient for the individual classifiers. As mentioned, this implies that the actual loss rate using the combined classifier will be less than that for using individual classifiers. Let y_i be the random variable describing the outcome of classifier i—either the score or the node of the classification tree—and if a particular consumer comes up with classifier values $y_1, y_2, \dots, y_n$, let $q(G|y_1, y_2, \dots, y_n) = q(G|\mathbf{y^n})$ be the probability such a consumer is good. Let $p(\mathbf{y^n}|G)$ and $p(\mathbf{y^n}|B)$ be the conditional probabilities that a good or bad consumer has classifier outcomes $y_1, y_2, \dots, y_n$. By Bayes's theorem,

$$q(G|\mathbf{y^n}) = \frac{p_G p(\mathbf{y^n}|G)}{p_G p(\mathbf{y^n}|G) + (1 - p_G)p(\mathbf{y^n}|B)}. \qquad (12.6)$$

Thus we need to estimate $p(\mathbf{y^n}|G)$. If one believed that the classifiers were stochastically independent, then one could assume that $p(\mathbf{y^n}|G) = \prod_{i=1}^{n} p_i(y_i|G)$, but in reality classifiers are likely to be correlated, so instead we define the n-dimensional conditional density $p(\mathbf{y^n}|G)$ as a product of $n - 1$ univariate conditional densities:

$$p(\mathbf{y^n}|G) = p(y_n|\mathbf{y^{n-1}}, G)p(y_{n-1}|\mathbf{y^{n-2}}, G) \cdots p(y_2|y_1, G)p(y_1|G). \qquad (12.7)$$

Then $p(\mathbf{y^n}|G)$ can be updated sequentially as

$$q(G|\mathbf{y^n}) = \frac{p(y_n|\mathbf{y^{n-1}}, G)q(G|\mathbf{y^{n-1}})}{k(y_n|\mathbf{y^{n-1}})}, \qquad (12.8)$$

where $k(y_n|\mathbf{y^{n-1}}) = p(y_n|\mathbf{y^{n-1}}, G)q(G|\mathbf{y^{n-1}}) + p(y_n|\mathbf{y^{n-1}}, B)(1 - q(G|\mathbf{y^{n-1}}))$. Thus all that is needed is to calculate $p(y_n|\mathbf{y^{n-1}}, G)$ and $p(y_n|\mathbf{y^{n-1}}, B)$.

Zhu, Beling, and Overstreet (1999b) suggested calculating these by transforming the y values so that they satisfy a normal distribution applying a regression equation to connect the transformed y_n with the $y_1, \dots, y_{n-1}$ and then transferring this relationship back to the y_ns. This has the advantage that the regression is done with normal distributions and that it can work whatever the distributions of $p(y_n|\mathbf{y^{n-1}}, G)$ and $p(y_n|\mathbf{y^{n-1}}, B)$.

Formally define z_k^i by $Q(z_k^i) = F(y_k|i)$, $k = 1, \ldots, n$, $i = G$ or B, where Q is the cumulative distribution function of the standard normal distribution. Obtain the conditional distributions $p(z_n^i|\mathbf{z}^{\mathbf{n}-1}, i)$ by applying a regression to these transformed values so that

$$z_n^i = r_n^i + s_{n1}^i z_1^i + \cdots + s_{n,n-1}^i z_{n-1}^i + e_{n-1}^i i = G \text{ or } B,$$

where e_n^i is $N(0, (\sigma_n^i)^2)$. This implies that $p(z_n^i|\mathbf{z}^{\mathbf{n}-1}, i)$ has a normal distribution with parameters $N(r_n^i + \sum_{j=1}^{n-1} s_{nj}^i z_j, (\sigma_n^i)^2)$. Transforming this back to the original distribution gives

$$p(y_n|y^{n-1}, i) = \frac{p(y_n|i)}{\sigma_n^i \phi(Q^{-1}(F(y_n|i))} \phi\left(\frac{Q^{-1}(f(y_n|i) - r_n^i - \sum_{j=1}^{n-1} s_{nj}^i Q^{-1}(F(y_j|i))}{\sigma_n^i}\right), \tag{12.9}$$

where ϕ is the density function of the standard normal distribution.

This Bayesian procedure produces a sequence of combined classifiers of increasing predictive power but also, unfortunately, increasing complexity. One way to simplify this is to use the idea of sufficiency introduced in the previous section since if S_1 is sufficient for the classifier based on S_1 and S_2, then $p(G|s_1, s_2) = p(G|s_1)$ and one can ignore classifier 2. Zhu, Beling, and Overstreet (1999b) used this approach to combine a linear regression, a logistic regression, and three neural net–based scorecards. The error rates of the individual classifiers were between 12% and 17% on a holdout sample, while the combined classifier, having recognized by the sufficiency tests that two of the classifiers were not needed, had an error rate of 7.5% on that sample.

Another way to combine classifiers would be to average them in some sense—somewhat akin to the basic generic scorecard of section 7.2. A more subtle approach would be to jointly calibrate the classifiers so that each classified the error between the true class and the average of the other classes. This corresponds to saying that if $\mathbf{x}$ are the characteristics and Y_i, $i = 1, \ldots, n$, are the different classifiers, then one defines

$$q(G|\mathbf{x}) = \alpha + \sum_i g(\mathbf{x}, Y_i),$$

where $g(\mathbf{x}, Y_i)$ is the prediction by classifier Y_i of the probability of being good.

Mertens and Hand (1997) looked at a specific example with two classifiers. The first is a linear regression model with parameters $\mathbf{w}$, so $g(\mathbf{x}, \mathbf{w}, LR) = \mathbf{x}^{\mathbf{T}} \cdot \mathbf{w}$. The second is a classification tree with final nodes given by hyperrectangles R_l in the characteristic space and the lth one having a good rate of g_l, so $g(\mathbf{x}, R, CT) = \sum_{l=1}^{L} \chi_{R_l}(\mathbf{x})$. $\chi_{R_l}(\mathbf{x})$ is the indictor function, which is 1 if $\mathbf{x} \in R$ and 0 otherwise. One can then alternate between the tree and the linear regression, keeping the parameters for one fixed and adjusting the parameters of the other to minimize the mean square classification error. Suppose there are n consumers, g of whom are good, in the sample. If the kth one has characteristics $\mathbf{x}_k$ and a probability y_k of being good ($y_k = 0$ or 1), the algorithm first defines $g_0(\mathbf{x}_k, \mathbf{w}, LR) = \frac{g}{n}$ and $g_0(\mathbf{x}_k, R, CT) = \frac{g}{n}$. At iteration i,

(a) calculate $\mathbf{w}_i$ by fitting the model $g_i(\mathbf{x}_k, \mathbf{w}_i, LR)$ as the best linear regressor of the residuals $y_k - g_{i-1}(\mathbf{x}_k, R_{i-1}, CT)$;

(b) grow the tree $g_i(\mathbf{x}_k, CT)$ to fit the residuals $y_k - g_i(\mathbf{x}_k, \mathbf{w}_i, LR)$ and prune the tree under some agreed procedure to obtain the classification tree $g_i(\mathbf{x}_k, R_i, LR)$;

(c) calculate the mean square error and the optimal constant α. If necessary, repeat the procedure.

This is a procedure for combining a regression and a classification tree, but this approach could be used for any number and any combination of classifying approaches. There are several other approaches for combining classifiers, but little work has been done on this area in credit scoring to date.

12.5 Indirect credit scoring

All the credit-scoring methods discussed so far seek to directly predict the future default risk, but one of the current moves in credit scoring is to predict expected profit rather than this default risk. One way to do this is to predict the future values of other variables, like balance outstanding, which affect the profit as well as the default risk. One can then use the estimates of these intermediate variables to forecast the profit or the default risk indirectly. The differences are displayed in Figure 12.2.

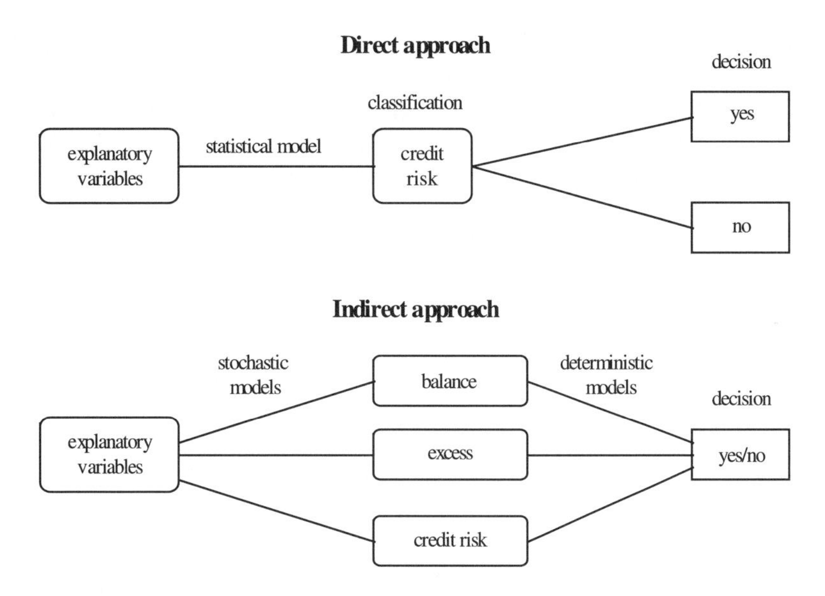

Figure 12.2. *Direct and indirect approaches to credit decisioning.*

There are several advantages to the indirect method. Since one is estimating a number of variables, one can get a better understanding of the way the repayment process develops; one can be more flexible in the definitions used to decide whether to grant credit, without having to recalculate the whole model; and validation of the model is easier to undertake. On the other hand, the more variables one estimates, the more errors that can be introduced into the system.

Li and Hand (1997) investigated this approach on a bank's current account data. They forecasted each component of a multicriterion definition of bad. Thus they used the data over the previous nine months to estimate the month-end balances and maximum balances, credit turnover, and amount the account is over the overdraft limit in months 9 and 10 in the future. These were the variables, $\mathbf{y}$, that they wanted to estimate. There were 29 predictor variables, $\mathbf{x}$, available, and they looked at three types of regression. The full model used lagged values of $\mathbf{x}$ and $\mathbf{y}$ to predict future $\mathbf{y}$; the predictor model used only lagged $\mathbf{x}$ values to predict $\mathbf{y}$, and the lagged model just used lagged values of **y** to predict future **y**. The definition of default taken by the bank involved either being £500 over the overdraft limit if the account was active, £100 over the overdraft limit if the account was inactive for one month, and £100 overdrawn if the account was inactive for two or more months. Thus having the predictions of the **y**, one could calculate whether the account was bad under this definition. This was compared with the logistic regression approach where each customer was directly classified as good or bad depending on their behavior in the ninth month. This was called the direct approach. The results they obtained are given in Table 12.2, where ER is the error rate and BR is the bad rate among those accepted under each approach.

Table 12.2. *Comparison of errors in the direct and indirect methods.*

Model	Full	Predictor	Lagged	Direct
ER	0.171	0.170	0.167	0.170
BR	0.162	0.160	0.163	0.163

These are very similar results, so Li and Hand (1997) then performed a number of simulation experiments. Although one might expect the performance of the indirect methods to be superior in that they used the exact definition of what is a bad, which is quite complex, this did not turn out to be the case. Only when one was able to classify the intermediate variables very accurately was the indirect method clearly superior, and when the cost of misclassifying a bad as a good was high, the direct method was clearly superior. Thus once again this approach needs some development before it can be used with confidence in practice, although given its flexibility it is clearly worth undertaking the development.

12.6 Graphical models and Bayesian networks applied to credit scoring

The indirect method outlined in the last section suggests that it is useful to model the connection between the variables that describe a consumer's performance and the attributes of the consumer. This is the consequence of moving from credit granting based purely on default risk to credit granting based on the much more complex objective of profit. Graphical modeling is a set of statistical tools that has proved useful in modeling the connections between variables. Recently, several authors (Whittaker 1990, Sewart and Whittaker 1998, Hand, McConway, and Stanghellini 1997, Chang et al. 2000) suggested that graphical models could prove useful in building models in the credit scoring context.

A graphical model of a situation consists of a number of vertices, each of which represents a continuous or discrete variable of interest. Some of these will be the application variables, like residential status or age; others can be outcome variables, like balance after 12 months or number of missed payments; others can be exogenous variables, like unemployment rate or interest rates; while some may be unobservable, like capacity to repay. The vertices are linked by edges, which describe the probabilistic dependency between them. In particular, the lack of an edge between two vertices implies the two variables are conditionally independent given the rest of the variables.

Definition 12.2. *Let X, Y, Z be a set of variables; then X is conditionally independent of Y given Z if when p is either the conditional density function (continuous variable) or the probability function (discrete variable), then $p_{\mathbf{x}|\mathbf{y},\mathbf{z}}(\mathbf{x}|\mathbf{y},\mathbf{z})$ does not depend on y. This is* *equivalent to saying*

$$p_{\mathbf{x}|\mathbf{y},\mathbf{z}}(\mathbf{x}|\mathbf{y},\mathbf{z}) = p_{\mathbf{x},\mathbf{z}}(\mathbf{x},\mathbf{z}) \cdot p_{\mathbf{y},\mathbf{z}}(\mathbf{y},\mathbf{z}). \tag{12.10}$$

With this definition of vertex and edge (or nonedge), the relationship between the variables in the model can be displayed as a graph. Hence the name *graphical models*. If two variables are separated in the graph, i.e., you cannot get from one to the other except through another variable, then they are conditionally independent given this other variable. This is called the global Markov property.

For example, consider four variables W, X, Y, Z in which W is conditionally independent of Z given (X, Y) and Y is conditionally independent on Z given (W, X). This is represented in Figure 12.3.

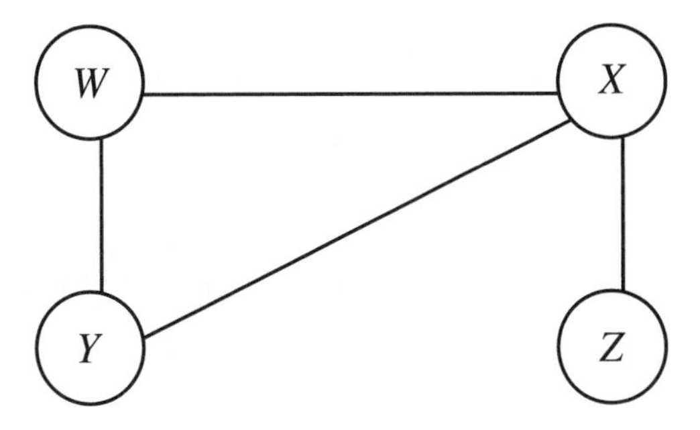

Figure 12.3. *A typical graphical model.*

This means that Z and W are separated by X, so Z is conditionally independent of W given X (and also Z is conditionally independent of W given Y only). Marginal indepedence is the normal definition of independence and says that $p_{xy}(x, y) = p_x(x)p_y(y)$. This is shown in Figure 12.4.

Figure 12.4. *A graphical model of marginal independence.*

We can also bracket variables together in groups called cliques if they are all adjacent to each other in the graph and are the largest group with this property. Thus if we added any other variable, we would not form a clique. Therefore, in Figure 12.3, $\{X, Y, W\}$ and $\{X, Z\}$ are cliques.

There are three ways in which these graphical models can be used in credit scoring. First, they illuminate the relationship between the factors that affect the behavior of borrowers. They model the interrelationships between the variables simultaneously without forcing any particular one to be distinguished as the outcome variable. Thus it may be easier to identify subpopulations with certain behavioral characteristics than by the standard credit-scoring methods, even those based on classification trees. Second, the graphical models can be used to predict risk or profit for each individual. Given the application variables, one can use the relationship to predict the other variables, including unknown ones like default risk. Thus it can be used in the same way as normal credit-scoring systems. Third, since in reality credit granting is a decision process, with different decisions being made at different times—to whom should we mail, who will respond, whom do we accept, who will then use the card, who will change to another lender, who will default—it is useful to understand when information will be known and what insights that information can give. Thus it is a way to relate the possible predictions to the decision-making process and so can be used to ensure that the decision-making process is well designed.

The graph connecting the variables can be constructed either by starting with the complete graph and asking which edges can be left out or by starting with a graph with no edges and adding them. In both cases, one is trying to compare two models, one with an edge in and one with it out. These models correspond to different statistical models depending on whether all the variables are discrete (categorical), all are continuous, or there is a mixture with some variables continuous and others discrete. The discrete variable models correspond to loglinear models; the continuous variable models become multivariate normal models if we assume that the distributions are normal. In this case, conditional independence corresponds to zero partial correlations. Graphical mixed models were introduced by Lauritzen and Wermuth (1989), and a software package, MIM, was developed by Edwards (1995) to estimate the parameters and calculate the graph.

We will sketch the calculations for the discrete case since it is so common in credit and behavioral scoring to translate continuous variables into categorical ones. Suppose there are three categorical variables X, Y, and Z and let $p_{jkl} = P\{X = j, Y = k, Z = l\}$. The most general loglinear model defines these probabilities by

$$\mathrm{Log}(p_{jkl}) = u + u_j^X + u_k^Y + u_l^Z + u_{jk}^{XY} + u_{jl}^{XZ} + u_{kl}^{YZ} + u_{jkl}^{XYZ}. \tag{12.11}$$

There is a relationship between independence of the variables X and Y and whether u^{XY} is zero. If Y is conditionally independent of Z, given X, then that is equivalent to $u_{kl}^{YZ} = u_{jkl}^{XYZ} = 0$. If there are N observations in the sample set and n_{jkl} are the numbers of observations where $X = j, Y = k, Z = l$, then this is a multinomial distribution. Thus the chance of this happening is

$$\left(\frac{N!}{\prod_{jkl} n_{jkl}!}\right) \prod_{jkl} p_{jkl}^{n_{jkl}}.$$

The log of the likelihood of this happening is

$$l_M = \log(L_M(p_{jkl}|n_{jkl})) = \log\left(\frac{N!}{\prod_{jkl} n_{jkl}!}\right) + \sum_{jkl} n_{jkl} \log(p_{jkl}). \tag{12.12}$$

Let l_{full} be the log of the maximum likelihood assuming the full model of (12.11) and let l_M be the log of the maximum likelihood assuming some submodel M (maybe with X conditionally independent of Y). The deviance G^2 is the likelihood ratio test, so $G^2 = 2(l_{\text{full}} - l_M)$. In the full model, the maximum likelihood estimates of p_{jkl} are $\frac{n_{jkl}}{N}$. If $\hat{p}_{jkl}^1$ are the ML estimates, assuming that M_1 and $\hat{m}_{jkl}^1 = N\hat{p}_{jkl}^1$ are the expected cell counts and $\hat{p}_{jkl}^2$ and $\hat{m}_{jkl}^2$ are the corresponding estimates under a different model M_2, $M_1 \subseteq M_2$, then the deviance difference d is given by

$$d = G_2^2 - G_1^2 = 2\left(\sum_{jkl} n_{jkl} \ln \hat{p}_{jkl}^2 - \sum_{jkl} n_{jkl} \ln \hat{p}_{jkl}^1\right) = 2\sum_{jkl} n_{jkl} \ln\left(\frac{\hat{m}_{jkl}^2}{\hat{m}_{jkl}^1}\right). \tag{12.13}$$

Under M_2, d is asymptotically χ_k^2, where k is the difference in the number of free parameters between M_2 and M_1. The alternative is to use the fact that G_1^2 is $\chi_{k'}^2$, where k' is the difference in the number of free parameters between M_1 and M_{full}.

Sewart and Whittaker (1998) gave some examples of the use of these techniques on credit scoring, and the following examples are based on these.

Example 12.1 (Is wealth related to default risk?). A sample of 5000 customers, was compared for wealth (poor or rich) and default status (yes or no). Table 12.3 gives the distributions. M_1 is the model where wealth and default status are independent, so the independent default rate is 2.4% ($\frac{120}{5000}$) and the poor rate is 50.2% ($\frac{2510}{5000}$). This gives expected numbers in each of the four cells of

$$2449.76 \left(2510 \times \frac{4880}{5000}\right), \qquad 60.24 \left(2510 \times \frac{120}{5000}\right),$$

$$2430.24 \left(2490 \times \frac{4880}{5000}\right), \qquad 59.76 \left(2490 \times \frac{120}{5000}\right).$$

Table 12.3. *Numbers in Example* 12.1.

		X		
		Not defaulted	**Defaulted**	**Total**
Y	**Poor**	2450	60	2510
	Rich	2430	60	2490
	Total	4880	120	5000

M_2 is the full model where wealth and default are connected and the expected numbers there are, of course, the actual numbers. There is one extra parameter in this model (u^{XY}). Hence

$$\begin{aligned} d &= 2\left(2450 \ln\left(\frac{2450}{2449.76}\right) + 60 \ln\left(\frac{60}{60.24}\right) + 2430 \ln\left(\frac{2430}{2430.24}\right) + 60 \ln\left(\frac{60}{59.76}\right)\right) \\ &= .002. \end{aligned}$$

$\chi_{0.95}^2 = .0039$ for one degree of freedom, which suggests that the M_2 model is not a significant improvement over the M_1 model. Hence wealth and default rate are marginally independent.

Example 12.2 (Are wealth and usage related to default risk? (Marginal independence does not imply conditional independence.)). The use made of the credit card by the 5000 customers in Example 12.1 is now added to obtain Table 12.4. Compare this time M_2, which is the full graph with usage, wealth, and default related, and M_1, the graph where default and usage are dependent and usage and wealth are dependent, but default and wealth are conditionally independent. (Remember that they were marginally independent originally.) Again under M_2 with the complete graph, the expected numbers in each cell are the actual numbers in this case. For M_1, we use the fact that estimating someone as a light user is .4844 (heavy .5156) and the estimates of the conditional probabilities are $\hat{p}(\text{default}|\text{light}) = \frac{12}{2422}$, $\hat{p}(\text{default}|\text{heavy}) = \frac{108}{2578}$, $\hat{p}(\text{poor}|\text{light}) = \frac{2220}{2422}$, and $\hat{p}(\text{poor}|\text{heavy}) = \frac{290}{2578}$. This leads to the estimates for the cell entries under M_1 given in Table 12.5.

Table 12.4. *Numbers in Example* 12.2.

		X		
Z	*Y*	**Not defaulted**	**Defaulted**	**Total**
Light use	**Poor**	2210	10	2422
	Rich	200	2	
Heavy use	**Poor**	240	50	2578
	Rich	2230	58	

Table 12.5. *Estimate for* M_1.

		X	
Z	*Y*	**Not defaulted**	**Defaulted**
Light use	**Poor**	2209	11
	Rich	201	1
Heavy use	**Poor**	277.85	12.15
	Rich	2192.15	95.85

There are two extra degrees of freedom between M_1 and M_2, i.e., u^{xy} and u^{xyz}, so the deviance difference (12.13) has a χ^2 distribution with two degrees of freedom. In this case,

$$\begin{aligned} d = 2 \Bigg\{ & 2210 \ln\left(\frac{2210}{2209}\right) + 10 \ln\left(\frac{10}{11}\right) + 200 \ln\left(\frac{200}{201}\right) + 2 \ln\left(\frac{2}{1}\right) + 240 \ln\left(\frac{240}{277.85}\right) \\ & + 50 \ln\left(\frac{50}{12.15}\right) + 2230 \ln\left(\frac{2230}{2192.15}\right) + 58 \ln\left(\frac{58}{95.85}\right) \Bigg\} \\ = {} & 90.5. \end{aligned}$$

The 95% significance value of the χ^2 distribution with two degrees of freedom is .103, so this is vastly over that and shows that the M_2 model is significantly better than the M_1 model. Hence the graphical models for Examples 12.1 and 12.2 would be as given in Figure 12.5.

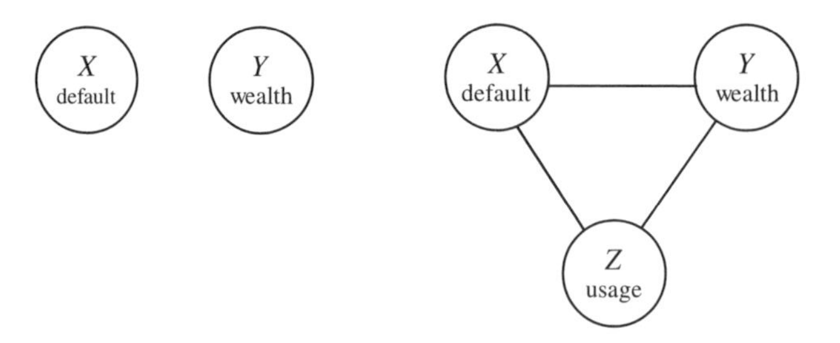

Figure 12.5. *Graphical models of Examples* 12.1 *and* 12.2.

Example 12.3. Another example of graphical models in credit scoring was studied by Hand, McConway, and Stanghellini (1997). They looked at 23,000 unsecured personal loans and investigated the graph of nine variables: AMOUNT (amount of loan); MARRY (marital status); RES (residential status); INCOME (disposable income); AGE (customer's age); INSURANCE (Did they take out insurance on the loan?); CURRAC (current account);

CCDEL (Has the applicant defaulted on a credit card?); and BAD (whether the loan became bad).

With nine variables, even if each variable has only two categories, that gives 512 cells, while if they have three categories each, there are 19,683 cells. Thus even with a sample of 23,000, one has to be very parsimonious in the number of attributes for each variable to avoid too many of the cells being empty. In this example, seven of the variables had two categories and two had three categories—2304 cells in total. The modeling started with the complete graph and tried to drop arcs and ended up with Figure 12.6.

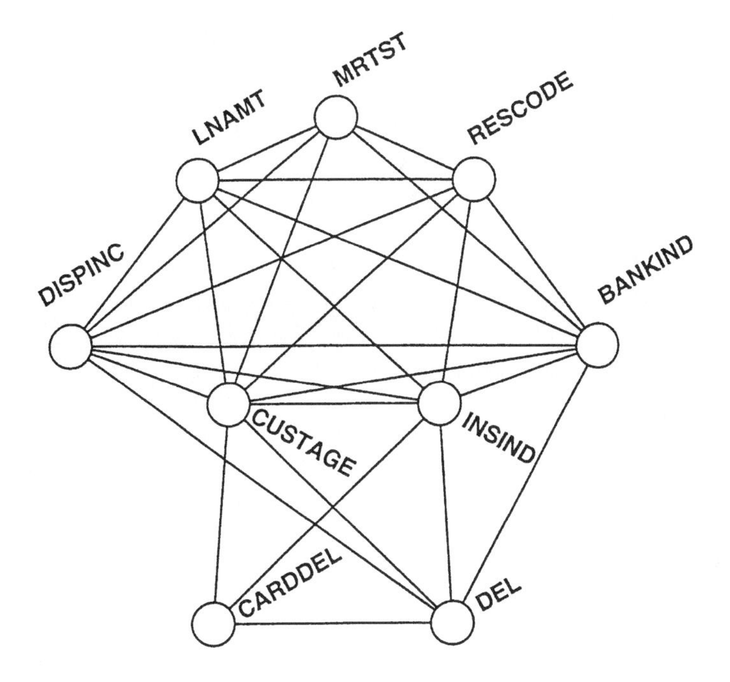

Figure 12.6. *Graphical model of Example* 12.3.

Only nine edges were dropped, but the graph can be decomposed into three groups. One is amount of loan, marital status, and residential position; the second is age, income, current account, and insurance; and the third is default on either credit card loan. What is interesting is that the first and third sets are separated by the second. Therefore, to estimate default likelihood, one does not benefit by knowing loan amount, marital status, and residential position if one knows age, income, insurance, and current account.

Thus far, we have considered graphical models where there is no direction to the edges because they represent conditional dependence between the variables. An undirected graph represents the joint distributions, but one can also define a joint distribution by a sequence of conditional probabilities. If this approach is taken, then one can put a direction on each arc of the graph to represent the way in which the conditioning is undertaken. One can also define a directed graph formally by saying that if $(X_1, \ldots, X_n)$ are a set of ordered variables,

$$p_{X_1,X_2,\ldots,X_n}(x_1, x_2, \ldots, x_n) = p_{X_1}(x_1)p_{X_2|X_1}(x_2|x_1) \cdots p_{X_n|X_{n-1},\ldots,X_1}(x_n|x_{n-1}, \ldots, x_1). \tag{12.14}$$

For $i < j$, an arrow is drawn from X_i to X_j unless $p_{X_j|X_{j-1},\ldots,X_1}(x_j|x_{j-1}, \ldots, x_1)$ does not depend on X_i; i.e., X_j is conditionally independent of X_i given $\{X_1, \ldots, X_{i-1}, X_{i+1}, \ldots, X_{j-1}\}$.

This directed graphical model is also called a Bayesian network because it was introduced to look at probabilistic expert systems where the emphasis is on Bayesian ideas of updating one's beliefs; see (Spiegelhalter et al. 1993) for a survey of the area. In these Bayesian networks, if there is a directed arc from vertex X_i to vertex X_j, then vertex i is a parent (direct predecessor) of node j. Let $P(j)$ be the set of parents of j; i.e., $P(j) = \{i|(i, j)$ is a directed arc$\}$. Similarly, $S(i)$ are the children (direct successors) of node i, so $S(i) = \{j|(i, j)$ is a directed arc$\}$. This enables us to define all the variables in the database that are relevant to the prediction of variable X_i. This is the Markov blanket surrounding node i. A Markov blanket of X_i in a Bayesian network is the subset of nodes

$$M(i) = (P(i) \cup S(i) \cup P(S(i)) - \{i\}; \tag{12.15}$$

i.e., it has the parents, the children, and the parents of the children of the variable X_i.

It follows from these definitions that if $p(.|.)$ is again a conditional probability or a conditional density, then we can show that

$$p(x_i|x_1, x_2, \ldots, x_{i-1}, x_{i+1}, \ldots, x \ldots) = p(x_i|\mathbf{x}_{M(i)}).$$

Chang et al. (2000) used the idea of a Bayesian network in the credit-scoring context in the following way. They took X_0 to be the variable describing risk of default (i.e., $G =$ good or $B =$ bad) and assumed that it had no predecessors but that the application variables $X_1, \ldots, X_p$ could be its children. They pointed out that with a suitable transformation, the score $S(\underline{X})$ for someone with application details x is equivalent to the log of the odds of being good to being bad. (This was the basis of the logistic regression models.) Applying this gives the relationship

$$s(x) = \log\left(\frac{q(G|x)}{q(B|x)}\right) = \log\left(\frac{p_G p(x|G)}{p_B p(x|B)}\right) = \log\left(\frac{p_G}{p_B}\right) + \log\left(\frac{p(x|G)}{p(x|B)}\right). \tag{12.16}$$

Chang et al. (2000) modified the definition of a clique to define cliques in directed graphs. They defined a successor clique C of the performance node to be a set of nodes at least one of whom is a successor of the performance node, and the set has the property that all children of nodes in the clique lie themselves within the clique; i.e., $S(C) \subseteq C$. This means that if we have two successor cliques C_1 and C_2, then

$$p(\mathbf{x}_{C_1 \cup C_2}|x_0) = p(x_{C_1}|x_0)p(x_{C_2}|x_0)$$

since no arcs reach from one clique to the next. This means that if there are K successor cliques $C_1, C_2, \ldots, C_K$ in the network,

$$p(\mathbf{x}|G) = p(\mathbf{x}_{C_1}|G)p(\mathbf{x}_{C_2}|G)\cdots p(\mathbf{x}_{C_K}|G) \quad \text{and}$$
$$p(\mathbf{x}|B) = p(\mathbf{x}_{C_1}|B)p(\mathbf{x}_{C_2}|B)\cdots p(\mathbf{x}_{C_K}|B).$$

Hence

$$s(x) = \log\left(\frac{p_G}{p_B}\right) + \sum_{k=1}^{K}\log\left(\frac{p(\mathbf{x}_{C_k}|G)}{p(\mathbf{x}_{C_k}|B)}\right) = s_0 + \sum_{k=1}^{K} s(C_k), \tag{12.17}$$

where

$$s(C_k) = \log\left(\frac{p(\mathbf{x}_{C_k}|G)}{p(x_{C_k}|B)}\right).$$

This means that the score splits into the sum of its clique scores. This result is useful in that there may be far fewer elements in a clique than the p original variables and so score calculation is easier. The structure of the cliques might also give insight into the economic and behavioral factors that most influence the score.

However, this decomposition into cliques may be less helpful than is first thought. Chang et al. (2000) applied their results to a sample of 7000 applicants for bank credit with 35 characteristics available on each. Starting with the model where each of the 35 characteristics was assumed to depend on the good-bad performance variable X_0 but to be conditionally independent of each other, a Bayesian network was built to find the Markov blanket of X_0 and the successor cliques. It produced a result where there were 25 nodes in the Markov blanket (so 10 variables were dropped altogether), and these split into 11 one-node cliques, 4 two-node cliques, and 2 three-node cliques. Thus the network got only slightly more complicated than the one that was taken initially.

These results suggest that more research is needed in developing the methodology for using graphical models in credit scoring. At present, it seems if you start with the complete graph, you will not get rid of many arcs, while if you start with a very sparse graph, you will not add many arcs. Clearly, however, the potential for this sort of approach to building profit-scoring models is considerable.

12.7 Survival analysis applied to credit scoring

Credit-scoring systems were built to answer the question, "How likely is a credit applicant to default by a given time in the future?" The methodology is to take a sample of previous customers and classify them into good or bad depending on their repayment performance over a given fixed period. Poor performance just before the end of this fixed period means that customer is classified as bad; poor performance just after the end of the period does not matter and the customer is classified as good. This arbitrary division can lead to less-than-robust scoring systems. Also, if one wants to move from credit scoring to profit scoring, then it matters when a customer defaults. One asks not if an applicant will default but when will they default. This is a more difficult question to answer because there are lots of answers, not just the yes or no of the "if" question, but it is the question that survival analysis tools address when modeling the lifetime of equipment, constructions, and humans.

Using survival analysis to answer the "when" question has several advantages over standard credit scoring. For example,

(a) it deals easily with censored data, where customers cease to be borrowers (either by paying back the loan, death, changing lender) before they default;

(b) it avoids the instability caused by having to choose a fixed period to measure satisfactory performance;

(c) estimating when there is a default is a major step toward calculating the profitability of an applicant;

(d) these estimates will give a forecast of the default levels as a function of time, which is useful in debt provisioning; and

(e) this approach may make it easier to incorporate estimates of changes in the economic climate into the scoring system.

Narain (1992) was one of the first to suggest that survival analysis would be used in credit scoring. Banasik, Crook, and Thomas (1999) make a comparison of the basic survival analysis approach with logistic regression–based scorecards and showed how competing risks can be used in the credit-scoring context. Stepanova and Thomas (1999, 2001) developed the ideas further and introduced tools for building survival analysis scorecards as well as introducing survival analysis ideas into behavioral scoring. This section is based on these last two papers.

Let T be the time until a loan defaults. Then there are three standard ways to describe the randomness of T in survival analysis (Collett 1994):

- survival function: $S(t) = \text{Prob}\{T \geq t\}$, where $F(t) = 1 - S(t)$ is the distribution function;
- density function: $f(t)$, where $\text{Prob}\{t \leq T \leq t + \delta t\} = f(t)\delta t$;
- hazard function: $h(t) = \frac{f(t)}{S(t)}$, so $h(t)\delta t = \text{Prob}\{t \leq T \leq t + \delta t | T \geq t\}$.

Two of the commonest lifetime distributions are the negative exponential, which with parameters λ has $S(t) = e^{-\lambda t}$, $f(t) = \lambda e^{-\lambda t}$, and $h(t) = \lambda$, and the Weibull distribution, which with scale λ and shape k has $S(t) = e^{-(\lambda t)^k}$, $f(t) = k\lambda t^{k-1}e^{-(\lambda t)^k}$, and $h(t) = k\lambda^k t^{k-1}$. The former has no aging effect in that the default rate stays the same over time; the latter is more likely to default early on if $k < 1$, is more likely to default late on if $k > 1$, and becomes the negative exponential distribution if $k = 1$.

In standard credit scoring, one assumes that the application characteristics affect the probability of default. Similarly, in this survival analysis approach, we want models that allow these characteristics to affect the probability of when a customer defaults. Two models have found favor in connecting explanatory variables to failure times in survival analysis: proportional hazard models and accelerated life models. If $\mathbf{x} = (x_1, \ldots, x_p)$ are the application (explanatory) characteristics, then an accelerated life model assumes that

$$S(t) = S_0(e^{\mathbf{w}\cdot\mathbf{x}}t) \quad \text{or} \quad h(t) = e^{\mathbf{w}\cdot\mathbf{x}}h_0(e^{\mathbf{w}\cdot\mathbf{x}}t), \tag{12.18}$$

where h_0 and S_0 are baseline functions, so the $\mathbf{x}$ can speed up or slow down the aging of the account. The proportional hazard assumes that

$$h(t) = e^{\mathbf{w}\cdot\mathbf{x}}h_0(t), \tag{12.19}$$

so the application variables $\mathbf{x}$ have a multiplier effect on the baseline hazard. One can use a parametric approach to both the proportional hazards and acceleration life models by assuming that $h_0(.)$ belongs to a particular family of distributions. It turns out that the negative exponential and the Weibull distributions are the only ones that are both accelerated life and proportional hazard models. The difference between the models is that in proportional hazards, the applicants most at risk for defaulting at any one time remain the ones most at risk for defaulting at any other time.

Cox (1972) pointed out that in proportional hazards one can estimate the weights $\mathbf{w}$ without knowing $h_0(t)$ using the ordering of the failure times and the censored times. If t_i and $\mathbf{x}_i$ are the failure (or censored) times and the application variables for each of the items under test, then the conditional probability that customer i defaults at time t_i given that $R(i)$ are the customers still operating just before t_i is given by

$$\frac{\exp\{\mathbf{w}\cdot\mathbf{x}_i\}h_0(t)}{\sum_{k\in R(i)}\exp\{\mathbf{w}\cdot\mathbf{x}_k\}h_0(t)} = \frac{\exp\{\mathbf{w}\cdot\mathbf{x}_i\}}{\sum_{k\in R(i)}\exp\{\mathbf{w}\cdot\mathbf{x}_k\}}, \tag{12.20}$$

which is independent of h_0.

There are two minor complications when applying these ideas to credit granting. First, the default data tend to be aggregated at a monthly level, so the default time is really discrete, not continuous, and this leads to lots of ties when there are a large number of customers who default or whose data is censored for other reasons in the same month. Cox (1972) suggested a linear log odds model as the discrete time equivalent or proportional hazards, where

$$\frac{h(t)\partial t}{1-h(t)\partial t} = \exp\{\mathbf{w}\cdot\mathbf{x}\}\left(\frac{h_0(t)\partial t}{1-h_0(t)\partial t}\right). \tag{12.21}$$

If d_i are the number of defaults (failures) at time t_i, let $R(t_i, d_i)$ be the set of all subsets of d_i customers taken from the risk set $R(i)$. Let R be any such subset of d_i customers in $R(t_i, d_i)$. Then let $\mathbf{s}_R = \sum_{r\in R}\mathbf{x}_r$ be the sum of the answers over the individuals in set R. Let D_i denote the set of the d_i of customers who actually defaulted at t_i and $\mathbf{s}_{D_i} = \sum_{r\in D_i}\mathbf{x}_r$. The likelihood function arising from model (12.21) is then

$$\prod_{i=1}^{K}\left(\frac{\exp(\mathbf{w}\cdot\mathbf{s}_{D_i})}{\sum_{R\in R(t_i;d_i)}\exp(\mathbf{w}\cdot\mathbf{s}_R)}\right). \tag{12.22}$$

Maximizing this is difficult because of the large number of sets that have to be summed over in the denominator, but Breslow (1974) and Effron (1977) have suggested simplifying approximations.

The results of applying Cox's proportional hazard model to a sample of 50,000 personal loans from a major U.K. financial institution are given in Figure 12.7. Two measures were used for the survival model based on application data:

(i) How good was the model at estimating which loans will default in the first 12 months, compared with a logistic regression scorecard with failure in the first 12 months as its definition of bad?

(ii) How good was the model at estimating, among those still repaying after 12 months, which ones will default in the second 12 months? This again is compared with a logistic regression built on those who have survived 12 months, with the definition of bad being default in the second 12 months.

Figure 12.7 shows that the proportional hazard model on each criterion is competitive with the logistic regression model built solely on that criterion. The left-hand graph (a) shows the ROC curves under criterion (i), and the right-hand graph (b) is the ROC curve under criterion (ii). This suggests that even as a measure of default risk on a fixed time horizon, proportional hazard models are very competitive.

The survival analysis approach has a number of extensions that can be used in credit lending modeling. There are other reasons apart from default why a customer may finish before the original intended term—moving to another lender, selling the item that the loan was used to purchase, taking out another loan. All these mean the lender does not make the profit that was expected, and so in their own way these loans are bad for the lender. The competing-risks approach to survival analysis allows us to build survivor function models for each of these risks separately. Consider a sample case where the loan can be defaulted on, paid off early, or paid to term, and let T_d and T_e be the lifetime of the loan until default and until early repayment. If T_m is the term of the loan, the number of months of repayment is $T = \min\{T_d, T_e, T_m\}$.

Just as we were able to estimate T_d previously, we could use the same analytic techniques to estimate T_e, the time until early repayment. One does not need to assume that T_d

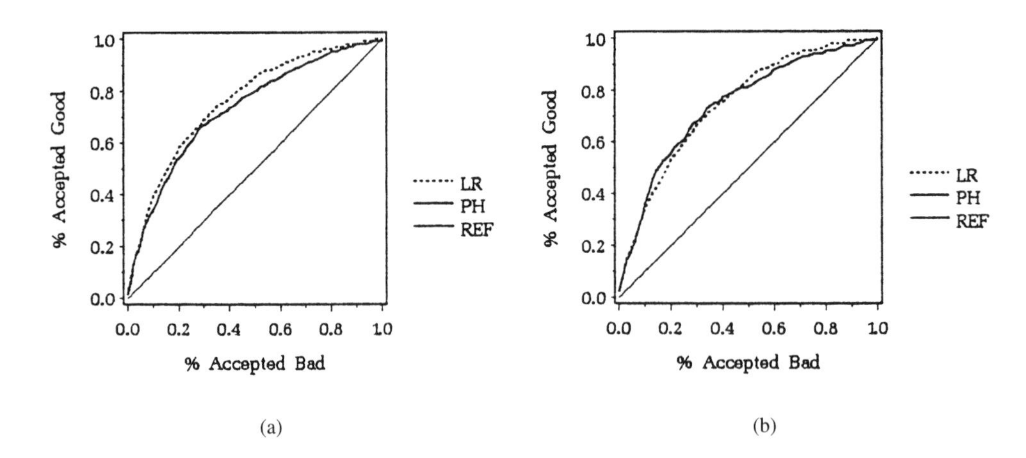

Figure 12.7. *ROC curves comparing proportional hazards and logistic regression models using default criterion.*

and T_e are independent variables, but as the calculations get complicated, if one does not (Collett 1994), it is usual to assume that is the case. If one repeats the exercise on the 50,000 personal loans and uses Cox's proportional hazard models to estimate early repayment both in the first year and then in the second year for those still repaying after 12 months, the proportional hazard model does not perform as well as the optimal logistic regression models, especially in the second year, where some of the results are worse than chance. Stepanova and Thomas (1999) showed that if one segmented the results on the term of the loan, the results are much better. Figure 12.8 shows the ROC curves for early repayment again in the first year and then in the second year for those still repaying after 12 months. Graph (a) is the ROC curve for this first criterion for loans with terms between months 12 and 24, and graph (b) is the ROC curve for the second criterion for such loans. Graphs (c) and (d) are ROC curves for these two criteria for loans between 24 and 30 months, and graphs (e) and (f) are the same ROC curves for loans with terms of more than three years. In the latter case, the survival approach works well. One of the interesting results of their analysis is that for early payments, it may be better to take t to be the number of months until completion of the term of loan, rather than the length of time since the loan was taken out. This is why segmenting on the term of the loan improved the results so dramatically.

One of the disadvantages of the proportional hazards assumption is that the relative ranking among the applicants of the risk (be it of default or early repayment) does not vary over time. This can be overcome by introducing time-dependent characteristics. So suppose that $x_1 = 1$ if the purpose of the loan is refinancing and 0 otherwise. One can introduce a second characteristic $x_2 = x_1 t$. In the model with just x_1 involved, the corresponding weight was $w_1 = 0.157$, so the hazard rate at time t for refinancing loans was $e^{0.157}h_0(t) = 1.17h_0(t)$, and for other loans, the hazard rate was $h_0(t)$. When the analysis was done with x_1 and x_2 involved, the coefficients of the proportional hazard loans were $w_1 = 0.32$, $w_2 = -0.01$. Thus for refinancing loans, the hazard rate at time t was $e^{0.32-0.01t}h_0(t)$ compared with others $h_0(t)$. Thus in month 1, the hazard from having a refinancing loan was $e^{0.31} = 1.36$ times higher than for a nonrefinancing loan, while after 36 months, the hazard rate for refinancing was $e^{-0.04} = 0.96$ of the hazard rate for not refinancing. Thus time-by-characteristic interactions in proportional hazard models allow the flexibility that the effect of a characteristic can increase or decrease with the age of the loan.

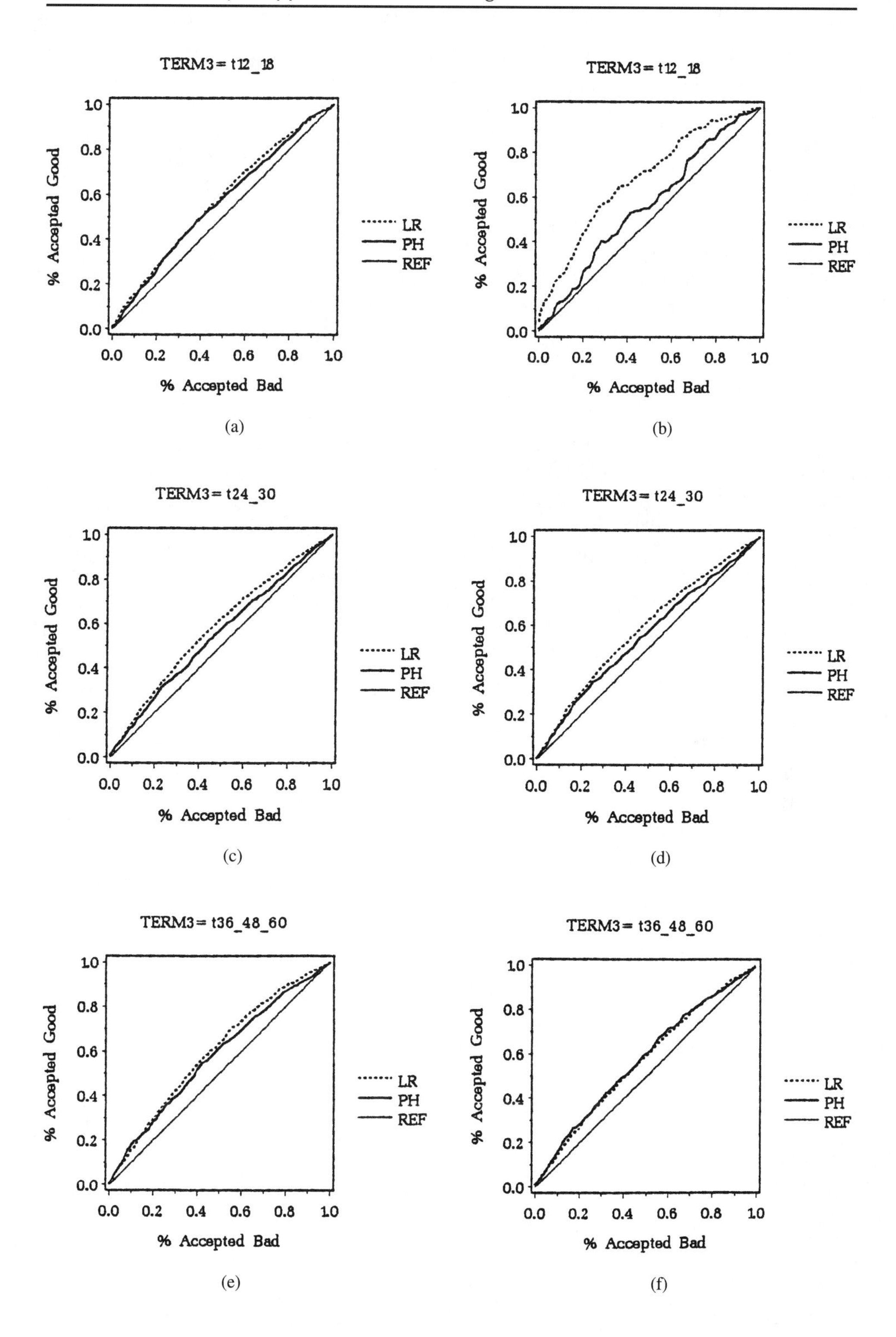

Figure 12.8. *ROC curves for early repayment segmented by term of loan.*

Survival techniques can also be applied in the behavioral scoring context, although a little more care is needed. Suppose that it is u periods since the start of the loan and $\mathbf{b}(u)$ are the behavioral characteristics in period u (which may include application characteristics). Then a proportional hazard model might say the hazard rate for defaulting in another t periods time, i.e., $t+u$ since the start of the loan, is $e^{\mathbf{w}(u)\cdot\mathbf{b}(u)}h_0^u(t)$. At the next period, $u+1$, the comparable hazard rate would be that for $t-1$ more periods to go, i.e., $e^{\mathbf{w}(u+1)\cdot\mathbf{b}(u+1)}h_0^{u+1}(t-1)$. Thus the coefficients $w(u)$ have to be estimated separately for each period u using only the data in the data set that have survived up to period u. As it stands, these coefficients could change significantly from one period to the next. One way to smooth these changes would be to make the behavioral score at the last period one of the characteristics for the current period. So suppose that $\mathbf{x} = (x_1, \ldots, x_p)$ were the application characteristics and $\mathbf{y}(u) = (y(u)_1, \ldots, y(u)_q)$ were the behavioral characteristics at period u. One defines a series of behavioral scores by $s(0) = \mathbf{w}\cdot\mathbf{x}$, where $e^{\mathbf{wx}}h_0^0(t)$ is the default rate hazard function at time 0. At time 1, the hazard function for time from now until default is $e^{\mathbf{w}_0^1 s(0)+\mathbf{w}(1)\cdot\mathbf{y}(1)}h_0^1(t)$, and define $s(1) = w_0(1)s(0) + \mathbf{w}(1)\mathbf{y}(1)$. Repeating this procedure leads to behavior scores at period u satisfying $s(u) = w_0(u)s(u-1) + \mathbf{w}(u)\cdot\mathbf{y}(u)$. Details of this analysis can be found in Stepanova and Thomas (2001).

Chapter 13

International Differences

13.1 Introduction

In this chapter, we explore a number of international differences in the behavior of borrowers and lenders and institutional arrangements affecting markets for credit. We begin by considering differences between countries in the volume of debt that households take on. Then we look at differences in credit bureau information between the U.K. and the U.S. Third, we review studies that have tried to explain differences in the choice of payment methods consumers use to buy goods and services. Fourth, we review some work that examined intercountry differences in scorecards, and finally we examine the rather unique position of the U.S. in the procedures available for debtors who declare themselves bankrupt and the possible affects this might have on some borrowers' behavior.

13.2 Use of credit

13.2.1 Consumer credit

Table 13.1 shows data for the U.S., the U.K., and Canada of consumer credit outstanding relative to GDP, consumer credit per capita of the population, and residential mortgage outstandings relative to GDP. While the U.S. has the largest consumer credit outstanding relative to GDP, Canada has the greatest stock of mortgage debt for its national output. In the U.S., there is \$4922 of consumer credit outstanding for every person in the country, but only £1727 (\$2862) in the U.K. Table 13.2 shows data for a larger number of countries for household debt outstanding as a percentage of GDP and for the net increase in debt outstanding as a percentage of GDP. These are all average figures for the periods shown in the table, which in most cases is 1992–1996. This period is determined by data availability and the attempt to span a full business cycle. Household debt is the sum of short-term and long-term loans owed by households. It includes consumer credit and mortgage debt. This table reveals great differences between countries. At the top of the list is the U.S., with household debt standing at 68.81% of GDP; Canada is not far behind at around 65%. Then come Spain, Germany, and France at between 54% and 42%, and then finally Italy at around 19% and Japan at a mere 11.98%. In terms of the average annual net increase in debt outstanding, the largest is again the U.S. at 4.4%, with Spain, Canada, and the U.K. not far behind at around 3%. Germany, France, and Italy have considerably lower increases at between 1% and 0.5%, and Japan actually had an average annual decrease, although it was very small. Of course, these figures cannot be regarded as precise because there are minor differences in the definitions used. But the overall impression is reasonably clear. The U.S. had the largest household debt outstanding and the largest annual net increase. Canada is very similar. Spain does not have as sizable a level of debt as the North American countries, but it has such a high rate of increase that its debt relative to GDP will increase further from the levels of comparable major European countries (with the exception of the U.K.). On the other hand, the stocks of consumer debt in Germany, France, Italy, and Japan not only are low relative to GDP but will stay low because of their low growth rates (unless there are dramatic changes in their growth rates). The immediate question arises as to why such differences exist.

Table 13.1. *International differences in the volume of credit,* 1998. (*Sources: Calculated from Financial Statistics, Office for National Statistics* (*U.K.*); *Federal Reserve Board, U.S.; Bank of Canada; and International Financial Statistics Yearbook, Vol.* 11, *International Monetary Fund,* 1999).

Country	Consumer credit outstanding/GDP (%)	Consumer credit outstanding per capita	Residential mortgage debt outstanding/GDP (%)
U.S.	15.65	US$4,922	52.63
U.K.	12.21	£1,727	54.53
Canada	17.67	C$5,223	44.53

Table 13.2. *Intercountry differences in household debt. The numerator consists of the sum of short-term and long-term household debt* (*or net increase in short-term plus long-term household debt*) *divided by the number of years in the period. The denominator is the mean GDP over the years in the period. All prices are current.* (*Sources: Calculated from OECD Financial Statistics, Part* 2: *Financial Accounts of OECD Countries, various issues, and International Monetary Fund, International Financial Statistics, January* 2000.)

Country	Period	Debt outstanding as percentage of GDP	Net increase in debt outstanding as percentage of GDP
U.S.	1992–1996	68.81	4.429
Germany	1992–1996	43.24	0.485
Canada	1992–1995	64.94	3.051
France	1992–1996	42.07	0.903
Spain	1992–1996	53.89	3.441
Italy	1992–1996	18.52	0.831
Japan	1992–1996	11.98	–0.001
U.K.	1992–1995	n.a.	3.323

Remember from Chapter 3 that the volume of credit observed depends on demand, supply, and the extent to which those who demand credit are unable to obtain it, that is, the extent of credit rationing. Therefore, differences in the volume of credit outstanding between countries will depend on differences in demand, in supply, or in the extent to which credit constraints apply between countries. This explanation for intercountry differences in the amount of debt outstanding was investigated by Jappelli and Pagano (1989), and we follow their findings here. Jappelli and Pagano examined the volume of debt as a proportion of consumer expenditure using data from around 1961 to around 1984 for seven countries: Sweden, the U.S., the U.K., Japan, Italy, Spain, and Greece. They used data for consumer credit that was available at that time on a consistent basis for these countries. However, while Jappelli and Pagano explained the rank ordering of debt outstanding in the 1960s, 1970s, and early 1980s, the rank ordering has, according to the figures in Table 13.2, changed. Thus in Jappelli and Pagano's study, the rank ordering was the U.S., the U.K., Japan, Italy, and Spain in descending order in terms of consumer plus mortgage debt outstanding as a percentage of consumption expenditure; in 1992–1996, the rank ordering in terms of short- plus long-term loans as a percentage of GDP was the U.S., Spain, Italy, and Japan (with no figures available for the U.K.). Unfortunately, we know of no other study that tried to rigorously explain intercountry differences in consumer debt. However, the factors that Jappelli and Pagano isolated to explain the ranking in their period may also explain the ranking in 1998; the difference may just be that the conditions that lead to a high ranking in their period now apply to different countries in the 1990s. In any event, we need to be cautious about taking Jappelli and Pagano's reasons as necessarily being applicable today.

First, we might expect that the greater the amount by which the rate of interest on loans exceeds that which lenders have to pay to gain funds (the interest rate wedge), the greater the amount of credit that lenders would wish to supply. However, Jappelli and Pagano found that in mortgage markets, there was no correlation between the wedge and volume of lending. For example, Sweden had a relatively high wedge but relatively low volume of loans, Italy had a high wedge and low volume of loans, and the U.S. had both a high wedge and a high loan loan volume.

Second, Jappelli and Pagano looked at the average percentage of house prices paid as down payment and at the percentage of home owners in younger age groups. They found that, generally speaking, in those countries where the down payments are a small percentage of a house price, home ownership occurs earlier. Thus they found that rationing in the market for mortgages is more common in Japan, Italy, and Spain than in the U.K., the U.S., and Sweden and that this, rather than differences in supply, may contribute to intercountry differences in mortgage debt outstanding.

Finally, they considered differences in demand. One factor that might explain intercountry differences in demand is differences in tax rates, with those countries offering greater tax deductions from interest payments having a greater demand, everything else constant. But this effect will depend on the marginal rate of tax in each country since that is the rate that borrowers would avoid paying. Jappelli and Pagano document the tax incentives to borrow in each sample country and consider a proxy for the marginal tax rate. They conclude that in five of the seven countries, there were no incentives to take consumer loans. However, the greater incentives in Sweden and the U.S. may partially explain the greater debt-to-consumption ratios observed there. In the case of mortgage loans, differences in tax regimens were also too small to explain differences in debt taken.

A second factor which may explain differences in demand is differences between countries in the age distribution and earnings profiles of their populations. To assess the effect of these, for each country, Jappelli and Pagano took the life cycle theory of consumption (see Chapter 3) to predict consumption at each age and the observed earnings profile to simulate the volume of debt-to-consumption ratio. This was then compared with the observed ratios. They concluded that there was little relation between the predicted and actual rankings of debt to consumption. For example, debt to consumption was predicted to be highest in Japan and lowest in the U.K., whereas in Japan the observed ratio was low and in the U.K. high.

Finally, preferences for debt may differ between countries. Unfortunately, in the absence of direct measures of tastes, Jappelli and Pagano can only consider differences in the proportion of consumers' expenditure on durables as indicative of tastes and they find no correlation between this and the debt-to-consumption ratio. Of course, this variable may be a very inaccurate indicator of tastes.

Overall, Jappelli and Pagano conclude that differences in the amount of credit rationing, especially in the markets for mortgages, explains intercountry differences in debt-to-consumption ratios.

13.2.2 Credit cards

Table 13.3 shows the number of credit and charge cards per head issued in Europe in 1997, where the numbers of cards were reported by Datamonitor (1998). The figures suggest that it is possible to distinguish between four groups. The very high users are the U.S. and the U.K. with 1,616 and 661 cards per 1000 people in the population, respectively. Belgium, the Netherlands, Spain, Sweden, and Germany have around 200 cards per 1000 of the population, and finally Italy and France have around 100 cards per 1000 of the population. The numbers of superpremium and premium cards, of corporate cards, and of standard cards per head are in the same rank order as the totals, with one or two exceptions. The exceptions are the very high number of premium and corporate cards per head in Sweden and the relatively high number of premium cards in Germany.

Table 13.3. *European credit and charge cards,* 1997 (*per* 1000 *people in population*).

Country	Superpremium + premium	Corporate	Standard	Total
U.S.	650.4	20.9	945.0	1616.3
U.K.	91.3	22.5	546.7	660.5
Belgium	53.0	6.9	197.4	257.3
Netherlands	38.3	9.4	195.9	243.5
Spain	26.5	4.3	212.0	242.8
Sweden	44.2	46.4	85.8	176.4
Germany	39.7	4.6	127.8	172.0
Italy	18.2	9.7	109.1	137.0
France	25.1	3.1	68.3	96.6

Fianco (1998) discussed some possible reasons for these differences given in the table. She argued that in the U.S., recent intense competition between card issuers has resulted in the acceptance of more risky applicants who previously would not have been accepted. In addition, extensive direct marketing mailshots and preapprovals were issued and interest rates lowered. However, delinquency rates on premium cards increased commensurately. The U.K. followed a pattern similar to that of the U.S., and U.S. card issuers who entered the U.K. market have taken large shares of the premium and superpremium markets.

Fianco also points out some interesting differences in the relative significance of fee and interest income in the premium and corporate sectors. Interest makes up 60% and 80% of income in the U.K. and the U.S., respectively, while in the rest of Europe, fee income makes up around 90% of income. The difference is caused by the greater use of revolving credit in the U.S. and the U.K. and the greater use of corporate cards as charge cards in Europe.

13.3 International differences in credit bureau reports

Details of the information contained in U.K. credit bureau reports are given in Chapter 8. In this chapter, we consider information in bureau reports from other countries.

13.3.1 The U.S.

In the U.S., there are more than 1000 credit bureaus, but almost all are either owned or under contract to one of three main credit reporting companies: Experian, TransUnion, and Equifax. Information contained in a U.S. credit report is similar to that in a U.K. report except that much more information about the characteristics of accounts opened by the person is recorded. Again, the precise items included vary between agencies, but a typical report would include the following information:

- Personal information. Name, address, former address, date of birth, name of current and former employers; Social Security number.
- Public record information. Any details of court judgments against the person, the amount, the balance, the plaintiff, and the current status; details of any declarations of bankruptcy and of tax liens.
- Credit accounts history. For each account, the account number and lender; the date it was opened; the type of account; credit limit; payment in dollars; whether the payments are up to date; the balance outstanding; and how many times the regular payment has been 30 days, 60 days, or 90 days overdue.
- Inquiries. The names of credit grantors that have asked for the credit report and the dates on which they did so.

Most adverse information about an account remains on a report for up to seven years from the date of the first delinquency on the account. The same applies to civil judgments, discharged or dismissed Chapter 13 bankruptcies, and tax liens from the date of payment. Chapters 7, 11, and 12 remain on the report for 10 years. Information on searches of your report remain for one to two years depending on the type of organization that carried out the search.

No credit reference reports in either country contain information on religion, race, or national origin. Credit reports do not include information on checking accounts or savings accounts.

In the U.S., access to a person's credit report is governed by the Federal Fair Credit Reporting Act of 1996, as are a person's rights if he is denied credit. In the U.K., access is governed by the Data Protection Act of 1998, and a person's rights are governed by the Consumer Credit (Credit Reference Agency Regulations) Act of 1999.

13.3.2 Other countries

The content and even availability of credit bureau reports varies greatly between countries. Jappelli and Pagano (2000) carried out a survey to establish the availability of bureau information among 45 countries. Here we briefly summarize some of their results for countries in the European Union (excluding Luxembourg). Jappelli and Pagano found that credit bureau data were available in all countries except France and Greece. In all countries, information on debts outstanding and on defaults was available except for Belgium, Denmark, Finland, and Spain, where only black data were available. They also found that public debt registers are available in all countries except Denmark, Finland, Ireland, the Netherlands, Sweden, and the U.K., although the amount of information varies between countries. Generally, if there is a public credit register, it contains information on defaults, arrears, total loans outstanding, and guarantees.

One would expect that in countries where there is no credit bureau data, then everything else being equal, the statistical discriminatory power would be rather lower than in countries where more information exists. However, there are no empirical studies of this issue.

13.4 Choice of payment vehicle

A small number of papers attempted to predict the method of payment most frequently used for different types of goods and services. Most of these papers relate to consumers in the U.S., although a few relate to consumers in other countries. These studies are difficult to compare because the alternative forms of payment that they analyze differ, as do the goods and services considered and the variables used to predict the customers' chosen methods. It may be that these differences, rather than the differences between the countries in which the consumers studied reside, explain any differences in payment method. This is a further opportunity for research.

The most recent study for the U.S., by Carow and Staten (1999), considers the choice between gas credit cards, cash, general-purpose credit cards, and debt as payment for gas. They found that consumers are more likely to use cash when their income is low, the purchaser is middle-aged, they have relatively little education, and the purchaser has few cards. Other credit and debit cards are more likely to be used when the purchaser is relatively young, is more educated, and holds more cards. Lindley et al. (1989), who considered purchases in 1983, also found that purchases of gas were more likely to be made using a credit card than by cash or checks when the buyer's income was higher. They also found the same result for purchases of furniture, household goods, and clothing, but they did not detect an influence of the buyer's age, nor for years of education, except in the case of clothing. For the purchase of household goods and of clothing, the chance of using a credit card was lower if the buyer owned his own house.

An interesting set of results relating perceived attributes to various characteristics of different forms of payment in 1980 was gained by Hirschman (1982) in his study of U.S. consumers. Hirschman attempted to predict the relative frequency of use of five methods of payment—cards, store cards, travel and entertainment cards, cash, and checks—for different types of goods and services. He found that, for example, the use of bank cards was associated with the perceptions of bank cards' high security, high prestige, speedy transfer time, higher acceptability, lower transaction time, the possession of a transaction record, and the greater ease of facilitating purchases by gaining debt.

For the U.K., Crook, Hamilton, and Thomas (1992b) used a sample of bank credit card holders to distinguish between those who, over a fixed period in the late 1980s, used the card and those who did not. They found that those 30 to 40 years old were more likely to use the card than other age groups and that those of higher income were more likely to use their cards, as were those who had had an account at the issuing bank for four to five years and those who had lived at the same address for under six months or for very many years. Of all the possible categories of residential status, those least likely to use their card were tenants in unfurnished accommodation.

The only published study for Australia is by Volker (1983), who distinguished between those who did not hold a credit card and those who both held a card and used it. The probability of possession and use of credit cards was found to be lower if the person was aged 16 to 19 years, 20 to 24 years, 45 to 54 years, or over 55 years. The probability was also lower for the skilled and unskilled occupational groups than for other occupational groups and for females than for males.

13.5 Differences in scorecards

In section 12.2, we considered generic scorecards versus those specific to a particular credit vendor. In this section, we consider a related issue: differences in scorecards between countries. Little is known about international differences in scorecards. However, with an increasing number of European countries adopting a common currency (the Euro), one might expect to see increasing numbers of applications for loans from overseas residents who are still within the same currency region. If both the lender and the borrower use the same currency, then fluctuations in exchange rates will have no effect on payments. Thus it would be in the interest of lenders for them to advertise for borrowers in other countries, especially if the lender can offer lower interest rates than lenders in the applicant's home country.

This raises the question of whether the same scorecard can be applied to applicants from foreign countries as to home applicants. The only publicly available study of this question is by Platts and Howe (1997), who considered data relating to applicants in the U.K., Germany, Greece, Belgium, and Italy. They found that different variables are collected in different countries, and initially they compared the discriminating power of each variable taken on its own by considering the weights of evidence for each variable in each country. They discovered that the same variable had very different predictive power in different countries. For example, not having a home telephone was very much more predictive of default in the U.K. than in any of the other countries; having a bad credit bureau report was more predictive of default in Italy, Germany, and the U.K. than in Greece; and being a tenant was more predictive in the U.K. than in any of the other countries.

Platts and Howe then built three types of scorecard. First, a separate scorecard for each of the five countries was built using, for each country, data only from that country and using variables that discriminated best for that country: the specific scorecard. The second and third scorecards used the same attributes (variables). The second consisted of a separate scorecard for each country but used data only for the home country of the applicant, thus allowing the scores from each attribute to differ between the countries and for different weights of evidence: the country scorecard. The third used data pooled across all applicants from the five countries: the European scorecard. Platts and Howe then compared the percentage improvement in bad debt when using each scorecard compared with that currently in use when the cutoff scores were chosen to maintain the same acceptance rate. Their results are shown in Figure 13.1.

Figure 13.1 shows that for all five countries, the country-specific scorecard performs better than the country scorecard, which in turn outperforms the European scorecard. In most cases, all the scorecards performed better than existing systems. The differences were smallest for Belgium but largest for the U.K. One possible explanation for the relatively poor performance of the European card in the U.K. is that in the U.K. there is a relatively large amount of credit bureau data. The greatest improvement in performance of the country scorecard over the European scorecard (both using the same attributes) was again in the U.K., with a 12% improvement in bad debt. The benefits to the other countries were very similar at between 4% and 6%. Platts and Howe concluded that these differences were sufficiently large that it was worth the additional cost of building country-specific cards rather than using one single generic card.

13.6 Bankruptcy

If an individual fails to keep up payments of loans, creditors typically pursue debt collection activities. In the U.S., creditors can, subject to federal regulations, garnish the debtor's

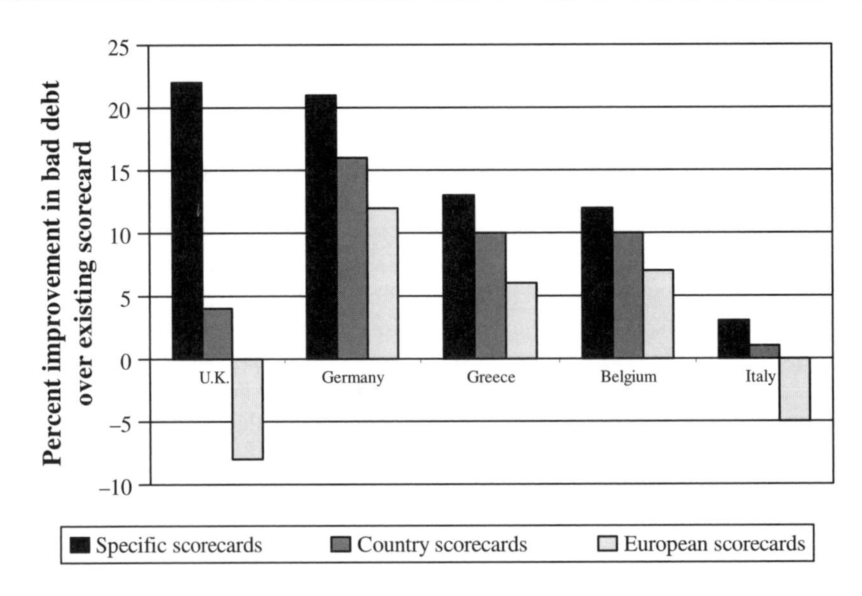

Figure 13.1. *Comparison of European, country, and specific scorecards.*

wages and take actions to repossess the debtor's property. Alternatively, debtors may declare themselves bankrupt. Unlike many other countries, the U.S. allows individuals to choose which of two types of bankruptcy to declare (under the Bankruptcy Reform Act of 1978): Chapter 7 procedures and Chapter 13 procedures.

Under a Chapter 7 filing, a debtor surrenders all of his nonexempt assets to the court. A plan is made to sell the assets and the proceeds to be distributed to creditors. Secured debt would be repaid first, for example, loans where houses or cars were offered as collateral. Unsecured loans would be repaid last. Unsecured include debt where the value of the collateral was less than the loan outstanding and also, usually, debt outstanding on credit cards and installment credit. Once the plan is agreed, the debtor is discharged from the requirement to repay almost all his debts. No further garnishment of wages or repossession of assets is allowed. The debts that are not waived include recent income tax and child support. Note also that if property is collateral for a loan, an exemption cannot prevent the creditor repossessing the property. The nature and value of exempt assets are stated in federal law but also vary considerably between states (see White 1998).

Under a Chapter 13 filing, debtors with regular income offer a plan to reschedule the payments for most of their debt over three to five years. One exception is mortgages, which cannot be rescheduled, although debtors can pay off mortgage arrears to prevent foreclosure. The debtor keeps all of his assets, and creditors are restrained from their efforts to collect debt. Often, debtors who file under Chapter 13 do not have to pay some part of their debt, although they must plan to repay as much unsecured debt as they would under Chapter 7. Chapter 13 plans require the approval only of the judge, not of the creditors, to be implemented.

There has been discussion as to how individuals might—and do—respond to these procedures. White (1998) considered several possible strategies an individual might adopt if he files under Chapter 7. The person may borrow relatively large amounts of unsecured debt (for example, installment credit or credit card debt) and either use it to buy consumption goods (such as holidays) and services that cannot be liquidated (or, if they can be sold, only for a small faction of their original cost) or, alternatively, use it to pay off nondischargeable debt. Debtors may then choose to declare themselves bankrupt. As a result, this recently acquired debt and older unsecured debt will all be written off by lenders. Indeed, Berkowitz and Hynes (1999) argue that "many debtors" file for bankruptcy so that they can discharge their nonmortgage debt to reduce their payments on these so as to have sufficient funds to keep repaying their mortgage. As evidence, Berkowitz and Hynes cite Sullivan, Warren, and Westerbrook (1989), who found that in a sample of 1529 bankruptcy cases in 1981, 40% of debtors had secured debt and 10% owned their own homes but did not declare a mortgage debt.

A second strategy White considered is for individuals to liquidate assets that are not exempt from seizure by debtors and use the proceeds to reduce the mortgage on their main home, which in many states under state law is at least partially exempt from seizure. This would enable the individual to retain a greater amount of assets after bankruptcy. A third strategy White considered exists if, having adopted the second strategy, the individual's mortgage is completely repaid but he still has some nonexempt assets. In this case, the individual would be able to retain even more assets after bankruptcy if he uses the proceeds to improve his residence or move to a more valuable residence.

Using data from the Survey of Consumer Finance (1992), White estimated the percentage of households in the U.S. as a whole and in eleven selected states who would benefit from these three strategies rather than declaring bankruptcy and not following any of these strategies. Her findings for the U.S. as a whole are that by not following one of the three strategies, 17% of households would benefit, whereas the percentage who would benefit from pursuing each of the strategies is 24% in each case.

The recent empirical evidence that examined the choice of chapters an individual makes is not so clear-cut. For example, Domowitz and Sartain (1999) found that the choice of chapter depended more on location, whether the person was married, employment, auto equity, home equity, and personal assets than on home exemption relative to debt, other exemptions relative to debt, the volume of credit card debt, or different types of secured or unsecured debt.

Chapter 14

Profit Scoring, Risk-Based Pricing, and Securitization

14.1 Introduction

Thus far, with a few exceptions, credit and behavioral scoring have been used to estimate the risk of a consumer defaulting on a loan. This information is used in the decision on whether to grant the loan (credit scoring) or extend the credit limit (behavioral scoring). In this chapter, we discuss how scoring can be used to make more sophisticated financial decisions, so we extend some of the topics outlined in Chapter 9. In section 14.2, we discuss how the accept-reject decision can be made explicitly in terms of maximizing the profit, even with a default-based credit score. We point out that inconsistencies in the scoring system can then lead to more complicated decisions than just one cutoff score. Moreover, the objective might be not just profit but could also involve the chance of losses being not too great and that the volume of loans accepted meets the lender's requirements. Section 14.3 describes a profit-based measure of scorecard accuracy that allows comparison of scorecards or cutoff levels purely in profit terms.

Section 14.4 presents a brief review of the ways one could try to build a profit-scoring system as opposed to a default-risk system. Although many organizations seek to develop such scorecards, it is fair to say that no one has yet perfected an approach to this problem. If profit is one's objective, then of course the lender's decision is not just, "Shall I accept the consumer or not?" but also, "What interest rate should I charge?" This leads to the idea of risk-based pricing, which has been very slow to come in credit cards or other forms of consumer lending. The reasons behind this hesitancy are discussed in section 14.5, which contains a simple model of risk-based pricing where the score can be used to set the interest rate.

Another area where the score could be used as a measure of general risk rather than just the default risk of the individual consumer is securitization. Groups of loans are packaged together, underwritten, and distributed to investors in the form of securities. The price at which the package of loans is sold must reflect the risk of nonrepayment, i.e., should be related to the credit scores of the individual loans in the package. In practice this link is not very transparent, and section 14.6 discusses the issue of securitization. The one area where securitization has been an overwhelming success is mortgage-backed securities in the U.S. This is the second-biggest bond market in the U.S. after Treasury bonds, so in the last section we look at how this market functions and why it has proved so much more successful in the U.S. than anywhere else.

14.2 Profit-maximizing decisions and default-based scores

Recall the model of the credit-granting process developed in section 4.2. The application characteristics $\mathbf{x}$ give rise to a score $s(\mathbf{x})$. (We drop the x dependency of the score at times for ease of notation.) p_G and p_B are the proportion of goods and bads in the population as a whole and so $o_p = \frac{p_G}{p_B}$ is the population odds. $q(G|s)$ ($q(B|s)$) is the conditional probability that a consumer with score s will be good (bad) and $q(G|s) + q(B|s) = 1$. Let $p(s)$ be the proportion of the population that has score s. Assume that the profit from a consumer is a random variable R, where

$$R = \begin{cases} 0 & \text{if the account is rejected,} \\ L & \text{if the account is accepted and becomes good (L is lost profit from ruling out a good),} \\ -D & \text{if the account is accepted and becomes bad (so D is the default amount).} \end{cases} \tag{14.1}$$

The expected profit per consumer if one accepts those with score s is

$$E\{R|s\} = Lq(G|s) - D(1 - q(G|s)) = (L + D)q(G|s) - D. \tag{14.2}$$

Thus to maximize profit, one should accept those with scores s if $q(G|s) \geq \frac{D}{D+L}$. Let A be the set of scores where the inequality holds; then the expected profit per consumer from the whole population is

$$E^*\{R\} = \sum_{s \in A} ((L + D)q(G|s) - D)p(s). \tag{14.3}$$

This analysis ignored fixed costs. As section 9.2 suggests, fixed costs encourage the lender to set lower cutoff scores than do those that maximize profits.

Notice that if the profits and default losses $L(s)$ and $D(s)$ were score dependent, the decision rule would be to accept if $q(G|s) \geq \frac{D(s)}{D(s)+L(s)}$. In that case, A may consist of several regions of scores, i.e., accept if the score is between 300 and 400, accept if between 500 and 550, and accept if over 750; reject otherwise. However, we assume that the profits and losses are independent of the score and that $q(G|s)$ is monotonically increasing in s. This means that $A = \{s|s \geq c\}$, where $q(G|s) \geq \frac{D}{D+L}$, so c is the cutoff score. In this case, define $F(s|G)$, $F(s|B)$ to be the probabilities a good or a bad has a score less than s:

$$\begin{aligned} E^*\{R\} &= \sum_{s \geq c} ((L + D)q(G|s) - D)p(s) = \sum_{s \geq c} (Lp_G p(s|G) - Dp_B p(s|B)) \\ &= Lp_G(1 - F(c|G)) - Dp_B(1 - F(c|B)) \\ &= Lp_G - Dp_B + (Dp_B F(c|B) - Lp_G F(c|G)). \end{aligned} \tag{14.4}$$

The first term on the right-hand side, $Lp_G - Dp_B$, is the profit if we accept everyone, so the second term is the profit that the scorecard brings. One can rewrite (14.4) in a different way to say how far away $E^*\{R\}$ is from the expected profit if there were perfect information $E\{PI\}$. With perfect information, one would only accept goods, so $E\{PI\} = Lp_G$. Hence

$$\begin{aligned} E^*\{R\} &= Lp_G - (Dp_B(1 - F(c|B)) + Lp_G F(c|G)) \\ &= E\{PI\} - (Dp_B(1 - F(c|B)) + Lp_G F(c|G)). \end{aligned} \tag{14.5}$$

This analysis follows the approach of Oliver (1993) very closely, and one of the points he makes is that terms like $q(G|s)$ and $F(c|G)$ have two interpretations. They can be the forecasted probabilities and they can be the actual proportions in the portfolio. Thus (14.4) and (14.5) give both the forecast performance profit and the actual performance profit, but these are two very different quantities although they have the same expression. It brings us back to some of the questions discussed in Chapter 7 on measuring the performance of a scorecard. If one thinks of a scorecard as a forecast, then the questions are how discriminating it is (how far from 0.5 $q(G|s)$ is) and how well calibrated it is (how different the forecast $q(G|s)$ is from the actual $q(G|s)$). Usually, the most discriminating forecasts are not well calibrated, and the well-calibrated forecasts are not usually the most discriminating.

In a subsequent paper, Oliver and Wells (2001) developed these models further to describe the different trade-offs that credit granters make in choosing their portfolios of loans by setting a cutoff. Let $F(s)$ be the proportion of scores below s; i.e., $F(s) = F(s|G)p_G + F(s|B)p_B$. Many lenders look at the trade-off of the bad acceptance rate (the percentage of the total bad population accepted) against the acceptance rate, i.e., $(1 - F(s|B))p_B$ against $1 - F(s)$. The actual bad rate, which is the percentage of those accepted who are bad, is the ratio of these two numbers, so

$$\text{Actual bad rate} = \frac{(1 - F(s|B))p_B}{1 - F(s)}. \tag{14.6}$$

This is called the strategy curve, and Figure 14.1 shows the strategy curve for a typical scorecard. The bold line shows the strategy curve for the scorecard with perfect information and hence perfect discrimination. Clearly, the nearer the scorecard strategy curve gets to that, the better it is.

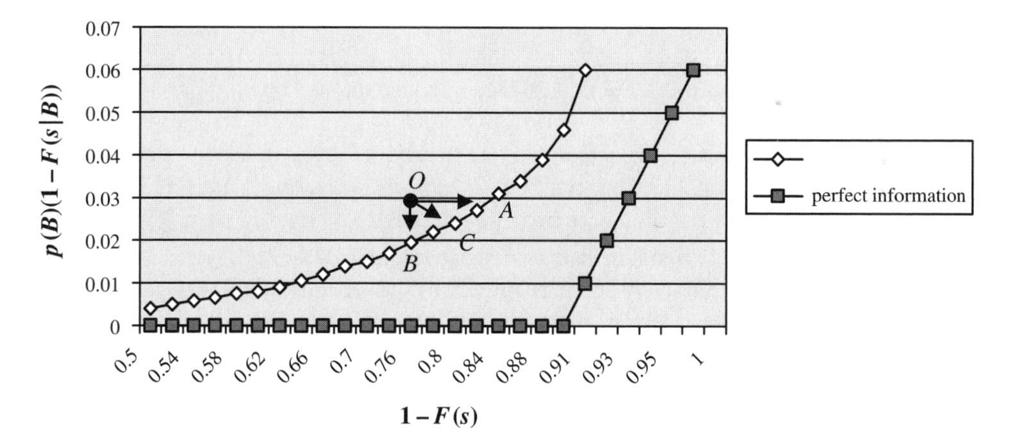

Figure 14.1. *Strategy curve.*

What often happens on introducing a new scorecard is that the existing operating policy gives a point O that is above the new strategy curve. The question then is where on the strategy curve one wants to go by choosing the appropriate cutoff. If one moves to A, then one keeps the bad acceptance rate the same but accepts more people, while moving to B would accept the same numbers but lower the bad acceptance rate and hence the bad rate. Moving to C would keep the bad rate the same and again increase the numbers accepted. The following example, based on Oliver and Wells (2001), investigates this.

Example 14.1. A bank's current policy O has an acceptance rate of 72%, a bad rate of 2.6%, and the population odds ($o = \frac{p_G}{p_B}$) of 10.3 to 1. The data for a new scorecard are as given in Table 14.1. It follows from the existing statistics that $p_G = \frac{o}{1+o} = \frac{10.3}{11.3} = 0.912$ and so $p_B = 0.088$. Since $1 - F(s) = 0.72$ (the acceptance rate), from (14.6) the bad acceptance rate

$$p_B(1 - F(s|B) = \text{(bad rate)} \cdot \text{(acceptance rate)} = 0.026 \cdot 0.72 = 0.0187.$$

Hence $1 - F(s|B) = \frac{0.0187}{0.088} = 0.213$. Since $1 - F(s) = (1 - F(s|G))p_G + (1 - F(s|B))p_B$, we can write

$$1 - F(s|G) = \frac{(1 - F(s)) - p_B(1 - F(s|B))}{p_G} = \frac{0.72 - 0.0187}{0.912} = 0.769.$$

Table 14.1. *Strategy curve details for new scorecard in Example* 14.1.

Marginal odds	7.2:1	7.8:1	8.3:1	8.9:1	9.6:1	11.0:1	12.5:1	15.1:1	20.6:1	30.2:1
$1 - F(s\|G)$	0.848	0.841	0.833	0.827	0.818	0.803	0.776	0.746	0.693	0.631
$1 - F(s\|B)$	0.233	0.218	0.213	0.193	0.181	0.166	0.144	0.126	0.097	0.069
$1 - F(s)$	0.793	0.786	0.778	0.771	0.762	0.747	0.720	0.691	0.640	0.581

We can use this table to calculate the statistics for the new possible points A, B, and C. At point A, the bad acceptance rate is maintained at 1.87%, so $1 - F(s|B) = \frac{0.0187}{0.088} = 0.213$. This is the point with marginal odds 8.3:1 in the table and corresponds to an acceptance rate of 77.8% and hence an actual bad rate of $\frac{0.0187}{0.778} = 0.024$, i.e., 2.4%.

At B, we keep the acceptance rate the same, so $1 - F(s) = 0.72$, which corresponds to the point with marginal cutoff odds of 12.5:1. Since for this point $1 - F(s|B) = 0.144$, the bad acceptance rate is $(0.144)(0.088) = 0.012672$, and hence the bad rate is $\frac{0.012672}{0.72} = 0.0176$, i.e., 1.76%.

Finally, if at C we want the actual bad rate to stay at 2.6%, we need to find the point where $\frac{0.088(1-F(s|B))}{1-F(s)} = 0.026$. Checking, this is the case at marginal odds of 7.2:1 when $1 - F(s|B) = 0.232$ and $1 - F(s) = 0.794$ since $\frac{0.088(0.232)}{0.794} = 0.0257$. Hence at this point the acceptance rate is 79.4% and the bad acceptance rate is 2.04%.

An alternative to the strategy curve, suggested by Oliver and Wells (2001), is to plot the expected losses against the expected acceptance rate, but all that this does is multiply the scale of the Y-axis in Figure 14.1 by D, the cost of a default. More interesting is to plot the expected losses against the expected profit, where the profit is $L(1 - F(s)) - (L + D)F(s|B)p_B$, so the new X-axis is at an angle to the original one in Figure 14.1. This leads to Figure 14.2.

This is intriguing in that one can get the same expected profit in at least two ways. The points on the lower part of the curve have higher cutoff scores, so less bads are accepted; those on the higher part of the curve correspond to lower cutoff scores with higher numbers of bads accepted. The efficient frontier of this curve is the lower part from C to D. These give points with an expected profit and an expected loss, in which the former cannot be raised without the latter also being raised.

If a lender is at present operating at point O, then again he can move onto the new scorecard curve either by keeping the bad acceptance rate the same, i.e., A, or by keeping the acceptance rate the same (which would move to point B on the curve). In this case, one would suggest that the move to A is less sensible because it is not an efficient point and one could have the same expected profit with lower expected losses.

One can obtain the efficient frontier in a more formal way as follows. One is seeking to minimize the expected loss with lower bound P^* on the expected profit:

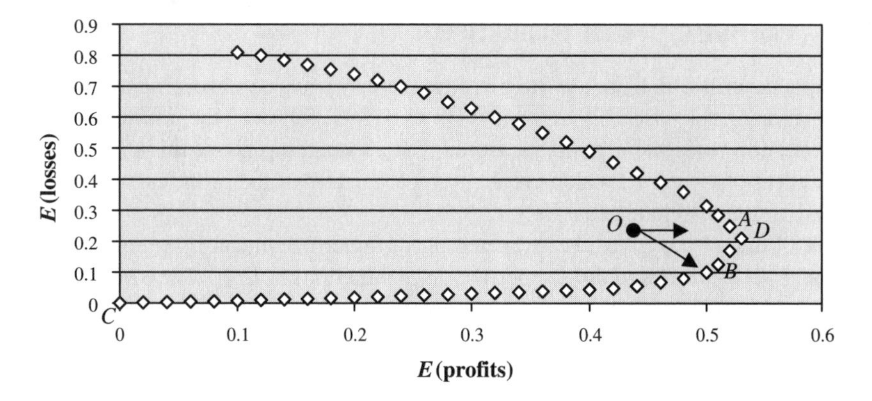

Figure 14.2. *Expected losses against expected profits.*

$$\begin{array}{ll} \text{Min}_s & Dp_B(1 - F(s|B)) \\ \text{subject to} & Lp_G(1 - F(s|G)) - Dp_B(1 - F(s|B)) \geq P^*. \end{array} \tag{14.7}$$

One can solve this using the Kuhn–Tucker optimality conditions for nonlinear optimization, i.e., the derivative of (objective + λ(constraint)) = 0 at a minimum, to get the condition

$$\begin{array}{r} Dp_B f(s|B) - \lambda(Lp_G(1 - f(s|G)) - Dp_B(1 - f(s|B))) = 0 \quad \text{and} \\ \lambda(Lp_G(1 - F(s|G)) - Dp_B(1 - F(s|B) - P^*)) = 0, \quad \lambda \geq 0, \end{array} \tag{14.8}$$

where $f(s|B)$ is the derivative of $F(s|B)$ with respect to s. Then the shadow price λ satisfies

$$\lambda = \frac{Dp_G f(s^*|B)}{Lp_G f(s^*|G) - Dp_B f(s^*|B)} = \frac{1}{\frac{o^*}{\bar{o}} - 1} > 0, \tag{14.9}$$

where $o^* = \frac{p_G f(s^*|G)}{p_B f(s^*|B)}$ is the marginal odds at the cutoff score and $\bar{o} = \frac{D}{L}$ is the optimal odds for the unconstrained problem. Notice that this implies that $o^* > \bar{o}$—the odds for the constrained problem should be higher than those for the unconstrained problem.

Some lenders add another constraint, namely, that there should be a lower bound on the numbers accepted. This leads to the following extension of (14.7), where the total profit is being maximized, N is the number in the population, and N_0 is the minimum number that must be accepted:

$$\begin{array}{ll} \text{Min}_s & DNp_B(1 - F(s|B)) \\ \text{subject to} & N(Lp_G(1 - F(s|G)) - Dp_B(1 - F(s|B))) \geq P^*, \quad N(1 - F(s)) \geq N_0. \end{array} \tag{14.10}$$

This splits into three cases depending on what happens to the multipliers λ and μ of the profit constraint and the size constraint at the solution point. If the profit constraint is exactly satisfied and the size constraint is a strict inequality ($\mu = 0$), then the result is similar to (14.9). If the profit constraint is a strict inequality ($\lambda = 0$) and the size constraint is an equality, then the number N_0 accepted by this constraint gives the cutoff limit by $1 - F(c) = \frac{N_0}{N}$. Third, if N_0 is very large, then there will be no feasible cutoff that meets both constraints.

14.3 Holistic profit measure

The criteria described in Chapter 7 to measure the performance of scoring systems were all discrimination or risk based. Hoadley and Oliver (1998) followed the profit approach of the previous section to define a profit measure. The clever thing about this measure is that it is independent of D and L, which are the two parameters of the profit calculations that are always difficult to estimate accurately.

Recall that (14.3) gives the expected profit under a scorecard with cutoff c, and the profit under perfect information (or perfect discrimination) is Lp_G. Then the holistic profit is defined as

$$\text{HP} = \frac{\text{expected profit under scorecard with cutoff } c}{\text{expected profit under perfect discrimination}} = 1 - F(c|G) - \frac{Dp_B}{Lp_G}(1 - F(c|B). \tag{14.11}$$

If instead of defining the cutoff c we define the marginal odds at the cutoff, $o^* = \frac{q(G|s)}{q(B|s)} = \frac{D}{L}$ and the population odds $o_0 = \frac{p_G}{p_B}$. Then we can rewrite (14.11) as the holistic profit scan $h(o^*)$, which is the holistic profit in terms of the marginal cutoff odds o^* by

$$h(o^*) = 1 - F(s(o^*)|G)| \left(\frac{o^*}{o_0}\right) (1 - F(s(o^*)|B)). \tag{14.12}$$

Using the data of Table 14.1, we get the holistic profit scan in Figure 14.3.

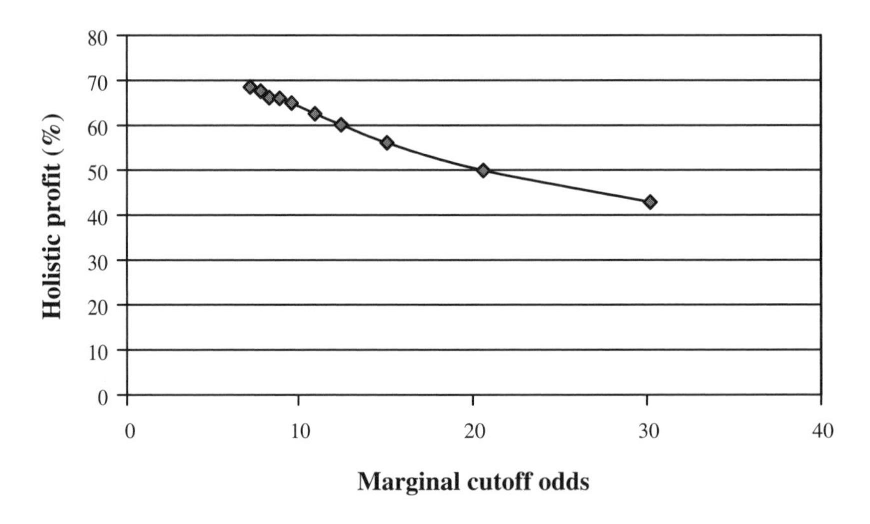

Figure 14.3. *Holistic profit scans.*

One has to be careful not to read too much into this graph. It is not the case that the best cutoff is where the curve is a maximum. It depends on the values of L and D. What the graph does is give a view of how the scorecard will perform compared with the perfect discriminating scorecard as the ratio $\frac{D}{L} = o^*$ varies.

One can also use holistic profit functions to assess how sensitive this profit is to choice of suboptimal cutoffs. Suppose that $o^* = \frac{D}{L}$; then one can draw the holistic profit function as

$$h(s|o^*) = 1 - F(s|G) - \left(\frac{o^*}{o_0}\right)(1 - F(s|B)). \tag{14.13}$$

This has its maximum when s is such that the marginal odds at the score are o^* and the curves tend to be fairly flat in the region close to this score; see Figure 14.4.

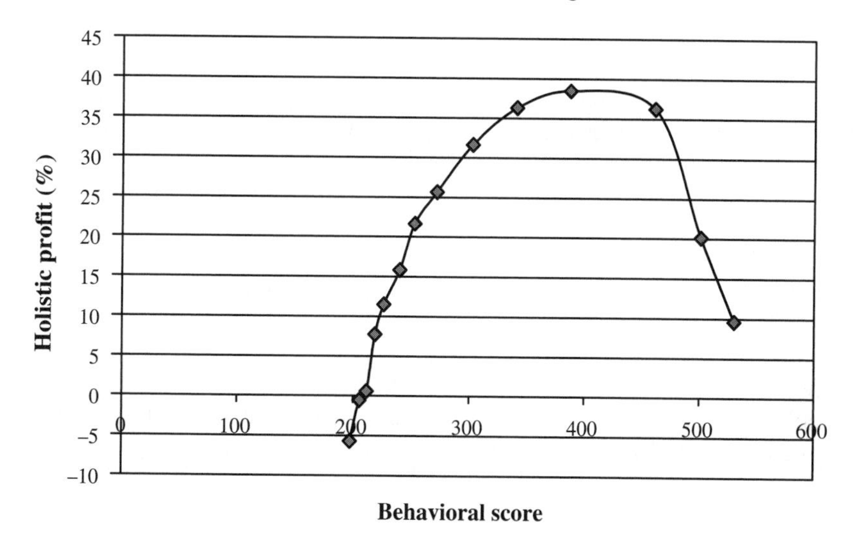

Figure 14.4. *Holistic profit if* $\frac{D}{L} = 20$.

If $o^* = o_0$, then (14.13) reduces to $F(s|B) - F(s|G)$, which is the Kolmogorov–Smirnov distance or, to put it another way, how far the ROC curve is away from the diagonal.

14.4 Profit-scoring systems

As mentioned at several points in this text—for example, in section 10.6—one way to deal with profitability in consumer lending is to change the objective of the scoring system from minimizing default to maximizing profit. Initially, one may feel that all is required is a change in the definition of good in the scoring techniques, and some organizations have gone along this path. However, whereas default rates are affected by acceptance decisions, credit limit decisions, and default recovery decisions, profits are affected by many more decisions, including marketing, service levels, and operation decisions, as well as pricing decisions. Thus moving to profit scoring implies that these techniques should help a whole new range of decisions—in fact, almost all the decisions a retailer or retail bank may be involved in.

Even if the impact of profit scoring is wider than firms first imagine, surely there should be no difficulty in implementing the change. However, a number of problems arise, which is why it is taking organizations so long to move to fully blown profit-scoring systems. First, there are data-warehousing problems in ensuring the accounts include all the elements that make up the profit. Even in credit card organizations, this has proved difficult in that the credit card company gets a certain percentage of each purchase made on the card paid back to it by the retailer—the merchant service charge. This charge varies considerably between the types of purchases, and the companies have had to revamp their systems so that this information can be readily accessed. Similarly, the retail part of an organization often writes off all or a fixed percentage of any bad debt a customer incurs and never checks how much of it is actually recovered subsequently by the debt-recovery department. These examples suggest that profit scoring requires a fully integrated information system in the organization. One needs the information on all the customers' transactions (and maybe a whole family's transactions) and accounts collated together to calculate the customers' profitability to the firm; hence the push to data warehousing by companies so that all this information is kept and is easily accessible. This could lead to legal problems because the use of personal information for reasons other than those for which it was originally collected is frowned on by legislators in many countries.

The advent of data-mining techniques—see (Jost 1998) for its uses in credit scoring—means that the technical problems of analyzing such vast amounts of data are being addressed. However, there are still major problems in developing models for profit scoring. Over what time horizon should profit be considered, given that we do not wish to alienate customers by charging high prices now and then losing future profit as they move to another lender? Also, profit is a function of economic conditions as well as the individual consumer's characteristics. (See Crook et al. (1992a) for the effect economic conditions have between one year and the next.) Thus it is even more important to include economic variables into profit scoring than it was in credit scoring. Profit is dependent on how long a customer stays with a lender, and so one wants to know how long customers stay and whether they default or just move their custom elsewhere. Thus one needs to estimate attrition rates as part of profit scoring.

Last, two difficulties affect which methodology to choose. Should one look at the profit on each product in isolation or look at the total profit over all possible products? The former means that one could decide not to offer a customer a new credit card because he does not use it enough. This refusal may offend a customer so much that his profitable home loan is moved to another lender. Going for total profit, on the other hand, ignores the fact that the decision on which products a customer takes is the customer's decision. He can cherry-pick as well and may refuse the product where the firm felt it would make the most profit from him. Second, there is the problem of censored data. In a sample of past transactions, the total profit for current customers will not be known but only the profit up to the date that the sample history finished.

So what approaches are being tried? They can be classified into four groups. One approach is to build on the existing scorecards, which estimate default, usage, acceptance, and attrition, and try to define the profit for groups of the population segmented according to their scores under these different measures. Oliver (1993) was one of the first to suggest this and looked at what decision rules should be used if one has a transaction profit score and a default score. Fishelson-Holstine (1998) described a case study where one tried to segment according to two types of profit. A bank runs a private-label credit card for a retailer. The retailer wants to increase profits by sending cards to people who will use it to buy more in their stores, while the bank wants the credit card operations of the customer to be profitable. By using a demographically based segmentation tool, details of the retailers sales, and the credit card transaction database, groups were identified who were profitable for both. The idea of estimating intermediate variables, outlined in section 12.5, is an obvious step toward estimating profit. One should try to estimate intermediate variables like balance outstanding and purchases and use these to estimate the final outcome. This approach is related to the graphical networks and the Bayesian learning models of section 12.6.

A second approach is to mimic the regression approach of credit scoring by trying to describe profit as a linear function of the categorical application form variables. Almost all the data will be censored in that the total profit is not known, but there is a body of literature on regression with censored data (Buckley and James 1979), and although the censoring that occurs in credit scoring has not yet been dealt with, research in this area is continuing (Lai and Ying 1994).

The third approach is to build on the Markov chain approaches to behavioral scoring outlined in section 6.4 and develop more precise stochastic models of customer behavior. Cyert, Davidson, and Thompson's (1962) original model was used to model profit in a one-product case, and these approaches have proved very successful in estimating debt provisioning for portfolios of customers with the same product. If one extends the ideas to the profit a customer brings when one has to look at the several products he has or may possibly have with the lender, the problems become one of data availability and computational power rather than of modeling difficulty. One runs up against the curse of dimensionality that arises when one seeks to use Markov chains to model complex real situations. However, a number of techniques are proving very successful on other areas to overcome this problem, including aggregation-disaggregation and parallel computing (Thomas 1994).

Fourth, one could recognize that the survival analysis approach of section 12.7 is another way to begin to address profit by estimating how long the loan will run before unexpected—and hence usually unprofitable—events occur. The techniques of survival analysis—proportional hazard models and accelerated life models—could then be used to estimate the long-term profit from a customer given only the experience in the first few months or years. Narain (1992) was the first to suggest that one could use this analysis on credit-scoring data, while the paper by Banasik, Crook, and Thomas (1999) showed that one could also use the idea of competing risks from reliability to get good estimates of when borrowers will default and when they will pay off early, thus incorporating default and attrition in the same analysis. Proportional hazards and accelerated life approaches are useful ways to think about how economic affects can be introduced into profit models. By taking the characteristic variables in proportional hazards and accelerated life to describe the economic conditions as well as the characteristics of the borrower, one can build a model that allows for the speeding up in default rates that occurs in poor economic conditions.

Profit-scoring systems seem more difficult to obtain than might have been first thought, but the prize for success is enormous. A profit-scoring system would provide a decision support aid that has the same focus throughout the different decision-making areas of the organization. It provides an excellent way to take advantage of all the new data on consumer behavior that has become available in the last few years with electronic point-of-sales equipment and loyalty cards.

14.5 Risk-based pricing

One of the real surprises in consumer credit over the 50 years of its history is how long it has taken lenders to start to price the cost of the loan to the risk involved. For most economic goods, the supplier produces a good at a price to clear the market, yet in the credit industry there are usually funds available for consumer credit—it being so much less risky than other forms of lending—and yet there are whole segments of the population who cannot get the credit they want. This is because normally instead of adjusting the price of credit, which is the interest rate they charge, lenders decide on a fixed interest rate (price) and use credit scoring to decide to whom to lend at this price. There have always been different groups of lenders targeting different parts of the credit market: naïvely, one could say gold cards are for the least risky group, normal credit cards for the next group, store cards for a slightlier risky group, and door-to-door collectors (tallymen in Old English) for the group even more at risk. Even so, there are groups who fall between or outside these bands and who find it hard to get the credit they want. Within each product group, the price tends to be fixed across most lenders. For example, it is only in the past decade that one has had subprime lending coming into the credit card market or some financial institutions offering very attractive rates to their own best customers; see Crook (1996) and Jappelli (1990), who identify the characteristics of those unable to gain the credit they wish.

This slowness in trying to relate price to risk is surprising in consumer lending because credit scoring is an ideal technique for setting risk-based prices or interest rates, as the following example shows.

Suppose one has a scoring system with p_G, p_B, $p(s|G)$, $p(s|B)$, $p(s) = p(s|G)p_G + p(s|B)p_B$, $q(G|s) = \frac{p(s|G)}{p(s)}$, $q(B|s) = \frac{p(s|B)}{p(s)}$ defined as in section 14.2. Assume that the interest rate charged, i, is a function of the credit score s, so $i(s)$. Let the cost of default be D, and we assume this is independent of the interest rate but that the profit on a good customer, $L(i)$, does depend monotonically on the interest rate i charged. Then for each score s, one has to decide whether to accept consumers at that score and, if so, what interest rate to charge. One is interested for a score s in maximizing

$$\max_i\{(L(i)q(G|s) - Dq(B|s))a_s(i), 0\}. \tag{14.14}$$

This is clearly maximized when $L(i)$ is maximized, and this silly result shows what is missing in that not all the potential customers with a credit score s will accept a loan or credit card when the interest rate being charges is i. So let $a_s(i)$ be the fraction of those with credit score s who will accept interest rate i, while $q(G|s,i)$, $q(B|s,i)$ are the fraction of acceptors with score s when the interest rate is i who are good or bad. Adverse selection suggests that these conditional probabilities do depend on i because as the interest rate increases, it is only the people who cannot get credit elsewhere who are accepting, and these will increasingly be poorer risks even if their credit score has not changed. Clearly, $a_s(i)$ is decreasing in i, and what we want is to maximize

$$\max_i\{(L(i)q(G|s,i) - Dq(B|s,i))a_s(i), 0\}. \tag{14.15}$$

Differentiating with respect to i and setting the derivative to 0 to find the maximum gives that the optimal interest rate for a score s satisfies

$$\begin{aligned} -L'(i)q(G|s)a_s(i) + (L(i)q'(G|s,i) - Dq'(B|s,i))a_s(i) \\ = (L(i)q(G|s) - Dq(B|s))a_s'(i). \end{aligned} \tag{14.16}$$

In a specific case, suppose that $a_s(i) = e^{-\alpha(s)(i-i^*)}$, i.e., everyone accepts an interest rate i^* and the subsequent dropoff is exponential, and that there is no effect of interest rate on the fraction of goods who accept. Let $L(i) = \frac{R}{(1+i)^T} - \frac{R}{(1+i^*)^T}$; i.e., there is one payment of R at time T with the interest charged being i, while the real cost of capital is i^*. Then (14.14) becomes

$$\begin{aligned} \frac{-R}{(1+i)^{T+1}}q(G|S)e^{-\alpha(s)(i-i*)} &= -\left(\frac{R}{(1+i)^T}q(G|s) - Dq(B|s)\right)\alpha(s)e^{-\alpha(s)(i-i*)} \\ \text{or} \quad \frac{q(G|s)}{(1+i)^T}\left(\alpha(s) + \frac{1}{1+i}\right) &= Dq(B|s)\alpha(s). \end{aligned} \tag{14.17}$$

Solving this gives the interest rate to be charged for those with credit score s. One can do similar calculations for all types of loan products.

One reason that risk-based pricing will become more prevalent for individual loans is that the idea of risk-based pricing of portfolios of loans is already with us in the concept of securitization. The last two sections of this chapter look at this issue.

14.6 Securitization

Securitization is the issuing of securities in the finance market backed not like bonds by the expected capacity of a corporation to repay but by the expected cash flows from specific assets. The term *securitization* is usually applied to transactions in which the underlying assets are removed from the original owner's balance sheet, called "off-balance sheet securitization," although in some European countries there is a system of "on-balance sheet securitization." The most widely used assets in securitization are residential mortgages, credit card receivables, automobile loans, and other consumer purchases—the very areas where credit scoring is used to assess the credit risk of the individual loan.

Securitization begins with a legal entity known as a special-purpose vehicle (SPV) set up by three parties—the sponsor, the servicer, and the trustee. The sponsor, who is often the original owner of the assets (the originator), but need not be, sells the receivables to the SPV, which is often called the issuer. The only business activity of the issuer is to acquire and hold the assets and issue securities backed by the assets, and the servicer is the administrator of this securitization. Often the sponsor is also the servicer, but there are third parties who also specialize in this area. To properly administer the securities, the servicer must be able to collect on delinquent loans and recover on defaulting loans. It must be able to repossess houses in the case of mortgage assets, dispose of collateral in the case of car loans, and generate new receivables in accordance with underwriting standards in the case of credit card loans. The trustee acts as an intermediary between the issuer and the investors in the event of default and ensures the orderly payment of the interest and the principal by the issuer to the investors. For the publicly offered securities, credit-rating agencies will assign a rating to the SPV based on the possible credit problems with the assets and the legal structure of the operation.

Securitization began in the 1970s to promote residential mortgage finance in the U.S., which is still one of the largest securitization markets. In the 1980s, the techniques developed were applied to an increasingly wide range of assets in the U.S. One would have expected these financial instruments to become worldwide very quickly. There has been considerable progress in Canada, the U.K., and Australia, which have legal and regulatory systems similar to the U.S. However, in many countries, the existing laws have prevented securitization, although some countries have started to change their rules to allow wider securitization. Even today, the most startling feature of securitization is the difference between the size of the asset-backed security market and the mortgage-backed security markets in the U.S. and those in the rest of the world.

It is vital for securitization that one can separate the assets from the originating firm so that the investors bear the risk only on a clearly defined existing pool of loans that meet specific criterion. They are not exposed to other risks that the originator might have, like the geographic concentration of its total portfolio (loans from several originators are put into the same SPV to overcome that), changes in the default rate of future customers, or losses on other of the originators assets. The assets can be transferred from the sponsor to the issuer in three ways:

(a) Novation or clean transfer. The rights and the obligations of the assets are transferred to the issuer. This is how most retail store cards are securitized. However, it requires the consent of all parties, including the borrower of the original loan, since consumer protection laws may be violated otherwise. In effect, a new contract is written between the original borrower and the issuer of the securities.

(b) Assignment. The original borrower keeps paying the originator, who in turn pays the money to the issuer. This is the norm for finance loans for automobile purchase. In this case, the issuer cannot change the rules of the original agreement with the borrower.

(c) Subparticipation. There is a separate loan between the issuer (the subparticipant) and the sponsor which the latter pays back out of the money repaid by the borrowers. The issuer has no rights over the original borrowers and so has a double credit risk of whether the original borrower and the sponsor will default.

The advantages of securitization are manifold. Borrowers can either borrow from financial institutions or raise funds directly in the capital market. It is usually cheaper to raise money by the sale of securities on the capital market than to borrow from banks. The cost of securitization must be less than the difference in the interest rates paid on the securities and on their loan equivalents. Therefore, securitization is typically used for loans where the credit risk is easy to assess (perhaps because there is lots of historic information) and where monitoring is adequate. Investors are willing to pay for the greater liquidity and credit transparency that the securities bring over loans. It is also the case that securitization allows originators who are not highly credit rated to get access to funds at better rates. Essentially, the originators' loans may be less risky than they are, and so by separating out their loans, the originators can create SPVs that have higher credit ratings than they do and so borrow money more cheaply. Banks also like securitization because it increases liquidity and because originating and then securitizing new loans increases their profit by raising the turnover rather than the volume of their assets. The biggest boost to securitization in the U.S. was the lack of liquidity in the banking sector associated with the asset quality problems of the savings and loan associations in the crisis of the mid 1980s. Securitization is also a way banks avoid having to hold capital to cover the risk of the loans defaulting, as they are required to do under the Basle rules.

There are reasons for not securitizing as well. Clearly, if a ready supply of cheap money is available from normal methods, there is less need for the hassle of securitization. There is also reluctance by some institutions to damage their special relationships with the original borrowers that creating an SPV involves. One has to be careful when securitizing the quality part of a loan portfolio to get the high credit rating that what remains is not so high risk and unbalanced that the whole organization is at risk. However, the main problem to securitization is still the financial cost and the need for skilled people to undertake the project. This is more so in Europe, where there has been a development in one-off exotic types of securitization rather than the standard securitizations that are the bulk of the U.S. market.

Another problem with securitization is that the credit risk of the portfolio of loans may be too great to be easily saleable. To reduce that risk, issuers will offer credit enhancement, which is usually of one of four types:

(a) Third-party enhancement. An external party like an insurance company will guarantee against the risk of nonpayment.

(b) Subordination. There are different sorts of investors, some of whom get priority in the case of difficulties in repayment in exchange for not getting such good rates of return.

(c) Overcollateralization. The assets are of greater value than is needed to support the agreed payments to the investors.

(d) Cash collateral accounts. Cash is held in a deposit account to be paid out if there a shortfall in the cash received from the receivables.

The second and even more vital part of reducing risk is the role of the ratings agencies, who will provide ratings of the credit (i.e., default) risk of the SPV. The ratings agency examines the historical performance of the receivables and uses simulations based on difficult economic conditions in the past to see how likely the loans are to default. They also examine the external credit enhancements and usually apply the weak-link principle, meaning that a security cannot be rated higher than the rating of the provider of the enhancement. The ratings given can be one of about a dozen levels going from AAA to C, for example.

So what is the relationship between credit scoring and securitization? There are two areas in which credit scoring can be used. The first is deciding what portfolio of loans to put together. Sometimes the portfolio is composed of loans all of the same type, for example, automobile purchasers of the same type of automobile, but one could also put together loans with the same likelihood of default, i.e., the same good:bad odds. Notice that we did not say the same score band because the loans could have come from a number of lenders with very different scorecards, but the odds ratio is the connection between them all. There is very little published work describing how sponsors put together their loan portfolios (probably because it is a very lucrative area), but one would assume that using the score is a sensible way to proceed.

There are some caveats, however, in that the credit score of two different accounts does not describe how those accounts are correlated in terms of risk. It could be that the borrowers have very similar characteristics and so might be vulnerable to the same economic or personal pressures that lead to defaulting, or they could be quite different, and although they have the same overall default risk, the underlying pressures are independent. Clearly, there is need for much more research to identify how credit-scoring techniques can be extended to help identify correlations.

The second and related area is how the ratings agencies can use the credit scores to come up with their overall ratings. One would expect a credit score giving the default risk of each loan in the portfolio would be of considerable help in determining the overall risk of default in the portfolio. However, the problem that the credit score does not describe correlations between the risks to the individuals still holds, and one may need to develop more creative approaches to get the credit score of a portfolio of loans. The methods the ratings agencies use are not widely disseminated, but again one feels there is room for considerable developments in this area. For more details of the whole of the securitization issues, there are good reviews by Lumpkin (1999), and for a view of the international scene, see an article in *Financial Market Trends* (1995).

14.7 Mortgage-backed securities

Mortgage-backed securities were the first of the consumer loans to be securitized and are still far and away the biggest market. In the mid 1990s, almost three-quarters of the home mortgage lending in the U.S. was financed through securitization. As of the end of 1998, the volume of outstanding mortgage-backed securities in the U.S. was $2.6 trillion, accounting for roughly half of all outstanding mortgage credit. The amount of other asset-backed securities in the U.S. was $284 billion. Compared with this, the mortgage-backed securities market in other countries is much smaller. Even in the U.K., where activity has grown considerably, nowhere near this percentage of the market is covered by securitized products.

The reasons for this unevenness are regulatory, historical, and related to the financial markets in the different countries. Historically, the U.S. housing market has long fixed-rate mortgages with financing provided by banks and savings and loan associations. In the 1970s, interest rate volatility heightened the mismatch between the short-term deposit rates and the long-term repayment rates. Thus federal agencies started to guarantee residential mortgages and then issued them as collateral securities to investors. These agencies—the Government National Mortgage Association (Ginnie Mae), the Federal Home Loan Mortgage Corporation (Freddie Mac), and the Federal National Mortgage Association (Fannie Mae)—were responsible for most mortgage-backed securitization. Also, the legal framework in the U.S. allowed for SPVs to be set up as trusts, while other countries have had to change their laws to allow them. It is also worth noting that in the late 1980s, U.S. banks started shifting away from traditional lending to fee-based sources of income. Securitization was the key component in this strategy, and only recently have banks in other countries started moving more to this strategic approach.

One of the major differences between mortgage-backed securities and asset-backed securities is that the chance of default is probably even smaller in the former than in the latter. However, the chance of prepayment or early repayment is much higher since the average mortgage loan term is more than 25 years but the average duration of a mortgage is less than 7 years. Prepayment has a negative effect on the investor as they were expecting to have their money tied up for longer maturities than turns out to be the case. So in terms of scoring, this means that scoring for early repayment is as important as scoring for default. Quite sophisticated models have been built to estimate the risk of prepayment of mortgage-backed securities in terms of changes in the interest rate, but these tend not to involve the early repayment scores or the default scores of the underlying accounts. In this area, the survival analysis models of section 12.7 as well as the Markov chain modeling of section 6.4 may prove alternatives to the interest-based models that are currently most commonly used.

References

H. T. ALBRIGHT (1994), *Construction of a Polynomial Classifier for Consumer Loan Applications Using Genetic Algorithms*, Working Paper, Department of Systems Engineering, University of Virginia, Charlottesville, VA.

E. I. ALTMAN (1968), Financial ratios, discriminant analysis and the prediction of corporate bankruptcy, *J. Finance*, 23, 589–609.

T. W. ANDERSON AND L. A. GOODMAN (1957), Statistical inference about Markov chains, *Ann. Math Statist.*, 28, 89–109.

P. BACCHETTA AND F. BALLABRIGA (1995), *The Impact of Monetary Policy and Bank Lending: Some International Evidence*, Working Paper 95.08, Studienzentrum Gerzensee, Gerzensee, Switzerland.

S. M. BAJGIER AND A. V. HILL (1982), An experimental comparison of statistical and linear programming approaches to the discriminant problem, *Decision Sci.*, 13, 604–611.

J. BANASIK, J. N. CROOK, AND L. C. THOMAS (1996), Does scoring a subpopulation make a difference?, *Internat. Rev. Retail Distribution Consumer Res.*, 6, 180–195.

J. BANASIK, J. N. CROOK, AND L. C. THOMAS (1999), Not if but when borrowers default, *J. Oper. Res. Soc.*, 50, 1185–1190.

M. S. BARTLETT (1951), The frequency goodness of fit test for probability chains, *Proc. Cambridge Philos. Soc.*, 47, 86–95.

G. BENNETT, G. PLATTS, AND J. CROSSLEY (1996), Inferring the inferred, *IMA J. Math. Appl. Business Indust.*, 7, 271–338.

J. BERKOWITZ AND R. HYNES (1999), Bankruptcy exemptions and the market for mortgage loans, *J. Law Econom.*, 42, 809–830.

B. S. BERNANKE (1986), Alternative explanations of money-income correlation, *Carnegie-Rochester Ser. Public Policy*, 25, 49–100.

B. S. BERNANKE AND A. S. BLINDER (1988), Is it money or credit, or both, or neither?, *Amer. Econom. Rev. Papers Proc.*, 78, 435–439.

B. S. BERNANKE AND A. S. BLINDER (1992), The federal funds rate and the channels of monetary transmission, *Amer. Econom. Rev.*, September, 901–921.

B. S. BERNANKE AND M. GERTLER (1995), Inside the black box: The credit channel of monetary policy transmission, *J. Econom. Perspectives*, 9, 27–48.

H. Bierman and W. H. Hausman (1970), The credit granting decision, *Management Sci.*, 16, 519–532.

D. Biggs, B. De Ville, and E. Suen (1991), A method of choosing multiway partitions for classification and decision trees, *J. Appl. Statist.*, 18, 49–62.

C. M. Bishop (1995), *Neural Networks for Pattern Recognition*, Oxford University Press, Oxford.

M. Blackwell (1993), *Measuring the Discriminatory Power of Characteristics, Proceedings of Credit Scoring and Credit Control* III, Credit Research Centre, University of Edinburgh, Edinburgh, Scotland.

Board of Governors of the Federal Reserve System (1992), *Survey of Consumer Finance*, Board of Governors of the Federal Reserve System, Washington, DC.

M. Boyle, J. N. Crook, R. Hamilton, and L. C. Thomas (1992), Methods for credit scoring applied to slow payers, in *Credit Scoring and Credit Control*, L. C. Thomas, J. N. Crook, and D. B. Edelman, eds., Oxford University Press, Oxford, 75–90.

L. Breiman, J. H. Friedman, R. A. Olshen, and C. J. Stone (1984), *Classification and Regression Trees*, Wadsworth, Belmont, CA.

N. Breslow (1974), Covariance analysis of censored survival data, *Biometrics*, 30, 89–99.

J. Buckley and I. James (1979), Linear regression with censored data, *Biometrika*, 66, 429–436.

N. Capon (1982), Credit scoring systems: A critical analysis, *J. Marketing*, 46, 82–91.

K. A. Carow and M. Staten (1999), Debit, credit or cash: Survey evidence on gasoline purchases, *J. Econom. Business*, 51, 409–421.

G. G. Chandler and D. C. Ewert (1976), *Discrimination on Basis of Sex and the Equal Credit Opportunity Act*, Credit Research Center, Purdue University, West Lafayette, IN.

K. C. Chang, R. Fung, A. Lucas, R. Oliver, and N. Shikaloff (2000), Bayesian networks applied to credit scoring, *IMA J. Math. Appl. Business Indust.*, 11, 1–18.

S. Chatterjee and S. Barcun (1970), A nonparametric approach to credit screening, *J. Amer. Statist. Assoc.*, 65, 150–154.

G. A. Churchill, J. R. Nevin, and R. R. Watson (1977), The role of credit scoring in the loan decision, *Credit World*, March, 6–10,

R. T. Clemen, A. H. Murphy, and R. L. Winkler (1995), Screening probability forecasts: Contrasts between choosing and combining, *Internat. J. Forecasting*, 11, 133–146.

J. Y. Coffman (1986), *The Proper Role of Tree Analysis in Forecasting the Risk Behavior of Borrowers*, MDS Reports, Management Decision Systems, Atlanta, GA, 3, 4, 7, 9.

D. Collett (1994), *Modelling Survival Data in Medical Research*, Chapman and Hall, London.

A. W. Corcoran (1978), The use of exponentially smoothed transition matrices to improve forecasting of cash flows from accounts receivable, *Management Sci.*, 24, 732–739.

D. R. Cox (1972), Regression models and life-tables (with discussion), *J. Roy. Statist. Soc. Ser. B*, 74, 187–220.

J. N. Crook (1989), The demand for retailer financed instalment credit: An econometric analysis, *Managerial Decision Econom.*, 10, 311–319.

J. N. Crook (1996), Credit constraints and U.S. households, *Appl. Financial Econom.*, 6, 477–485.

J. N. Crook (1998), Consumer credit and business cycles, in *Statistics in Finance*, D. J. Hand and S. D. Jacka, eds., Arnold, London.

J. N. Crook (2000), The demand for household credit in the U.S.: Evidence from the 1995 Survey of Consumer Finance, *Appl. Financial Econom.*, to appear.

J. N. Crook, R. Hamilton, and L. C. Thomas (1992a), The degradation of the scorecard over the business cycle, *IMA J. Math. Appl. Business Indust.*, 4, 111–123.

J. N. Crook, R. Hamilton, and L. C. Thomas (1992b), Credit card holders: Users and nonusers, *Service Industries J.*, 12 (2), 251–262.

R. M. Cyert, H. J. Davidson, and G. L. Thompson (1962), Estimation of allowance for doubtful accounts by Markov chains, *Management Sci.*, 8, 287–303.

S. Dale and A. Haldane (1995), Interest rates and the channels of monetary transmission: Some sectoral estimates, *European Econom. Rev.*, 39, 1611–1626.

C. Darwin (1859), *On the Origin of Species by Means of Natural Selection, or the Preservation of Favoured Races in the Struggle for Life*, John Murray, London.

Datamonitor (1998), *Analysis Profitability in Prestige Sectors*, Report DMFS0376, Datamonitor, London.

R. H. Davis, D. B. Edelman, and A. J. Gammerman (1992), Machine-learning algorithms for credit-card applications, *IMA J. Math. Appl. Business Indust.*, 4, 43–52.

M. H. De Groot and E. A. Ericksson (1985), Probability forecasting, stochastic dominance and the Lorenz curve, in *Bayesian Statistics*, J. M. Bernado, M. H. De Groot, D. V. Lindley, and A. F. M. Smith, eds., North–Holland, Amsterdam, 99–118.

V. S. Desai, D. G. Convay, J. N. Crook, and G. A. Overstreet (1997), Credit scoring models in the credit union environment using neural networks and genetic algorithms, *IMA J. Math. Appl. Business Indust.*, 8, 323–346.

V. S. Desai, J. N. Crook, G. A. Overstreet (1996), A comparison of neural networks and linear scoring models in the credit environment, *European J. Oper. Res.*, 95, 24–37.

Y. M. I. Dirickx and L. Wakeman (1976), An extension of the Bierman-Hausman model for credit granting, *Management Sci.*, 22, 1229–1237.

I. Domowitz and R. L. Sartain (1999), Incentives and bankruptcy chapter choice: Evidence from the Reform Act 1978, *J. Legal Stud.*, 28, 461–487.

L. M. Drake and M. J. Holmes (1995), Adverse selection and the market for consumer credit, *Appl. Financial Econom.*, 5, 161–167.

L. M. DRAKE AND M. J. HOLMES (1997), Adverse selection and the market for building society mortgage finance, *Manchester School*, 65 (1), 58–70.

D. DURAND (1941), *Risk Elements in Consumer Instalment Financing*, National Bureau of Economic Research, New York.

D. B. EDELMAN (1988), Some thoughts on the coding of employment categories, *Viewpoints*, 12 (4), San Rafael, CA.

D. B. EDELMAN (1997), Credit scoring for lending to small businesses, in *Proceedings of Credit Scoring and Credit Control* V, Credit Research Centre, University of Edinburgh, Edinburgh, Scotland.

D. B. EDELMAN (1999), Building a model to forecast arrears and provisions, in *Proceedings of Credit Scoring and Credit Control* VI, Credit Research Centre, University of Edinburgh, Edinburgh, Scotland.

D. EDWARDS (1995), *Introduction to Graphical Modelling*, Springer-Verlag, New York.

B. EFRON (1977), The efficiency of Cox's likelihood function for censored data, *J. Amer. Statist. Assoc.*, 72, 557–565.

R. A. EISENBEIS (1977), Pitfalls in the application of discriminant analysis in business, finance and economics, *J. Finance*, 32, 875–900.

R. A. EISENBEIS (1978), Problems in applying discriminant analysis in credit scoring models, *J. Banking Finance*, 2, 205–219.

R. A. EISENBEIS (1996), Recent developments in the application of credit scoring techniques to the evaluation of commercial loans, *IMA J. Math. Appl. Business Indust.*, 7, 271–290.

EQUAL CREDIT OPPORTUNITY ACT (1975), *United States Code*, Title 15, Section 1691 et seq.

EQUAL CREDIT OPPORTUNITY ACT AMENDMENTS OF 1976 (1976), *Report of the Committee on Banking Housing and Urban Affairs, 94th Congress*, U.S. Government Printing Office, Washington, DC.

S. S. ERENGUC AND G. J. KOEHLER (1990), Survey of mathematical programming models and experimental results for linear discriminant analysis, *Managerial Decision Econom.*, 11, 215–225.

S. FIANCO (1988), Platinum players, *European Card Rev.*, October, 26–28.

FINANCE AND LEASING ASSOCIATION (2000), *Guide to Credit Scoring* 2000, Finance and Leasing Association, London.

FINANCIAL MARKET TRENDS (1995), Securitisation: An international perspective, *Financial Market Trends*, 61, 33–53.

H. FISHELSON-HOLSTINE (1998), Case studies in credit risk model development, in *Credit Risk Modeling*, E. Mays, ed., Glenlake Publishing, Chicago, 169–180.

R. A. FISHER (1936), The use of multiple measurements in taxonomic problems, *Ann. Eugenics*, 7, 179–188.

E. Fix and J. Hodges (1952), *Discriminatory Analysis, Nonparametric Discrimination, Consistency Properties*, Report 4, Project 21-49-004, School of Aviation Medicine, Randolph Field, TX.

C. Forgy (1982), Rete: A fast algorithm for the many pattern/many object pattern match problem, *Artif. Intell.*, 19, 17–37.

N. Freed and F. Glover (1981a), A linear programming approach to the discriminant problem, *Decision Sci.*, 12, 68–74.

N. Freed and F. Glover (1981b), Simple but powerful goal programming formulations for the discriminant problem, *European J. Oper. Res.*, 7, 44–60.

N. Freed and F. Glover (1986a), Evaluating alternative linear programming models to solve the two-group discriminant problem, *Decision Sci.*, 17, 151–162.

N. Freed and F. Glover (1986b), Resolving certain difficulties and improving classification power of LP discriminant analysis formulations, *Decision Sci.*, 17, 589–595.

M. Friedman and A. J. Schwartz (1963), *A Monetary History of the United States*, 1867–1960, Princeton University Press, Princeton, NJ.

H. Frydman (1984), Maximum likelihood estimation in the Mover-Stayer model, *J. Amer. Statist. Assoc.*, 79, 632–638.

H. Frydman, J. G. Kallberg, and D.-L. Kao (1985), Testing the adequacy of Markov chains and Mover-Stayer models as representations of credit behaviour, *Oper. Res.*, 33, 1203–1214.

K. Fukanaga and T. E. Flick (1984), An optimal global nearest neighbour metric, *IEEE Trans. Pattern Anal. Mach. Intell.*, PAMI-1, 25–37.

N. C. Garganas (1975), An analysis of consumer credit and its effects on purchases of consumer durables, in *Modelling the Economy*, G. Renton, ed., Heinemann Educational Books, London.

G. D. Garson (1998), *Neural Networks: An Introductory Guide for Social Scientists*, Sage, London.

M. Gertler and S. Gilchrist (1993), The role of credit market imperfections in the transmission of monetary policy: Arguments and evidence, *Scand. J. Econom.*, 95 (1), 43–64.

J. J. Glen (1999), Integer programming models for normalisation and variable selection in mathematical programming models for discriminant analysis, *J. Oper. Res. Soc.*, 50, 1043–1053.

F. Glover (1990), Improved linear programming models for discriminant analysis, *Decision Sci.*, 21, 771–785.

D. E. Goldberg (1989), *Genetic Algorithms in Search Optimization and Machine Learning*, Addison–Wesley, Reading, MA.

B. J. Grablowsky and W. K. Talley (1981), Probit and discriminant functions for classifying credit applicants: A comparison, *J. Econom. Business*, 33, 254–261.

D. J. HAND (1981), *Discrimination and Classification*, John Wiley, Chichester, U.K.

D. J. HAND (1997), *Construction and Assessment of Classification Rules*, John Wiley, Chichester, U.K.

D. J. HAND AND W. E. HENLEY (1993), Can reject inference ever work?, *IMA J. Math. Appl. Business Indust.*, 5, 45–55.

D. J. HAND AND W. E. HENLEY (1997), Statistical classification methods in consumer credit, *J. Roy. Statist. Soc. Ser. A*, 160, 523–541.

D. J. HAND AND S. D. JACKA (1998), *Statistics in Finance*, Arnold, London.

D. J. HAND, K. J. MCCONWAY, AND E. STANGHELLINI (1997), Graphical models of applications for credit, *IMA J. Math. Appl. Business Indust.*, 8, 143–155.

A. HARTROPP (1992), Demand for consumer borrowing in the U.K., 1969–90, *Appl. Financial Econom.*, 2, 11–20.

S. HAYKIN (1999), *Neural Networks: A Comprehensive Foundation*, Prentice–Hall International, London.

W. E. HENLEY (1995), *Statistical Aspects of Credit Scoring*, Ph.D. thesis, Open University, Milton Keynes, U.K.

W. E. HENLEY AND D. J. HAND (1996), A k-NN classifier for assessing consumer credit risk, *Statistician*, 65, 77–95.

E. C. HIRSCHMAN (1982), Consumer payment systems: The relationship of attribute structure to preference and usage, *J. Business*, 55, 531–545.

B. HOADLEY AND R. M. OLIVER (1998), Business measures of scorecard benefit, *IMA J. Math. Appl. Business Indust.*, 9, 55–64.

P. G. HOEL (1954), A test for Markov chains, *Biometrika*, 41, 430–433.

K. HOFFMANN (2000), Stein estimation: A review, *Statist. Papers*, 41, 127–158.

J. H. HOLLAND (1968), *Hierarchical Description of Universal and Adaptive Systems*, Department of Computer and Communication Sciences, University of Michigan, Ann Arbor.

J. H. HOLLAND (1975), *Adaptation in Natural and Artificial Systems*, University of Michigan Press, Ann Arbor.

M. A. HOPPER AND E. M. LEWIS (1991), Development and use of credit profit measures for account management, *IMA J. Math. Appl. Business Indust.*, 4, 3–17.

M. A. HOPPER AND E. M. LEWIS (1992), Behaviour scoring and adaptive control systems, in *Credit Scoring and Credit Control*, L. C. Thomas, J. N. Crook, and D. B. Edelman, eds., Oxford University Press, Oxford, 257–276.

D. C. HSIA (1978), Credit scoring and the Equal Credit Opportunity Act, *Hastings Law J.*, 30, 371–448.

R. JAMBER, S. JAV, D. MEDLER, AND P. KLAHR (1991), The credit clearing house expert system, in, *Innovative Applications of Artificial Intelligence* 3, R. Smith and C. Scott, eds., AAAI Press, Menlo Park, CA.

T. JAPPELLI (1990), Who is credit constrained in the U.S.?, *Quart. J. Econom.*, 105, 219–234.

T. JAPPELLI AND M. PAGANO (1989), Consumption and capital market imperfections, *Amer. Econom. Rev.*, 79 (5), 1088–1105.

T. JAPPELLI AND M. PAGANO (2000), *Information Sharing, Lending and Defaults: Cross-Country Evidence*, Working Paper 22, Centre for Studies in Economics and Finance, Dipartimento di Scienze Economiche, Università degli Studi di Salerno, Salerno, Italy.

E. A. JOACHIMSTHALER AND A. STAM (1990), Mathematical programming approaches for the classification problem in two-group discriminant analysis, *Multivariate Behavioural Res.*, 25, 427–454.

R. W. JOHNSON (1992), Legal, social and economic issues implementing scoring in the U.S., in *Credit Scoring and Credit Control*, L. C. Thomas, J. N. Crook, and D. B. Edelman, eds., Oxford University Press, Oxford, 19–32.

A. JOST (1998), Data mining, in *Credit Risk Modeling*, E. Mays, ed., Glenlake Publishing, Chicago, 129–154.

J. G. KALLBERG AND A. SAUNDERS (1983), Markov chain approach to the analysis of payment behaviour of retail credit customers, *Financial Management*, 12, 5–14.

A. K. KASHYAP, J. C. STEIN, AND D. W. WILCOX (1993), Monetary policy and credit conditions: Evidence from the composition of external finance, *Amer. Econom. Rev.*, 83 (1), 78–98.

S. R. KING (1986), Monetary transmission through bank loans or bank liabilities?, *J. Money Credit Banking*, 18 (3), 290–303.

G. J. KOEHLER AND S. S. ERENGUC (1990), Minimising misclassification in linear discriminant analysis, *Decision Sci.*, 21, 63–85.

P. KOLESAR AND J. L. SHOWERS (1985), A robust credit screening model using categorical data, *Management Sci.*, 31, 123–133.

T. L. LAI AND Z. L. YING (1994), A Missing information principle and M-estimators in regression analysis with censored and truncated data, *Ann. Statist.*, 22, 1222–1255.

S. L. LAURITZEN AND N. WERMUTH (1989), Graphical models for association between variables some of which are qualitative and some quantitative, *Ann. Statist.*, 17, 51–57.

Y. LE CUN (1985), Une procedure d'apprentissage pour resau a seuil assymetrique, in *Proceedings of Cognitiva* 85: *À la Frontiere de l'Intelligence Artificielle, des Sciences de la Connaissance et des Neurosciences*, CESTA, Paris, 500–604.

K. J. LEONARD (1993A), Detecting credit card fraud using expert systems, *Comput. Indust. Engrg.*, 25, 103–106.

K. J. LEONARD (1993B), A fraud alert model for credit cards during the authorization process, *IMA J. Math. Appl. Business Indust.*, 5, 57–62.

K. J. LEONARD (2000), *The Development of a Just-in-Time Risk Scoring System for "Early Delinquent Account" Management*, Working Paper, Department of Health Administration, University of Toronto, Toronto, ON, Canada.

E. M. LEWIS (1992), *An Introduction to Credit Scoring*, Athena Press, San Rafael, CA.

H. G. LI AND D. J. HAND (1997), Direct versus indirect credit scoring classification, in *Proceedings of Credit Scoring and Credit Control* V, Credit Research Centre, University of Edinburgh, Edinburgh, Scotland.

J. T. LINDLEY, P. RUDOLPH, AND E. B. SELBY JR. (1989), Credit card possession and use: Changes over time. *J. Econom. Business*, 127–142.

A. D. LOVIE AND P. LOVIE (1986), The flat maximum effect and linear scoring models for prediction, *J. Forecasting*, 5, 159–186.

A. LUCAS AND J. POWELL (1997), Small sample scoring, in *Proceedings of Credit Scoring and Credit Control* V, Credit Research Centre, University of Edinburgh, Edinburgh, Scotland.

S. LUDVIGSON (1998), The channel of monetary transmission to demand: Evidence from the market for automobile credit, *J. Money Credit Banking*, 30 (3), 365–383.

S. LUMPKIN (1999), Trends and developments in securitisation, *Financial Market Trends*, 74, 25–59.

P. MAKOWSKI (1985), Credit scoring branches out, *Credit World*, 75, 30–37.

W. M. MAKUCH (1999), The basics of a better application score, in *Credit Risk Modeling, Design and Applications*, E. Mays, ed., Fitzroy Dearborn Publishers, Chicago, 59–80.

O. L. MANGASARIAN (1965), Linear and nonlinear separation of patterns by linear programming, *Oper. Res.*, 13, 444–452.

R. E. MARTIN AND D. J. SMYTH (1991), Adverse selection and moral hazard effects in the mortgage market: An empirical analysis, *Southern Econom. J.*, 1071–1084.

E. MAYS (1998), *Credit Risk Modeling*, Glenlake Publishing, Chicago.

B. J. A. MERTENS AND D. J. HAND (1997), *Adjusted Estimation for the Combination of Classifiers*, Department of Statistics, Trinity College, Dublin.

Z. MICHALEWICZ (1996), *Genetic Algorithms + Data Structures = Evolution Programs*, Springer-Verlag, Berlin.

M. L. MINSKY AND S. A. PAPERT (1969), *Perceptrons*, MIT Press, Cambridge, MA.

F. MODIGLIANI (1986), Life cycle, individual thrift and the wealth of nations, *Amer. Econom. Rev.*, 76, 297–313.

J. H. MYERS AND E. W. FORGY (1963), The development of numerical credit evaluation systems, *J. Amer. Statist. Assoc.*, 58, 799–806.

B. NARAIN (1992), Survival analysis and the credit granting decision, in *Credit Scoring and Credit Control*, L. C. Thomas, J. N. Crook, D. B. Edelman, eds., Oxford University Press, Oxford, 109–122.

R. NATH, W. M. JACKSON, AND T. W. JONES (1992), A comparison of the classical and the linear programming approaches to the classification problem in discriminant analysis, *J. Statist. Comput. Simul.*, 41, 73–93.

R. NATH AND T. W. JONES (1988), A variable selection criterion in the linear programming approaches to discriminant analysis, *Decision Sci.*, 19, 554–563.

J. R. NEVIN AND G. A. CHURCHILL (1979), The Equal Credit Opportunity Act: An evaluation, *J. Marketing*, 42, 95–104.

R. M. OLIVER (1993), Effects of calibration and discrimination on profitability scoring, in *Proceedings of Credit Scoring and Credit Control* III, Credit Research Centre, University of Edinburgh, Edinburgh, Scotland.

R. M. OLIVER AND E. WELLS (2001), Efficient frontier cutoff policies in credit problems, *J. Oper. Res. Soc.*, 52, 1025–1033.

G. A. OVERSTREET AND E. L. BRADLEY (1996), Applicability of generic linear scoring models in the U.S. credit union environment, *IMA J. Math. Appl. Business Indust.*, 7, 271–338.

G. A. OVERSTREET, E. L. BRADLEY, AND R. S. KEMP (1992), The flat maximum effect and generic linear scoring models: A test, *IMA J. Math. Appl. Business Indust.*, 4, 97–110.

S. PARK (1993), The determinants of consumer instalment credit, *Federal Reserve Bank of St. Louis Economic Rev.*, November/December, 23–38.

D. B. PARKER (1982), *Learning Logic*, Invention Report S81-64, File 1, Office of Technology Licencing, Stanford University, Stanford, CA.

R. PAVAR, P. WANART, AND C. LOUCOPOULOS (1997), Examination of the classification performance of MIP models with secondary goals for the two group discriminant problem, *Ann. Oper. Res.*, 74, 173–189.

A. P. DUARTE SILVA AND A. STAM (1997), A mixed integer programming algorithm for minimising the training sample misclassification cost in two-group classifications, *Ann. Oper. Res.*, 74, 129–157.

G. PLATTS AND I. HOWE (1997), A single European scorecard, in *Proceedings of Credit Scoring and Credit Control* V, Credit Research Centre, University of Edinburgh, Edinburgh, Scotland.

R. POLLIN (1988), The growth of U.S. household debt: Demand side influences, *J. Macroeconomics*, 10 (2), 231–248.

M. L. PUTERMAN (1994), *Markov Decision Processes*, John Wiley, New York.

J. R. QUINLAN (1993), *C4.5: Programs for Machine Learning*, Morgan–Kaufman, San Mateo, CA.

A. K. REICHERT, C.-C. CHO, G. M. WAGNER (1983), An examination of the conceptual issues involved in developing credit scoring models, *J. Business Econom. Statist.*, 1, 101–114.

M. D. RICHARD AND R. P. LIPMAN (1991), Neural network classifiers estimate Bayesian a posteriori probabilities, *Neural Comput.*, 3, 461–483.

C. D. Romer and D. H. Romer (1990), New evidence on the monetary transmission mechanism, in *Brookings Papers on Economic Activity*, Brookings Institution Press, Washington, DC.

E. Rosenberg and A. Gleit (1994), Quantitative methods in credit management: A survey, *Oper. Res.*, 42, 589–613.

F. Rosenblatt (1958), The perceptron: A probabilistic model for information storage and organization in the brain, *Psychological Rev.*, 65, 386–408.

F. Rosenblatt (1960), *On the Convergence of Reinforcement Procedures in Simple Perceptrons*, Report VG-1196-G-4, Cornell Aeronautical Laboratory, Buffalo, NY.

P. A. Rubin (1990), Heuristic solution procedures for a mixed-integer programming discriminant model, *Managerial Decision Econom.*, 11, 255–266.

P. A. Rubin (1997), Solving mixed integer classification problems by decomposition, *Ann. Oper. Res.*, 74, 51–64.

D. E. Rumelhart, G. E. Hinton, and R. J. Williams (1986a), Learning representation by back-propagating errors, *Nature*, 323, 533–536.

D. E. Rumelhart, G. E. Hinton, and R. J. Williams (1986b), Learning internal representations by error backpropagation, in *Parallel Distributed Processing: Explorations in the Microstructure of Cognition*, Vol. 1, D. E. Rumelhart and J. L. McClelland, eds., MIT Press, Cambridge, MA.

D. E. Rumelhart and J. L. McClelland, eds. (1986), *Parallel Distributed Processing: Explorations in the Microstructure of Cognition*, Vol. 1, MIT Press, Cambridge, MA.

S. R. Safavian and D. Landgrebe (1991), A survey of decision tree classifier methodology, *IEEE Trans. Systems Man Cybernetics*, 21, 660–674

B. S. Sangha (1998), A systematic approach for managing credit score overrides, in *Credit Risk Modeling*, E. Mays, ed., Glenlake Publishing, Chicago, 221–244

J. Saunders (1985), This is credit scoring, *Credit Management*, September, 23–26.

P. Sewart and J. Whittaker (1998), Fitting graphical models to credit scoring data, *IMA J. Math. Appl. Business Indust.*, 9, 241–266.

G. M. Shepherd and C. Koch (1990), Introduction to synaptic circuits, in *The Synaptic Organization of the Brain*, G. M. Shepherd, ed., Oxford University Press, New York.

J. L. Showers and L. M. Chakrin (1981), Reducing revenue from residential telephone customers, *Interfaces*, 11, 21–31.

O. A. Smalley and F. D. Sturdivant (1973), *The Credit Merchants: A History of Spiegel Inc.*, Southern Illinois University Press, Carbondale, IL.

D. J. Spiegelhalter, A. P. Dawid, S. L. Lauritzen, and R. G. Cowell (1993), Bayesian analysis in expert systems, *Statist. Sci.*, 8, 219–283.

V. Srinivasan and Y. H. Kim (1987a), The Bierman-Hausman credit granting model: A note, *Management Sci.*, 33, 1361–1362.

V. Srinivasan and Y. H. Kim (1987b), Credit granting: A comparative analysis of classification procedures, *J. Finance*, 42, 665–683.

M. Stepanova and L. C. Thomas (1999), Survival analysis methods for personal loan data, in *Proceedings of Credit Scoring and Credit Control* VI, Credit Research Centre, University of Edinburgh, Edinburgh, Scotland.

M. Stepanova and L. C. Thomas (2001), PHAB scores: Proportional hazards analysis behavioural scores, *J. Oper. Res. Soc.*, 52, 1007–1016.

J. Stiglitz and A. Weiss (1981), Credit rationing in markets with imperfect information, *Amer. Econom. Rev.*, 71, 393–410.

A. C. Sullivan (1987), *Economic Factors Associated with Delinquency Rates on Consumer Installment Debt*, Working Paper 55, Credit Research Center, Krannert Graduate School of Management, Purdue University, West Lafayette, IN.

T. A. Sullivan, E. Warren, and J. L. Westerbrook (1989), *As We Forgive Our Debtors*, Oxford University Press, New York.

H. Talebzadeh, S. Mandutianu, and C. Winner (1994), Countrywide loan underwriting expert system, in *Proceedings of the Sixth Innovative Applications of Artificial Intelligence Conference*, AAAI Press, Menlo Park, CA.

L. C. Thomas (1992), Financial risk management models, in *Risk Analysis, Assessment and Management*, J. Ansell and F. Wharton, eds., John Wiley, Chichester, U.K., 55–70.

L. C. Thomas (1994), Applications and solution algorithms for dynamic programming, *Bull. IMA*, 30, 116–122.

L.C. Thomas (1998), Methodologies for classifying applicants for credit, in *Statistics in Finance*, D. J. Hand and S. D. Jacka, eds., Arnold, London, 83–103.

L. C. Thomas, J. Banasik, and J. N. Crook (2001), Recalibrating scorecards, *J. Oper. Res. Soc.*, 52, 981–988.

L. C. Thomas, J. N. Crook, and D. B. Edelman (1992), *Credit Scoring and Credit Control*, Oxford University Press, Oxford.

H.-T. Tsai, L. C. Thomas, and H.-C. Yeh (1999), Using screening variables to control default rate in credit assessment problems, in *Proceedings of Credit Scoring and Credit Control* VI, Credit Research Centre, University of Edinburgh, Edinburgh, Scotland.

H.-T. Tsai and H.-C. Yeh (1999), *A Two Stage Screening Procedure and Mailing Credit Assessment*, Working Paper, Department of Business Management, National Sun Yat-Sen University, Kaohsiung, Taiwan.

J. A. M. van Kuelen, J. Spronk, and A. W. Corcoran (1981), Note on the Cyert-Davidson-Thompson doubtful accounts model, *Management Sci.*, 27, 108–112.

P. A. Volker (1983), A note on factors influencing the utilization of bankcard, *Econom. Record*, September, 281–289.

H. M. Weingartner (1966), Concepts and utilization of credit scoring techniques, *Banking*, 58, 51–53.

H. WHITE (1989), Learning in artificial neural networks: A statistical perspective, *Neural Comput.*, 1, 425–464.

M. J. WHITE (1998), Why it pays to file for bankruptcy: A critical look at the incentives under U.S. personal bankruptcy law and a proposal for change, *Univ. Chicago Law Rev.*, 65 (3), 685–732.

J. WHITTAKER (1990), *Graphical Models in Applied Multivariate Statistics*, John Wiley, New York.

J. C. WIGINTON (1980), A note on the comparison of logit and discriminant models of consumer credit behaviour, *J. Financial Quantitative Anal.*, 15, 757–770.

E. F. WONDERLIC (1952), An analysis of factors in granting credit, *Indiana Univ. Bull.*, 50, 163–176.

M. B. YOBAS, J. N. CROOK, AND P. ROSS (1997), *Credit Scoring Using Neural and Evolutionary Techniques*, Working Paper 97/2, Credit Research Centre, University of Edinburgh, Edinburgh, Scotland.

H. A. ZIARI, D. J. LEATHAM, AND P. N. ELLINGER (1997), Development of statistical discriminant mathematical programming model via resampling estimation techniques, *Amer. J. Agricultural Econom.*, 79, 1352–1362.

H. ZHU, P. A. BELING, AND G. A. OVERSTREET (1999), A Bayesian framework for the combination of classifier outputs, in *Proceedings of Credit Scoring and Credit Control* VI, Credit Research Centre, University of Edinburgh, Edinburgh, Scotland.

H. ZHU, P. A. BELING, AND G. A. OVERSTREET (2001), A study in the combination of two consumer credit scores, *J. Oper. Res. Soc.*, 52, 974–980.

Index

absolute errors, 64
accelerated life models, 204
activation function, 71–79
agents, 33
allele, 79–84
aperiodic chains, 93
application scoring, 161
artificial intelligence, 63
assignment, 229
attributes, 41, 43
attrition, 4
augmentation, 141, 142

back-propagation, 73–76
backward pass, 73–76
balance sheet channel, 33–38
bank lending channel, 33–38
bank reserves, 33–38
bankruptcy, 5, 215–217
Bankruptcy Reform Act, 216
Basle capital accord, 230
Bayesian approach, 102, 105
Bayesian learning networks, 196
behavioral scoring, 1, 89–90, 208
Bernoulli random variables, 103, 106
beta distribution, 103, 189
bias, 65
binary variable, 65
bonds, 30
bootstrapping, 67, 107, 111, 188
branch-and-bound, 68
budget constraint, 23
business borrowing, 38

C5, 85
calibration of scorecards, 147–149
cash, 32
cash collateral accounts, 230
certificates of deposit, 38
Chapter 7, *see* bankruptcy
Chapter 13, *see* bankruptcy
characteristic, 12, 43, 124, 126, 139
    categorical, 132
    continuous, 132
    transactional, 126
checking account, 15
checks, 214
χ^2-statistic, 132
χ^2 test, 58
chromosome, 80
classification tree, 41, 53–59, 86, 87
clique, 197
coarse classifying, 121, 131
    monotone, 138
conditionally independent, 196
confusion matrix, 109, 115
conjugate family of priors, 103
Consumer Credit (Credit Reference Agency Regulations) Act, 213
consumer durables, 29
consumer installment loans, 40
correct alignment, 121
cost of misclassifying, 68
Council of Mortgage Lenders, 39
county court judgements, 12, 127
credit bureau, 12, 14, 42, 65, 86, 126–131
    reports, 12, 212–214
credit cards, 10, 212, 214
credit channel, 29, 33–38
credit constraints, 25–27, 210
credit enhancement, 230
credit extended, 19
credit limit, 91
    matrix, 91
credit rationing, *see* credit constraints
credit scoring, 1, 3
    indirect models, 195
    system, 12
crossover, 80–84
cross-validation, 110, 188

current account, 15
curse of dimensionality, 227
cutoff scores, 64, 121, 131, 145–147, 156

data mining, 2, 6
Data Protection Act, 16
debit cards, 214
debt management, 4
debt outstanding, 19, 40, 211
decision theory approach, 42
default rate, 39
defining length of a schema, 83
delinquency rate, 40, 212
delta rule, 74
  generalized, 75
demand
  elasticity of, 27
  for credit, 19, 22–28
  for money, 30
deposits, 32
development sample, 111
deviation error, 66
discount factor, 22
discounted total reward, 96
discriminant
  analysis, 42
  score, 69

elasticity of demand, 27
electronic funds transfer at point of sale, 6
entropy index, 58
Equal Credit Opportunity Acts, 4, 124
error function, 79
error rate
  actual, 108
  Bayes, 108
European scorecard, 215
exchange rate channel, 29
expected average reward, 96
expected loss
  actual, 109
  optimal, 109
expert systems, 63, 84–85
Exports Credit Guarantee Department, 177
external finance premium, 33
extrapolation, 141, 142

Federal Reserve Board, 32, 37
Fisher, 41
fitness, 80
flat maximum effect, 86
forward pass, 73
fraud
  prevention, 129, 171
  scoring, 4
Freedom of Information Act, 16

gender, 5
generalized delta rule, 75
generic score, 16
generic scorecards, 86, 129, 188
genetic algorithms, 64, 79–84, 88
Gini coefficient, 108, 116
Gini index, 57
global Markov property, 197
goodness of fit, 63
graphical models, 196
gross domestic product, 19–21, 39, 209
gross national product, 37

hazard function, 204
hidden layer, 72
hire purchase, 11
holdout sample, 59, 107, 108

impurity index, 56
income, 22–28
indifference curve, 23
indirect credit-scoring models, 195
inference engine, 84
inflation, 40
information statistic, 133
installment loans, 10
integer programming, 63, 68–70
interactions, 87
interest, 2
interest rate, 21–28
  channel, *see* money channel
interior deviation, 69
investment and saving curve, 30

jackknifing, 67, 69, 107, 111, 188
James–Stein estimate, 189

Keynesian-type model, 29–31
Kolmogorov–Smirnov statistic, 55, 107, 113
Kuhn–Tucker conditions, 223

leasing, 11
leave-one-out method, 107, 111
life cycle theory, 24
likelihood function, 79
linear discriminant
  analysis, 79
  function, 41
linear programming, 63–68, 87, 149
linear regression, 41, 86, 226
LM curve, 31, 34
logistic
  function, 71
  regression, 41, 50, 79, 86
long-run distribution, 93
Lorentz diagram, 115
loyalty cards, 6

Mahalanobis distance, 107, 113
Markov blanket, 202
Markov chain, 92, 227
  nonstationary, 98
  order, 101
  stationary, 98
Markov decision process, 96
maximum likelihood
  estimate, 98
  estimators, 52
MMD, 64
monetary policy, 37
money channel, 29
money multiplier, 34
money supply, 29–38
monitoring, 17, 152
moral hazard costs, 33
mortgage debt, 209
mortgage market, 27, 211
mortgage-backed securities, 231
mortgages, 11
mover-stayer, 101
mutation, 81

nearest neighbor, 60, 88
net present value, 22, 29, 34, 145, 181
neural networks, 63, 70–79, 85
  architecture, 76–77
  error functions, 78–79
  multilayer, 72–73
  single-layer, 70–72
neuron, 70
non-Markovity, 101
normalization, 69
novation, 229

Office of the Data Protection Registrar, 127
optimality equation, 96
outcome
  period, 91
  point, 90
overcollateralization, 230
overdrafts, 10
overrides, 17, 121, 144
  high-side, 144
  low-side, 144

pawn shops, 2
perceptron, 72–73
perfect information, 220
performance period, 89
persistent states, 93
personal disposable income, 27, 40
planned consumption, 29
posterior probability, 78
precautionary motive, 30
present value, 22
probit function, 52
production rules, 84
profit maximizing, 220
profit scoring, 219, 225
proportional hazard models, 204
  Cox, 204

race, 5
receiver operating characteristic curve, *see* ROC curve
recursive partitioning
  algorithm, 53
  approach, 41
referral band, 13
regression, 67
  linear, 41, 86, 226
  logistic, 41, 50, 79, 86
reject bias, 141
reject inference, 121, 141–144, 170
relative entropy criterion, 79
religion, 5
response scoring, 4, 170
retention scoring, 4
return, 146, 182

risk-based pricing, 13, 174–175, 227
ROC curve, 107, 115, 205, 206
rotation, 111

sample, 121
savings, 22
schemata, 81–84
  defining lengths of, 83
  theorem, 84
score, 12
scorecard, 12, 13
  calibration of, 147–149
  European, 215
  generic, 86
  monitoring, 152
  tracking, 152
screening models, 192
seasonal effects, 160
securitization, 229
segmentation, 7, 90
sensitivity (Se), 115
separating hyperplane, 64
sight deposits, 30, 32
simulated annealing, 64
small firms, 38
small samples, 110, 188
softmax, 79
Somer's concordance statistic, 134
special-purpose vehicle, 229
specificity (Sp), 115
splitting rule, 54
squashing function, *see* activation function
stopping rule, 54
strategy curve, 221
string, 79
subordination, 230
subparticipation, 230
subpopulations, 131
subprime lending, 227
sufficient, 190
super-fail, 13
super-pass, 13
supply of credit, 25–27
Survey of Consumer Finance, 217
survival
  analysis, 203, 227
  function, 204
symbolic empirical learning, 85

tallyman, 227
tanh function, 71
tax rates, 211
third-party enhancement, 230
three-way classification, 141, 143
threshold function, 71
tobit analysis, 53
tracking, 17, 152
training rate coefficient, 74
transfer function, *see* activation function
transient states, 93
transmission mechanism, 29, 32

unemployment, 37, 38, 40
unsecured loans, 216
usage scoring, 4
usury, 2

value iteration, 97, 105
vector autoregression models, 37

wealth, 28
Widrow–Hoff rule, 74
withdrawal, 144

XOR problem, 77